T0180552

Studies in Systems, Decision and Control

Volume 4

Series editor

Janusz Kacprzyk, Polish Academy of Sciences, Warsaw, Poland
e-mail: kacprzyk@ibspan.waw.pl

For further volumes:
http://www.springer.com/series/13304

About this Series

The series "Studies in Systems, Decision and Control" (SSDC) covers both new developments and advances, as well as the state of the art, in the various areas of broadly perceived systems, decision making and control- quickly, up to date and with a high quality. The intent is to cover the theory, applications, and perspectives on the state of the art and future developments relevant to systems, decision making, control, complex processes and related areas, as embedded in the fields of engineering, computer science, physics, economics, social and life sciences, as well as the paradigms and methodologies behind them. The series contains monographs, textbooks, lecture notes and edited volumes in systems, decision making and control spanning the areas of Cyber-Physical Systems, Autonomous Systems, Sensor Networks, Control Systems, Energy Systems, Automotive Systems, Biological Systems, Vehicular Networking and Connected Vehicles, Aerospace Systems, Automation, Manufacturing, Smart Grids, Nonlinear Systems, Power Systems, Robotics, Social Systems, Economic Systems and other. Of particular value to both the contributors and the readership are the short publication timeframe and the world-wide distribution and exposure which enable both a wide and rapid dissemination of research output.

Kofi Kissi Dompere

Social Goal-Objective Formation, Democracy and National Interest

A Theory of Political Economy Under Fuzzy Rationality

Kofi Kissi Dompere
Department of Economics
Howard University
Washington District of Columbia
USA

ISSN 2198-4182 ISSN 2198-4190 (electronic)
ISBN 978-3-319-38275-3 ISBN 978-3-319-05173-4 (eBook)
DOI 10.1007/978-3-319-05173-4
Springer Cham Heidelberg New York Dordrecht London

Printed on acid-free paper

Springer is part of Springer Science+Business Media (www.springer.com)

Dedication

To all economists and social theorists who are seeking new methodological paths
For explaining individual-collective decision-choice actions in any political economy;
To all members in a thinking society with cognitive independence
In thought, reasoning and practice of ideas with the understanding of the power of interdependence of ideas, its practices and history;
Where new ideas are locked in audacity, curiosity and creativity;
With a full recognition of the nature of knowledge-production system,
Where most ideas need continual corrections towards perfections as
Information-knowledge system expands with interdependence and unity of social decision-choice systems, and with the view that knowledge is self-excited and Self-corrective which is decision-choice determined.
In this respect, the monograph is dedicated to the generalist.

Acknowledgements

The growth and development of the diameter of individual cognition in the understanding of events of in political economies arise from the perception of individual and social exigencies. This diameter and perceptions that shape it not only define the circumference of one's acceptance of general social vision but are always influenced either negatively or positively by the impacts of social infrastructure, institutions that define limitations on collective and individual reasoning, global events and the dominant socio-scientific ideology which has taken hold of the society with a recognition that the universe is a knowledge laboratory for democratic participation. My cognitive diameter is not and cannot be an exception. Its basic foundation has been laid down and shaped by African politico-economic traditions which have expressions in pre-colonial, colonial and post-colonial periods. Its further development was made possible with my presence in Philadelphia and Washington D.C. in The United States. The social exigencies from these periods and places in their interactive modes have provided me some important analytical unity to reflect on political economy in general. The lengthening or shortening of this diameter that affects the circumference of my analytical vision is motivated by a clear distinction that principal-agent cost-benefit conditions of human social organisms offer us in our contemporary nation states within the circumference of human weakness. The diameter has been refined in length by my critical encounter with fuzzy logic that allowed me to handle elements generated by the principles of opposites that view the universe in terms of dualities with continuum and unity from the African thought system. Here I am grateful to the cognitive traditions of the Nile valley, Dogon people and Akans. My development and refinement of fuzzy paradigm and its laws of thought, and their applications to the content of political economy own a gratitude to all working on the problems of fuzzy systems, and to all the authors who appear in the reference list of this monograph.

I wish to express my thanks to all my friends who, critics and admirers who have given me encouragements and emotional support. I am grateful to Akuafo Hall at the University of Ghana for its accommodations when I was editing and refining my views. Special thanks to Ms. Jasmine Blackman for editorial suggestions on the first draft. Controversial ideas and terminologies are intentional and intentionally directed to restructure the paradigm of thinking to account for new explanations in the theory on political economies. I accept all responsibilities for errors that may logically arise.

Preface

The neo-neoclassicals seem to follow the method of J.B. Clark that is, enunciating some proposition such as: What a social class gets is under natural law, what it contributes to the general output of industry and then hastily shoveling some assumptions underneath it to try to make it stand up.

Joan Robinson: Economic Heresies, 1973 [R15.43, p. xii]

In the general mass of notions and sentiments that make up anideology those concerned with economic life play a large part, and economics itself (that is the subject as it is taught in universities and evening classes and pronounced upon in articles) has always been partly a vehicle for the ruling ideology of each period as well as partly a method of scientific investigation.

The purpose of studying economics is not to acquire a set of ready-made answers to economic questions, but to learn how to avoid being deceived by economists.

Joan Robinson: Economic: Economic Philosophy, [R15.40, p. 1].

In economics as in anatomy the whole is much more that the sum of the parts. This is certainly so when the parts are in support of each other or in conflict with each other or are shaped by the fact of their common existence.

John Kenneth Galbraith, Economics and the Public Purpose, [R1123 p.ix].

A government, to perform its tasks successfully needs the supports of an organized majority committed to a programme of action. And to give the people an option, there must be a similarly organised opposition capable of forming an alternative government

F.A. Hayek, New Studies in Philosophy, Politics, Economics and History of Ideas, [R15.20,p.159]

I: The Monograph

This monograph is the first in the series of my treatment on the problems in the political economy where social decision-choice actions are framed in terms of a democratic collective decision-choice system under a given information constraint. The motivation in the development of this monograph stems from a basic discontent with the neoclassical theory of production and distribution where the underlying institutional forces are locked in the assumption of *all things being equal* and taken as given. The institutions of economics, politics and law are hidden under quantities of output, factors and prices that avail themselves to the classical paradigm with its laws of thought and mathematics for exact rigid determination. The transformations of institutional quality that create changing environments of decision-choice actions and social information that affects human decision-choice actions are irrelevant to the theories of consumer and producer decision-choice actions. In every social system, any decision-choice agent must be seen as three in one, which is identified as a social decision-choice agent. The social decision-choice agent is a composite of economic, political and legal decision-choice agents whose actions are confronted by an interdependent system of the three structures of economics, politics and law under extreme synergetic complexity with a continually changing information structure.

In this respect, the notion of free to choose is merely ideological to justify a particular social order. Every freedom operates in a constrained space that varies from one social formation to another. The constrained space is induced by costumes, traditions, rules, regulations, laws and social goals and objectives defined around social vision and the national interest. Such a decision-choice agent is both producer and consumer in one whose decision-making power is continually under the threat of other powers and the power of the social set-up that he or she belongs to. Given the freedom to exercise his or her decision-making power, the decision-choice elements available for production and consumption are institutionally and ideologically constrained by the social goal-objective formation in support of the national interest and social vision.

In this respect, it is not only necessary but imperative to understand the economics, politics and law of the social goal-objective formation as an integrated part of the theory of political economy and as the underlying forces of the market system of decision-choice actions for production, distribution and consumption. The concept of choice is only meaningful in specifying the given underlying social institutions of individual and collective decision actions. One of the essential aspects of this monograph is the view that individual and collective decision-choices are defined within a social goal-objective set, national interest and social vision that together induce the legal structure from the political structure to govern decision-choice actions in all the structures and sectors. The other important and new aspect of this monograph is the utilization of fuzzy paradigm, its logic and mathematics in the analyses of individual preference ordering and the aggregation into the collective decision-choice actions. In this way, the theory of fuzzy sets is introduced as the mathematics of vagueness and inexactness to deal with the penumbral regions of choice. The essential point is that the fuzzy paradigm provides us with channels to introduce effective mathematical structures into the examination of socio-political and coalition games and processes that account for vagueness, inexactness, qualitative transformations and subjective judgments in relation to the principal-agent duality which forms the foundation of decision-choice actions.

II: The Objective of the Monograph

The monograph is about the study and explanation of collective behavior in the social decision-choice space involving social goal-objective formation that must deal with the problems of multi-persons, multi-attributes, multi-criteria and conflicts in the individual preferences under individual-community duality. The decision-choice actions are taken in an environment of uncertainty and amplification of uncertainty within a democratic decision-choice system. The uncertainty is defined by a defective information structure composed of fuzziness and incompleteness. The amplification of uncertainty is defined by a deceptive information structure composed of misinformation and disinformation. The social goal-objective formation is in support of the national interest and social vision that must also be determined. All individual and collective conflicts and decision-choice actions are seen in terms of cost-benefit rationality in the presence of a defective information structure in addition to a deceptive information structure.

The cost-benefit rationality of social decision-choice actions in any political economy is always related to the selection of social goals and objectives that will support an implicitly or explicitly stated national interest and social vision. Hopefully, the cost-benefit rationality and the social vision are established under a defined cost-benefit rationality. The national interest is mapped onto the national vision that is used in defining the framework for organizing appropriate institutions to implement the elements in the goal-objective set. When it is established, the elements in the goal-objective set are stated in terms of policy and transmitted through institutions. In this respect, the study of the political economy under cost-benefit rationality may be decomposed into theories of cost-benefit identification systems, cost-benefit computable systems and social goal-objective formation systems. These theories are integrated as a unified system but may be examined individually in a relational format and then connected to resource allocations, production of goods and services, income distribution and consumption relative to the established social goal-objective set which is under continual transformations as time passes and individual and collective preferences change. The theory of economic behavior and the legal behavior are governed by cost-benefit rationality under quality-quantity duality.

III: Points of Entry and Departure

The entry point of this monograph is the examination of the problems of social goal-objective formation, national interests and social vision in political economies under given and evolving information structures, where the individual private decisions are motivated around the elements in the social goal-objective set. Unlike the traditional economic theories, there is a departure, where information is neither carried by prices as in the Walrasian system nor carried by output as in the Marshallian system but by costs and benefits that may be directly or indirectly related to prices and output quantities. Another point of departure is that the decision-choice theoretic structure does not work with a perfect information structure but with defective and deceptive information structures that render the classical paradigm with its logic and mathematics wanting. Furthermore, the market structure and the corresponding industrial organizations are governed and defined within the national interest and social vision. The monograph, thus, presents a theoretical framework where the formation of social goal-objective set is seen as a collective decision-choice problem whose solution affects the individual and collective actions in the social production-consumption space where the classical paradigm of reasoning presents some logical difficulties. The mechanism for selection is guided by pure democratic principle under majority rule in the collective decision-choice system. The collective decision-choice process is discussed in terms of a democratic collective decision-choice system whose characteristic structure is presented.

The deceptive information structure is made up of misinformation and disinformation characteristic sets and maintained by instruments of mass distraction through ideological and material sub-systems. The defective information structure is defined by information vagueness and information incompleteness that create a fuzzy decision-choice space for the goal-objective selections which are defined as games and coalition formations and the power to decide. Another point of departure of the monograph, relative to the tradition, is the position that the types of available commodities allowable for production and consumption in the social set-up are institutionally constrained by the goal-objective set that will support the national interest and social vision. Freedom of decision-choice actions and market institutions are constrained by the institutions that are constructed to support the national interest and social vision.

IV: The Toolbox for Reasoning, Analyses and Syntheses

The analytical tools used in the formulation of the problem, abstraction of the solution and in the understanding of the democratic collective decision-choice problem of the goal-objective formation are primary category, derived category, opposites, duality, trinity, continuum, conflicts, relational unity, pyramidal logic and fuzzy rationality. A further point of departure is the use of fuzzy paradigm in reasoning that allows the cost-benefit and true-false characteristics to simultaneously exist in one decision-choice element, where one cannot select the benefit and neglect the cost, or do away with the costs and receive the benefits.

V: The Organization of the Monograph

Chapter One presents the essential structure of the problem. Chapter Two is used to develop the theory of social goal-objective formation under a democratic collective decision-choice system and approximate reasoning through the technique of relevance-irrelevance duality under the principle of fuzzy continuum. The reader is introduced to the set of relevant fuzzy numbers for the analysis under the principles of opposites, duality, relational continuum, unity and trinity. In Chapter Three, the social goal-objective formation is seen as a socio-economic game in maximizing decision-making power over the political space, where the theoretical structure of the game of the social goal-objective formation is presented under different participation conditions of the members of the society. The game is played among the decision-making core, the public-sector advocates and the private-sector advocates. From this game structure, combined with a deceptive information structure and the instruments of mass distraction, discussions are made regarding the dilemma in the democratic collective decision-choice system and non-consensus democracy, where the voting process is defined as an instrument of resolving conflicts in the collective decision-choice space without violence. In Chapter Four the conditions of the conflicts in the democracy are presented through coalition formations under conditions of individual preferences and the presence of principal-agent phenomenon.

The coalition formations are specified on the micro-level and macro-level. At the micro-level, the individual preferences are such that by benefit-cost implications, they are classified into groups of private-sector advocates and public-sector advocates with an independent group whose decision-choice behavior alters the strength and the decision-making power of either the public-sector or private-sector advocates. At the macro-level, the coalition formations are among the decision-making core, the private-sector advocates and the public-sector advocates. The decision-choice game, viewed in terms of voting, is defined in terms of the power for the provision of goods and services of the society and social decision-making power which are then related to Marxian and Schumpeterian political economies at extremes, where the concept of private-public-sector efficiency frontier as a trade-off relation is introduced to link them at an increasing cost of private-public-sector transformations.

Chapter Five is used to conclude the monograph on the political economy of goal-objective formation under fuzzy rationality operating with the toolbox of cost-benefit duality, relevance-irrelevance duality, continuum, conflict, and relational unity where the relevance-irrelevance duality is mapped onto the cost-benefit duality in a continuum. Here, the principles of fuzzy laws of thought are used to construct and examine negotiated equilibria of the formation of the social goal-objective set in the political economy through the methods of fuzzy optimization and soft computing under defective and deceptive information structures. The solution set is then used to examine the debate of the relative validity of the appropriate size of the government over the private-public sector duality and its efficiency frontier.

The size of any government is argued to be in relation to national interests, social vision and the supporting social goal-objective set that in turn affects the choice of possible institutional configuration and resource allocation that may be constructed to incorporate the private and public sector ideologies. Again, it is useful to add another point of entry into the theory of the political economy. This point of entry into the political economy is that democracy is defined in terms of institutions to reconcile conflicts in individual preferences in the collective decision space, where individual preferences are allowed to count in all aspects of national life through the establishment of a principal-agent vehicle in national life. In this respect, a democratic social organization contains an important dilemma where the will of the majority may not be followed at the effective operations of the deceptive information structure, instruments of mass distraction and the preferences of the decision-making core (the agent) that may substantially deviate from the principal (the people) in order to establish non-democratic governance within a democratic social formation. The destruction of democracy may be accomplished with instruments that include the practice of non-transparency under the principles of fear, public safety, national security, governmental information classification and privacy violations. The principal is disadvantaged where the agent (government) operates in secrecy but monitors the privacy information of the members of the principal.

Prologue

To understand and appreciate the theory that is being advanced, and the framework for the development of this monograph, some reflections may be useful in the development of the form of the analysis on society and democracy. This developmental framework relates to the problems of collective decision-making with national interest and social vision under a principal-agent structure, where the concept of individual decision-choice sovereignty counts, but the individual sovereignties may assume differential weights in terms of preference ranking. The differential weights of the individuals are abstracted from the distribution of social decision-making power in the social set-up. The weights are defined in the social power spectrum that houses the aggregate sovereignty and are related to individual freedom in the production-distribution space. Individual sovereignty is equivalent to the power to decide and to choose. The weight of the individual power to decide and choose lies between zero and one that is created under some rules of decision-choice behavior. The practice of the individual and collective decision-choice process takes place through an institutional configuration. The preferred decision-choice outcomes of the national interests, social vision and the needed goal-objective set help to establish the required institutions and market or non-market system for the politico-economic decision-choice activities. These individual-collective decision-choice problems may be solved through the market institutions or non-market institutions. It is the perceived greater individual freedom and greater flexibility in weight distribution over the individual sovereignties that provide ideological preferences for market solutions of the collective-decision choice problems as compared to non-market solutions.

I. Reflections on Social Systems

To follow the orientation of this monograph and the subject under discussion, it is useful to reflect on the nature of social systems as human organizations for resource utilization in support of life. Every social system may be seen as composing of three structures in unity. They are the economic structure, political structure and the legal structure. Each of these structures contains its own problems and institutions through the problems are solved by decision-choice and policy actions. Together they constitute the institutional trinity of the social set-up. These three structures are integrated by institutions and organizations which must be managed and administered for social decisions to create stability and progress in all aspects of individual and collective lives. The social administration and management of these institutions take place in the political structure that holds the social decision-making power to create them as well as shape the nature of the legal structure. The legal structure holds the power to regulate conflicts in preferences and establishes social stability and the integrity of the collective behavior in all three structures as they interact to establish the identity of the social system.

The economic structure is the material foundation. It holds the decision-making power to provide goods and services to satisfy the needs and wants regarding food, shelter and collective defense which are broadly defined to include education, medical and others services for the material growth and individual and collective welfare improvements. Thus, the *social power* is the union and interrelation of economic, political and legal powers. The three structures are related to three sets of decision-choice actions in economics, politics and law which together constitute the general decision-choice space in which human social life draws its existence and demands. The general social decision-choice space, involving the individual and collective activities, is partitioned into private decision-choice space and collective decision-choice space. The power to make private decisions is vested in the individual while the power to make social decisions is vested in the political structure in which all members have the opportunity to participate. The individual decision-making sovereignty in the social set-up is the aggregate of their private and public decision powers that are accorded by the social institutional arrangements.

There is a relationship between the private and social decision powers and this relationship, induced by social distribution, is under manipulation from the power of the legal structure controlled by the power vested in the political structure that can be used to effect the individual and collective decision-choice actions. In this respect, all decision-choice actions are constrained not only by resources but by the social and ideological framework with conditions of rules of enforcement within the society. Any element in the private decision-choice space may, by the actions in the political structure, be moved into the collective decision-choice space through the activities in the legal structure. Similarly, an element in the collective decision-choice space may likewise be moved into the private decision-choice space through the politico-legal process. This politico-legal process constantly shapes the preferences in the two spaces through its effects on the social cost-benefit distribution, while the politico-legal process is automatically directed by the conflicting preferences in the individual-collective duality.

The social power-distribution, its internal relationship to sub-powers and applications are ultimately related to the elements in the social goal-objective set that may be established in the political economy. The form of this established social goal-objective set affects the nature and the dynamics of social institutions and their interactions, nature of ownership, production, industrial structure, competitive structure and strategic games in relation to the Darwinian type economy, income-wealth distribution and consumption of all members. Social policies, no matter how they are established, are related to the elements in the social goal-objective set. Goods and services are produced according to social policies in support of the elements in the social goal-objective set. The social policies are transmitted through social institutions, as well as affect the path and structure of private decision-choice space, all of which are related to the conditions of the political economy of the social set-up. The study of the political economy is such that, the economic outcomes and their progress may be viewed in terms of the study of social goal-objective formation and collective interest, and how they are supported by a provision of goods and services of the society in either competitive or non-competitive settings under a cost-benefit structure.

The study of the social goal-objective formation is an approach to the study of the political economy through the theory of collective decision-choice actions with multi-objective characteristics for any given information structure and institutional configuration. Such a study may proceed from different forms of collective decision-choice systems between the extremes of dictatorship and consensus with a relational continuum that reflects the underlying institutional arrangements. This monograph is devoted to the study of collective decision-choice problems with multi-objective conditions and conflicting individual preferences involving the formation of the social goal-objective set under a democratic collective decision-choice system with a majoritarian principle. The neo-classical economic theory of the behaviors of decision-choice agents in the political economy assumes static and given institutions in all political, legal and economic structures, and in fact without interactions. Additionally, the approach of the neo-classical logical system is to impose a set of assumptions about information structure in order to allow its methods of theoretical and empirical investigations to be compatible with the environment of the classical paradigm with its exact information representation, its exact laws of thought and the corresponding mathematics for exact rigid determination. The generalization of the static problem into the dynamic problem carries with it these restrictive assumptions on both information and institutional structures. In terms of pyramidal geometric representation, we have the problems of allocation, production and distribution, on which is imposed the decision-choice system composed of power, the individual and the collective freedoms. The center is the socio-economic imputations of the decision-choice system in relation to allocation-production-distribution decisions in the social set-up and how these decisions are affected by the individual-collective freedoms. The relational structure is presented as a cognitive geometry in Figure 1.1.1.

In this neo-classical thought process, quality of the social set-up is held constant where we only deal with quantitative dispositions of decision-choice items of individual private decision-choice agents under the assumed institutional configurations in the institutional trinity. This approach is then used to impose a *rigid relationship* between capitalism and democracy under an assumed concept of the supremacy of the private ownership of the means of productive capital and freedom of individual preferences over the collective decision space of the social set-up as the natural principle of social organization that provides an efficiency of social resource usage and production of a good life for all. The restraints on freedom and the role of the political and legal orders are never brought into the analysis. The inconsistencies of this approach are hidden mostly in the sometimes

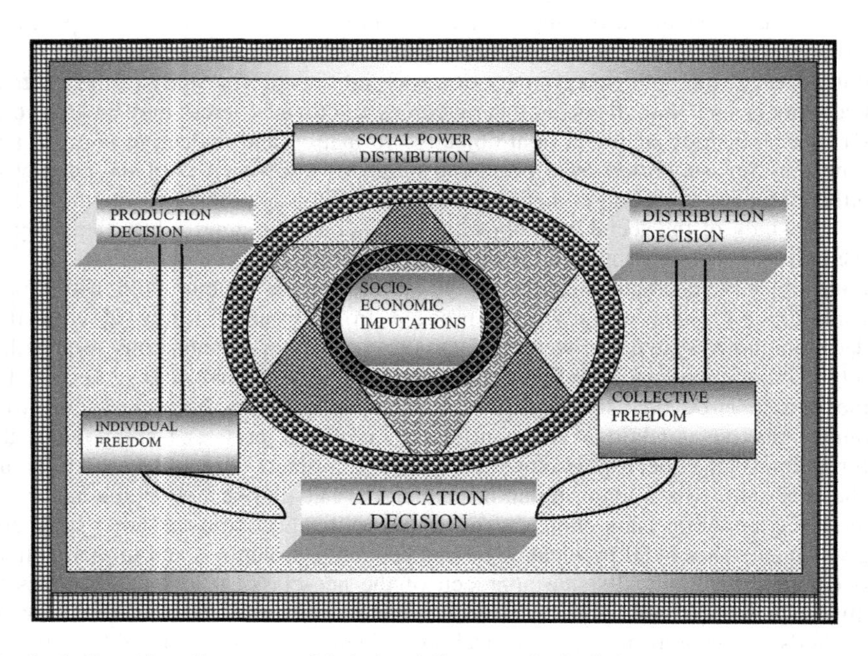

Fig. 1. A Cognitive Geometry of Relational Structure in Defining the Environment for Social Power distribution, in relation to the Individual and Collective Freedoms for Allocation, Production and Distribution Decision-choice system

unhelpful mathematical abstractions of the classical nature by suppressing qualitative dispositions at any given time and over periods. The implicit institutional assumptions in addition to the assumption of exact symbolic information representation in both the neo-classical and neo-Keynesian theories on the economy are very essential and extremely sensitive to the analytical results that may be obtained and the applications that may be required of them in terms of policies and actions.

The social allocation problems involve freedom, justice, costs, benefits and power or sovereignty as seen in the individual-community duality with a relational continuum in the three structures of politics, law and economics. These allocation problems translate into a set of social decision-choice problems that must be solved in any political economy. The decision-choice problems may be stated as 1) the national vision-interest decision-choice problem, 2) the goal-objective-formation decision-choice problem and 3) the allocation-production-distribution decision-choice problem. Problems 1 and 2, given the economic structure, reveal the *qualitative disposition* of the political economy. They are solved in the politico-legal structures and relate to freedom-justice-sovereignty questions, the solutions of which must be translated into the distribution of the social cost-benefit configuration at each point of the political economy. These two problems lead to the development of the theory of *national interest-vision formation* and the theory of *social goal-objective formation*. Problem 3, given the politico-legal structures, reveals *the quantitative disposition* of the political economy. It is solved in the economic structure, the solution of which must be translated into social cost-benefit distribution in the political economy. The problem 3 leads to the development of theories of microeconomics and macroeconomics. These decision-choice problems are relationally connected. Their solutions provide the identity and integrity of the political economy. The cognitive geometry of the analytic-synthesis process is provided in Figure 2 in support of Figure .1.

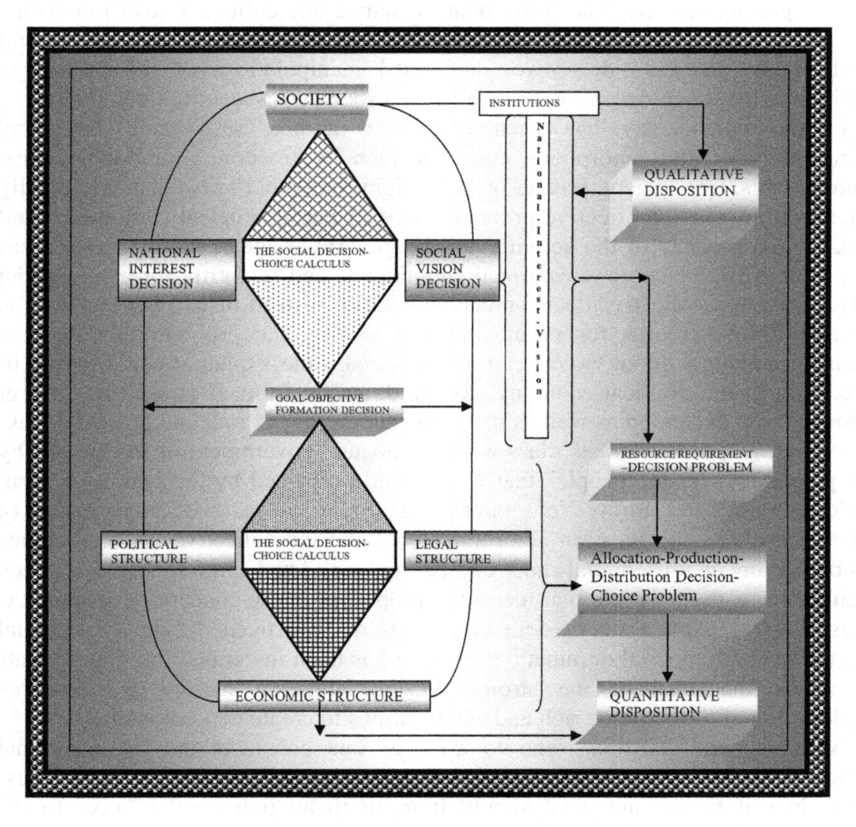

Fig. 2. A Cognitive Geometry of Decision-Choice Problems in the Political Economy

II: On the Concept of Democracy and Social Systems

One of the approaches in resolving conflicts in individual preferences and solving the collective and individual decision-choice problems is democracy, among others such as one-man imposition and general consensus. Most people have come to believe in the goodness of democracy in collective decision-choice systems. Here, a problem arises as to what democracy connotes. In respect of the organization for social formation and power distribution for provision of goods and services, the concept of democracy has acquired many implicit meanings with an amorphous general understanding and claims without any understanding of its relationship to the problem of social decisions and conflicts in the individual preferences. Democracy has been transformed from its role as a social algorithm in resolving conflicts in collective decision-choice actions to become an instrument of propaganda that is integrated into the attack arsenals of some nations in the Western Hemisphere. It has lost its material content and importance in the social decision-making process that can help human progress, justice and social stability. This loss is universal that presents varying degrees of oppression under various political covers. The instrument of democratic collective decision-choice making is effectively used by the radicals, the conservatives and the liberal intellectuals alike as deceptive ideological manipulations of the masses in addition to mass-self deception for control and social cost-benefit manipulations.

In fact, democracy has been placed in a box surrounded by many complementary and amorphous concepts such as freedom, individual rights, human rights, free society, private-property rights and other sweet and emotionally sounding terms as an effective protective belt of the ideological camouflage and lack of transparency of the activities of the governing agent. Its popular use has lost its intrinsic value, organizational significance and scientific force in conflict resolutions in a collective decision-choice system. Instead of its powerful role as an algorithmic process for organizing and solving the problem of collective decision-making without violence, it has acquired useless qualitative properties of deception in the political economy. At the domestic front, it is used as a cover through mass deception as well as mass distraction for the rich and powerful class to exploit the poor and weak class with the slogan "government of the people, by the people and for the people" that is constantly repeated by the masses without understanding its utility or oppressive significance. At the level of internationalism, the Western algorithm in solving the collective decision-choice problems under differential and competing individual preferences has been claimed to be the only one that meets the properties of the concept of democracy. This Western explication of democracy was not practiced during the colonial aspirations and imperial dominations. Now it has been integrated into the arsenals of attack machines of the strong Western imperial nations for resource exploitation of the resource rich and weak nations to create neo-colonial states.

A number of questions tend to arise as one considers the organizational structure of the decision-choice system of the social set-up. What is democracy, what does it mean, and what should it mean to an individual relative to the collective? How does democracy relate to the other popular ideas that have crept into the popular minds and language? What kind of institutional configuration will make possible the practice of democracy with an acceptance of common meaning? These questions bring us into the conflict zone of individual-community duality in terms of the political economy of social decisions with respect to the provision and distribution of goods and services in the social set-up. How much freedom and restraint should the individual have over the collective decision-choice actions, and how much restraint and freedom should the collective have over the individual in the exercise of freedom and restraint? In this respect, the concept of democracy must be defined and explicated to make it analytically and scientifically useful in order to deal with these questions and to relate them to the management and administration of the society composed of the institutional trinity and inputs of multi-person preferences for individuals and the collective. We must keep in mind that the democratic collective decision-choice system is the algorithmic foundation in solving individual-collective decision-choice problems in the three structures of economics, politics and law. The characteristics of democracy as algorithmic foundations relate to freedom, justice and social stability. The meanings of freedom, justice and social stability are social cost-benefit dependent. The understanding of the framing of the social decision-choice problems and the search for the solutions by the algorithms of democracy, requires us to have clear notions of its concept and content and the applications that may be required of them for any given information structure. It further requires of us to have the understanding of the costs and benefits in the application of democracy where the costs and benefits are inseparable, in that every benefit has a cost support and every cost has a benefit support in the decision-choice space.

Let us specify some definitions of the concept of democracy that are claimed to be consistent with forms of human organization under decision-choice problems in the individual-community duality that holds the conflicts within individual preferences on one hand, and between individual and collective preferences from which the principles of law-and-order, freedom-and-justice and the rule-of-law find their ways to define the direction of organizational quality of the relational form of individuals and the collective on the other hand.

III: Some Available Definitions of Democracy

Definition I

Democracy is a way of life where individuals in the society have equal opportunity to participate in the affairs of the state through collective decision-choice actions that affect the individual and the social collectivity.

Definition II

Democracy may be viewed in terms of a theory of government and the relationship between the governors and the governed where the governors constitute the agent and the governed constitute the principal.

In definition one, democracy as a way of life is empty but acquires some content when linked to collective decision-choice actions. It has a problem as to what constitutes equal participation. Is it equal in terms of preferences and hence utility, or is it equal in terms of cost-benefit balances? In definition II, there is an interpretation where the governors may be taken to owe obedience to the *will of the people*. Here, a question arises as to what is the will of the people, and how is this *collective will* established. Alternatively, the definition may project a set of conditions where the individuals have the opportunity and right to establish the will of the people, the national interest and social vision that may be followed by the governing unit (the agent) irrespective of how it is established. The will of the people, the social vision or the national interest may be taken as established by a *voting process* with either a consensus or by the majority, or some decision-choice calculus that is provided by a set of established rules and coded in the legal structure. The collective will, the social vision and the national interest may also be imposed by a powerful social group which may contain one or more individuals. In this respect, a question arises. What are the relationships among the collective will, social vision and the national interest, and how are they related to the individual and the collective socio-economic conditions and the sovereignties?

It will become clear that the notion of *a government by the people, of the people and for the people* has merely an ideological substance but not material substance in concept and practice of the management and administration of the social set-up. In all social set-ups, history reveals that the social goal-objective set, the national interest and social vision are put together by a small powerful elite and imposed on the social set-up; and then sold to the masses through ideological propaganda and mass deception irrespective of whether the governance is democratically or non-democratically structured. The social vision, national interest and the corresponding social goal-objective set may reflect the elite interest but not necessarily the interest of the general society. To follow and appreciate the contextual argument being offered in this monograph, one must distinguish between government and governance. The government is a holding place and is neutral to the members of the society while the governance is always transitional and is not neutral. Its neutrality depends on the character of the governing class in its behavior to fairness, freedom, equality and justice.

Definition III:

Democracy may be viewed in terms of citizens' participation in the decision-making process of the state composed of the governmental system and sub-systems by means of which political sovereignty is vested in and exercised by the citizens in the three structures of economics, law and politics.

This definition is amorphous; it includes all aspects of the state without indicating the process and the rules of participation. It, however, points to an important concept of decision-making which must be associated with the actions that affect the cost-benefit configurations of the individuals and the collective in participation. There are many variations of these definitions of the concept of democracy. All the definitions attempt to explicitly or implicitly incorporate, in one form or the other, the collective decision-choice phenomenon with individual relations into the political sovereignty and the collective. Most of them fail to acknowledge the problem that gives rise to the concept and practice of democracy. The problem that gives rise to the decision-choice principle of democracy or its possible alternatives finds expression in the conflict-power *relationality* in the social set-up and how the conflicts in the individual preferences may be reconciled in the social decision-choice space under the prevailing *information structure*.

This problem is not unique to any particular social set-up. Different societies have designed different methods to solve the conflict problem over the collective decision-choice space. The nature of the solution is linked to freedom-and-justice on one hand and law-and-order on the other hand, which is all defined in the political structure that imposes the institutions of law and regulations. For the purpose of the development of this monograph, it will be useful to present an analytical decision-choice explication of the concept of democracy. It will also be useful to distinguish between the *concept of democracy* and the *practice of democracy* in order to appreciate the interactive roles of information, decision and power in the social space and where the information structure presents itself as *defective information sub-structure* and *deceptive information sub-structure* to create differential advantages in favor of some segments in the society. These differential advantages further intensify the conflicts in the individual preferences within the collective and provide the continual energy of the forces of institutional changes and social transformations. The forces of institutional changes and social transformations of any society are accomplished by nonviolent means of applying social pressures and waging conflicts for resolving power differences to accomplish social vision and national interest, even though violence sometimes erupts to create social instability.

IV: Society and Its Organizational Form in the Political Economy

Let us take a look at what a society, as an organization, is essentially made up of in order to relate it to the problem of relational continuum in the principal-agent duality in the social decision-choice space. Every society, no matter how small or big, may be viewed as held together by three building blocks of *economic structure*, *political structure* and *the legal structure*. The economic structure, composed of material production, distribution and consumption relations, forms the foundation of material survival of the society. It is the real life support of the society and its members. There are no political and legal structures without the economic structure. The economic structure is thus concerned with the material relations of the members of the society for survivability. Here, we speak of *economic power* and its distribution for the control of the economic decision-choice system. The political structure, composed of the power production, distribution and consumption relations, forms the fundamental decision-making authority of the society. The political structure is thus concerned, from the viewpoint of organization, with the power relations of the members of the society regarding individual and collective decisions that affect the direction of individual actions in all relevant spaces, in addition to the history of the individual within the collective as well as the collective itself. Here, we speak of *political power* and its distribution. The legal structure, composed of production, distribution and consumption of judicial relations, forms the fundamental conditions of justice and social conditions of just order by the established rules in relation to acceptable relational interactions among the individuals and between the individual and the collective in the three structures, to define and ensure freedoms and fairness.

It is here that social injustice may be amplified through enforcements as crafted from the political structure. In the legal structure we speak of *legal power* and its distribution. The political structure, viewed through the lenses of the legal structure, is thus concerned, from the viewpoint of a social organism, with the relations of justices among the individuals and the collective that affect the conflicts, peace, freedom and harmony in the social set-up for stability in the unity of decision-choice behavior in the political, legal and economic structures. It is the special position which is held by the political structure and its direct or indirect control over both the economic and legal structures that gives contents and meaning to the statement that political power is an inescapable prerequisite to economic and social power, and hence places a nerving importance on the slogan *seek yee first the political kingdom* and the decision-choice sovereignty in designing the path of history is opened to you [R15.33]. How does democracy emerge out of the social organism, what does democracy mean to the members of the collective and what substance does it acquire for the members and the collective in their lives? The answers to these questions are extremely important in the understanding of events and outcomes in the political economy regarding poverty, social unrest, injustices and many social ills and goods.

The meaning and the relevance of democracy must be seen in terms of power distributions and uses as well as their impact on the individual and social decision-choice process that imposes costs and benefits on the individuals and the collective. The power to make decisions within the economic structure that affects the individual and the collective lives is actually designed from the nature of the legal structure. Thus economic power that one holds is the result of the events in the legal structure. The design of the rules, regulations and laws that hold the three structures together as well as regulate the individual and collective behaviors is done in the political structure. The harvesting of the cost-benefit characteristics of the design is in the economic structure. The decision-making is an intrinsic part of human life in all three structures. In every society, the elements of the decision-making come to us as 1) those that are *individual-specific* in terms of personal decisions 2) those that are *collective-specific* in terms of social decisions and 3) those that are simultaneously individual and collective in form.

The degrees of potential participation by either the individual or the collective in the social decision-choice space are defined in the political structure as well as set by the power distribution in the political structure. The manner, in which the political structure is constructed, given the individual preferences, ranges from a) complete opportunities for individual participation as the primacy of organization of decision-choice actions to b) complete opportunities for the collective as the primacy in the organization of decision-choice actions in the political economy. It may be noted that the opportunities are merely potential and must be translated into practice through some institutional arrangements which may either facilitate or hinder effective participation in terms of equal sovereignty or access. Such equality in participation is designed in the political structure, codified and vested in the legal structure as one of the rules or laws of constraint on individual sovereignties.

The social system is integrated by three structures to present an organic social decision-choice space that is partitioned into three decision-choice sub-spaces. The three sub-spaces are: 1) the sub-space of completely individual decision-choice actions, 2) the sub-space of completely collective decision-choice actions, and 3) the sub-space of joint individual-collective decision-choice actions which must be regulated in terms of decision-making power and the sovereignty to exercise the power. The second and third portions of the partition will be placed in the category of the second. In the completely individual decision-choice space, the individual decisions affect the decision maker with very little or no social externalities as analytically conceived. There are, however, some individual decisions whose outcomes may have important indirect effects on the collective welfare. These externalities create second order conflicts between the individual and the collective freedoms as well as impose demand on the social monitoring of individual decision-choice actions. It is the existence of distributions of sovereignties and the corresponding cost-benefit configuration that provide justification for the study of externalities in decision-choice systems.

Given the general decision-choice space, how should the social decision-making power be distributed among the individuals and between the individual and the collective, and to whom should the power to make the collective decision be vested? The question simply boils down to the question about the nature of power distribution and relations in the individual-community duality for collective decision-making which finds expression in the politico-legal structures under a designed power-distribution regime. In a sense, the study of the political economy is the study of power distribution and power relations in the decision-choice structures of the institutional trinity of economics, politics and law, and how these relations are transmitted into cost-benefit relations in the welfare space in which individuals and the collective find living expression in preferences. It is these three decision-choice sub-spaces that impose on the decision-choice system the nature of the private-public sector distribution of institutional formation that may be consistent with the ideological position in the social set-up. The completely private decision-choice space is associated with perfect capitalism and unlimited individual sovereignty. The completely collective decision-choice space is associated with perfect socialism with unlimited collective sovereignty.

V: The Individual-Community Duality and the Social Decision-Making Power

The answers to the questions raised above are provided by the nature of institutions of participation that are either agreed upon by the collective, or imposed on the individuals and the collective from the legal structure, or have evolved over a period of time as part of the culture the origins of which are currently and exactly untraceable into the distant past. The individual decision-making and the collective decision-making reside in a relational continuum of a duality with dynamic conflicts that provide the evolving social institutions with energy for transformations at any given information regime. The purpose of the transformation has always been continual attempts to reconcile the conflicts of the individual-collective duality in the social decision-choice space with or without violence. In other words, the reconciliation is to define the boundaries of social power distribution regarding the individual decision-making and the extent to which the society can interfere on one hand and the collective decision-making and the degree to which an individual participation is allowed by the collective on the other hand. This involves the problem of freedom and justice and the solution to the primacy problem of the individual-community freedoms in the three structures of economics, politics and law.

The degree of freedoms and justice that are socially allowed to the individual depends on the solution to the power-distribution problem which always generates conflicts in the individual-community preferences over the collective decision-choice space, as well as contributes to instabilities and social revolts due to possible unfairness in the social cost-benefit distribution. The social instabilities and revolts are transformational always seeking to change the existing institutional arrangement in support of the power distribution in the three structures. In this respect, individual freedom resides in the collective freedom while the collective freedom finds expression in the individual freedom within the individual-community duality of the political economy. The decision-choice questions involve the individual-community *primacy problem* of the fundamental ethical principles in the sense that: should the individual freedom be prior and emphasized over that of the collective freedom, or should the collective freedom be prior and emphasized over the individual freedom in the social decision-choice space to create a balance in the social power space with a proportional distribution of power in the continuum? Every solution to the individual-community conflict is a temporary one that generates new conflicts in the duality. In this respect, every temporary solution awaits to be dislodged by another conflict resolution through changes in and shifts of the institutional configuration and the corresponding value system.

The resolution of the conflicts, with or without violence in the individual-community duality in freedom and preferences, has lead to the evolution of institutional configurations that range from an emphasis on the primacy of the individual freedom and power in individual decision-making space and full participatory opportunities in the collective decision-making, and to the primacy of the collective decision-making power and freedom in the collective decision space with defined opportunities for the degrees of participation in the collective and individual decision-making process. The former comes under the *fundamental ethical postulate of individualism* that gives rise to the nature of the evolving social institutional configuration as well as reveals the structure of the social formation to ensure that the decision-making power is vested in the individual who must compete with other individuals for a voice in the political economy. The created institutional configuration must be compactable with this principle where sovereignty is completely vested in the individual. The latter comes under the *fundamental ethical postulate of collectivism* which provides a guidance for the nature of the evolving social institutional configuration that will reveal the basic structure of the social formation where the decision-making power is fully vested in the collective regarding social decisions with constrained freedoms for the individual private decisions, to the extent to which the outcomes of such decisions do not negatively impact on the collective in some substantial way. The created institutional configuration, in this respect, must be compactable with this principle where sovereignty is completely vested in the collective.

In all these situations, we are confronted with the proportional distribution of the degrees of individual and collective freedoms and power in the social decision-choice space involving the activities in the economic, political and legal structures that form the institutional trinity of the social set-up. The principle governing the structure of the political economy is such that, the fundamental sovereignty rests with the individual in social formation organized under the ideological and philosophical principle of individualism, while the fundamental sovereignty rests with the collective in social formation organized under the ideological and philosophical principle of collectivism. The analytical position adopted in this monograph is that all social activities whether individual or collective are decision-choice actions that bring about transformations with costs and benefits, the distributions of which generate perceived conditions of fairness and justice that are mapped onto the conflict space of human action. The cost-befit relative structures change the welfare configurations of the social states. How these social cost-benefit configurations are distributed among the members are established by the power distribution and corresponding power relations that generate justice and injustice. It is here that the concepts, which ranging from dictatorship to unanimity, find expression; and this expression is in relation to the nature of power distributions, power relation and the institutions that maintain them for liberty and subjugation. In this respect, we may speak of dictatorship-unanimity duality with a relational continuum. Within the dictatorship-unanimity duality are defined infinite possibilities of democratic institutional arrangements with varying degrees of individual-collective freedoms where such freedoms and the corresponding fairness and justice take their meanings in the perception space which is then mapped into the action space. In other words, in social practice reality is of less importance than perception that gives rise to social actions.

Given the institutional trinity of the social set-up of the political economy, a number of questions tends to arise. What should the social vision and national interests be, how are they established, by whom, and in relation to what freedoms? How should the national interests and the social vision be related to the individuals and the collective? Given the national interest and the social vision, what goals and objectives must the society establish in support of them, how are these goals and objectives created, by whom, and whose interest must these goals and objectives serve? Given the establishment of the social goal-objective set, how should the national resource endowment relate to allocation, production and distribution? Finally, given the social vision, national interest, the goal-objective set and national resource endowment, how must the social power for decision-making in the economic, political and legal structures be distributed to ensure a balance of freedom and justice between the individual and the collective to produce fairness among the individuals, and what institutional configuration must be established to ensure the ability to exercise the decision-making powers?

These questions and the problems that they engender must be faced, answered and solved by any society irrespective of the characterization imposed on it. The analysis and synthesis of the understanding of the solutions constitute the main corpus of the *theory of the political economy.* It is the manner in which these questions and problems are answered and solved by a society that defines its character and form in relation to power, freedom, justice and social oppression. Every set of answers and solutions is a temporary one waiting to be destroyed by its own internal conflicts and be dislodged with a new answer-solution set. A search for answers to these questions and others becomes a problem of looking for solutions to the problem of power distribution in the society in terms of how power is shared in the social decision-choice space. The same search for answers has been the source of internal conflicts, struggles and institutional transformations with never-ending processes and continual temporary equilibrium-disequilibrium processes that establish the social history and the corresponding culture as generated by the corresponding outcomes of the social decision-choice activities in the society.

The rules and regulations that have been created, the institutions that have been constructed through which these rules and regulations are administered and the outer shell composed of ideology, culture and beliefs have adjectival names for different social formations where such social formations are in competitive conflicts due to the fundamental value principle that is imposed on the society from within the individual-collective duality. The point, here, is that names used in characterizing social formations find expression, meaning and material content in the social decision-choice space as seen from the social distribution of decision-making power in the society either in a closed system of a political economy or in an open-system of a political economy. The expressions, meanings and material contents of the social form are enhanced and or, given the institutional configuration or arrangements through which the social decision-making power is distributed and practiced. The distribution and practice take place within the ideological and belief system that guides the creative-destructive process of the institutional arrangements to provide temporary resolutions of individual-collective freedoms under the culture of human thought. Given this background, and on the basis of resolution of conflicts in individual and collective preferences, a decision-choice definition of democracy that is appropriate to the task of the monograph is offered.

Definition IV

Democracy is an arrangement of the elements in the institutional configuration, power-distributional form and power-relational structures that establish general and specific processes and algorithms for resolving individual conflicts without violence and coercion among the conflicting preferences of the members of the social collectivity in the social decision-making space that encompasses economic, political and legal structures, in which the costs and benefits of decision-choice outcomes affect the individual and the collective existence as well as the social history in an organized social system.

As defined, democracy is seen in terms of an institutional arrangement to resolve conflicts in individual preferences to create collective preferences for ranking social states through the distribution of social decision-making power under a defined degree of individual participation under the principle of cost-benefit perceptions that relate to fairness and freedom. For example, one person one vote points to the concept of equal weights. Democracy is an institution of social calculus for collective decision-making without violence or coercion in the economic, political and legal structures. In this frame, we speak of a *democratic social decision-choice system* of the collectives, societies and states. The notions of without violence and coercion in the definition are very important as they point to a voluntary participation within a given set of rules and legal form in terms of peaceful resolutions of the conflicts in the individual and group preferences (see [R2] [R2.5] [R7.16]).

Individual participation in the social decision-choice process takes place through vote-casting. A vote-casting is an expression of non-violence and social stability that must be protected by the collective. The prevention of individual vote-casting of any form or any restriction imposed is an expression in favor of violence and social instability. The perfect democratic decision-choice system is complete consensus that corresponds to complete individual or collective freedoms where the individual freedoms coincide with the collective freedom in the preference space. There are three important forms that come out of the decision-making approach in defining democracy not simply as a form of a government but as a form of resolving conflicts in the social decision-choice space where the setting up of government is part of it. Firstly, we have democratic social formation that establishes its compatible institutional configuration and its protective belts where both the institutional configuration and the protective belts are constantly evolving and adjusting themselves to correct inefficiencies and undesirable outcomes that generate injustices and restrict freedoms. Secondly, there is the specification of the set of rules, regulations, culture of judicial and value-practices that indicate how the distribution of the social decision-making power must occur as well as must be enforced for practice. Thirdly, there is a form of institutional conditions and tools that must be there if the practices and the uses of the distributed decision-making power are to be successful in managing the social formation in states and over states of the social evolutions.

The decision-oriented definition of the concept of democracy has nothing directly to do with the definitional character of the social system of resource allocation, production and income distribution on the basis of type of ownership of the means of production. It, however, has some importance to the sovereignty question of the individuals and the government in terms of ownership of the decision-making power and the restraints on its exercise over the private and public domains. This sovereignty of exercising one's decision-making power is compatible with all ownership systems of regarding allocation of national resources, production of goods and services and the distribution of income for consumption. The conditions required to effect the stability and success of democratic social formation will affect, however, the acceptable design of the private-public sector combination of ownership of the means of production and provision of goods and services in the society. The concept of ownership of the means of production is an element of the set of conditions that allows effective exercise of social decision-making power that one has for participation. The degree of effectiveness of this exercise of the individual and collective sovereignties may be restrained from the political sector through the legal sector.

In this respect, one must take note of the concept of democracy and the conditions for its effective practice in creating a social preference order over the collective decision-choice space. Here, freedom of speech, association, and grouping; law and justice, rule of law, right to information, access to institutions, equal participation, responsibility, accountability, just distribution of costs and benefits of government and others are elements that enhance effective participation in democratic decision-choice systems but they cannot constitute the definitional concepts of the democracy. Similarly, classification of information on the basis of national security, public safety and others in addition to secrecy and lack of transparency in the governmental machinery tend to destroy democracy and enhance forms of dictatorship [R15.17]. There are countries that call themselves democratic but use the legal structures to disenfranchise segments of the population as if they do not exist. In most cases, some of the elements of enhancement in the practice of democracy have been equated to the concept of democracy. For example, vote-casting is a way of expressing preference through the exercise of the decision-making process. It does not by itself constitute the definitional content of the concept of democracy. It is simply a tool. Freedom of speech is an information sharing that may positively or negatively affect the resulting knowledge in support of the individual decision-making process and the resulting outcome of the collective decision under the tool of vote-casting in a democratic decision-choice system. It does not by itself constitute the content characteristic for explicating the concept of democracy. However, a restraint on the instruments that enhance the individual and collective participation reduces the degree of participation in the democratic collective decision-choice system and hence destroys its organizational power to resolve individual conflicts in the collective decision-choice space.

In the theoretical construct of either explanatory or prescriptive structure of the democratic decision-choice system in the political economy, there are two important assumptions that fix the form of the collective decision-choice actions. One is the assumption of the *degree of individual sovereignty* and the other is the assumption about the nature of *the information structure,* given the individual preferences. In the construct of the theory, one may assume equal participation or unequal participation defined in terms of degrees, ranging from zero to one, that are attached as weights on the individual preferences. Here, zero implies complete non-participation across all sovereignties and one applies to full equal participation. At the level of information input, one may assume perfect information or imperfect information which is here defined in terms of *defective* and *deceptive* information structures. An example of a theory of democratic collective decision-choice system is that of Arrow's with a resulting paradox [R2.5], [R7.15].

In general, the following four scenarios of participation may be presented: 1) Equal participation under perfect information structure; 2) Unequal participation under perfect information structure; 3) Equal participation under imperfect information structure; 4) Unequal participation under imperfect information structure. Each of these cases may require a different mathematical structure and logical reasoning in constructing the theory of the political economy. The weight-participation combinations produce zonal structures that are shown in Figure 3. The importance of the zonal areas cannot be underestimated in that the weight distribution and the information structure, given the individual preferences and the culture of decision-making, determine the social decision-choice outcomes in the political economy, as well as shape the path of the national history.

The prologue simply presents the nature of the set of problems in dealing with the social goal-objective formation, the national interests and the social vision through the framework of a democratic collective decision-choice system under defective-deceptive information structures. Under the *defective-deceptive information structure* the democratic collective decision-choice problem is formulated and solved with the toolbox fuzzy paradigm, composed of fuzzy logic and fuzzy mathematics with the corresponding fuzzy rationality. The fuzzy rationality is composed of fuzzy optimality and fuzzy conditionality [R15.13][R15.14]. The imposed analytical environment will account for unequal and equal conditions and information imperfections including the ideological regime of the social set-up. Here, a counsel from Joan Robinson is useful in that "*Any economic system requires a set of rules, an ideology to justify them, and a conscience in the individual which makes him strives to carry them out*" [R15.40, p13].

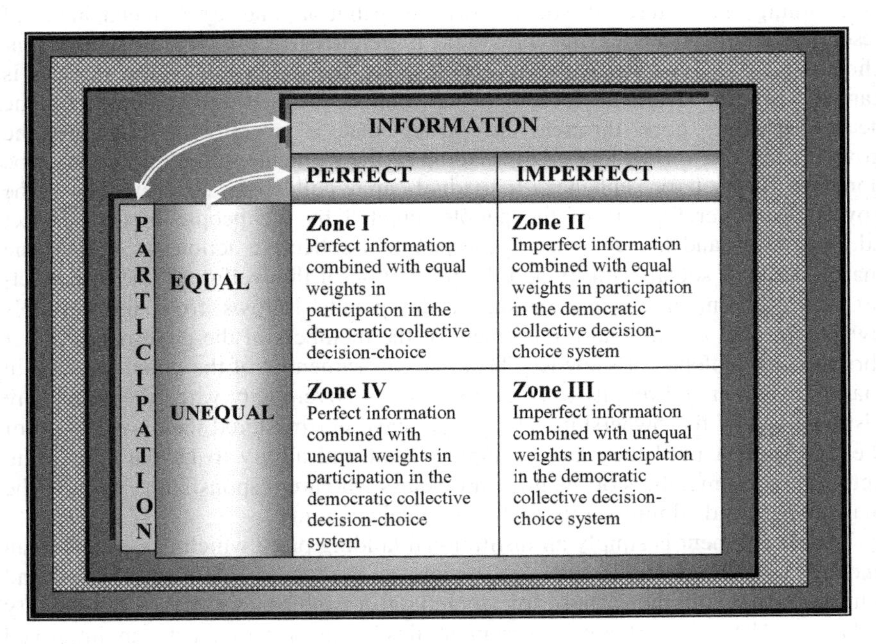

Fig. 3. The Participation-Information Structure in Democratic Collective Decision-Choice System

VI: A Simple Reflection on Government, Governance and the Theory of the Political Economy

The theory of the political economy is, here, defined as a theory of decision-choice systems with a distribution of decision-making power that is made up of an individual-collective combination. It is composed of three sub-theories of 1) the theory of national interest-vision formation, 2) the theory of social goal-objective formation and 3) the theory of allocation-production-distribution decision which is divided into interdependent conceptual sub-systems of a) a theory of production, b) a theory of distribution, and c) a theory of consumption. The theories of national interest-vision formation and social goal-objective formation are intimately connected with the theories of government and governance. These conceptual systems may be related to theories of democratic collective decision-choice and non-democratic collective decision-choice systems.

In the definitional structure and conceptual system of the theory of the political economy, government is explicitly distinguished from governance in both concept and social practice. Government is a permanent holding facility that houses a variable toolbox to be used by the social decision-making core (elected officials) to work within the social collective decision-choice system. The toolbox contains among other things 1) executing social power to make laws, rules and regulations and create institutions, 2) enforcement power, 3) a mechanism of determining social policy, 4) a mechanism of restricting behavior, 5) a mechanism of determining the nature of social-power distribution and 6) a mechanism of resolving conflicts among the individual preferences in the collective decision-choice space. Governance is a process where the execution of the process is carried with the available toolbox which can be refined and reshaped by the decision-making core through the power that is socially invested in the government. The refinement and reshaping of the elements in the toolbox may be done to restrict or enhance individual and collective sovereignties. The government, therefore, is of the people, created by the people in the service, administration and management of the people's collective actions as well as the maintenance of social stability and the promotion of the welfare of the social set-up. The government as a creation of the people follows from the people's ownership with a continual replacement of the members of the governance either through non-violence or violence. That the government is of the people is seen in mass movement and revolts to cease its inherent sovereignty when the people are dispossessed of the ownership. The people have not mastered their ownership of the government until they have complete right and ability to be the agent who determines the members of the governing class who are responsible to them in the management and administration of the political economy.

The government is simply an institutional holding place which is a means to an end of providing social services which is composed of management and administration of the affairs of society to establish individual-community balances. The value of any government lies in its use to create an improved welfare for the people, which is not possible under individual actions. Governance is a social decision-choice process assumed by the members of the social decision-making core to effect the administration and management of the inherent sovereignty of the government in the service of its owners, the people. The people have no control over the nature of governance if they have not mastered their collective sovereignty over the governing class and the collective ownership of the government. That they have not mastered their ownership of the government and have no ownership of the sovereignty is seen in the deterioration of freedom and justice and concentration of power in the hands of the governing class supported by special ownership class in the political economy. That the government is neutral and the governance is not is seen in the good-evil duality that is characteristic of the social set-up involving the social character of the governance ranging from law and order on one hand to freedom and justice on the other which are attributed to the government. The degree to which the democratic collective decision-choice system is allowed to work in any of the three structures of law, politics and economics is not of the government but of the governance. The fair application of laws, rules and regulations of the politico-legal structures to produce freedom, justice, law and order is of the governance. The people are the backbone of freedom and justice. It is by the people's effort, that oppression and subjugation are rooted. It is also by the relentless toil and heavy sweat from the people's existence that democratic institutions are created and efficiently managed. The people, therefore, are simply the reality of democracy, freedom and justice as the enduring qualitative characteristics of their culture and a way of collective life. Unquestionably, it is the people who suffer the depredation of injustices, abuses and lack of freedom that are important qualitative characteristics of their collective welfare. This collective welfare is compromised when the people fail at any moment of time to understand their sovereignty over the government and allow it to be taken over by a governing class. The people have no claim to sovereignty until they have mastered the costs and benefits of freedom. The claim to the ownership of sovereignty over the government has always produced conflicts in the principal-agent duality of every social system. It is these conflicts which generate forces of internal social evolution and revolution in the political economy for growth and development. In this respect, the political economy is self-exiting, self-correcting and self-learning social system which is driven by conflicting interests within the domain of justice-injustice duality with a relational continuum and unity for social goal-objective formations, national interests, and social vision under either the principles of non-democracy or democracy.

Contents

Chapter 1
The Problem of the Social Goal-Objective Formation in Democratic Societies of Collective Decision-Choice Systems

1.1 Introduction

Every theoretical construct must define the space of analysis. This space must contain the questions and problems that are of interest to cognitive agents. These problems and questions must relate to social set-ups and natural order. In this frame, theory is about problem solving, as well as a search for answers to questions. These problems and questions arise out of social exigencies which keep on creating new ones as existing problems get solved and existing questions get answered. Theories are, thus, not about purely intellectual activities for their own sake but directional attempts for explanation and prescription. These problems and questions are part of the initial information structure and its evolution for continual knowing. Every social formation is composed of the individuals and the collective in the social action space that must reveal the individual and collective behavioral actions in the decision-choice system. The study of the behavior of any society may proceed by studying the individual and collective choice-behavior and how the individual behaviors interact to produce the observed individual and collective outcomes. It is from the structure of the social set-up that attention must be paid to the individual-community duality as a conceptual foundation in dealing with the problems of explanation and prescription in the social system and its qualitative and quantitative dynamics as it interacts with resources for individual and collective survival.

All decisions are complex involving psycho-economic imputations of benefit and costs associated with well-defined and not well-defined preferences in the social decision-choice space. The boundaries of the social decision-choice space are vaguely fixed by culture, social norms and existing institutions that provide a general framework of decision-choice rationality as an acceptable guide for the individual and the collective behaviors. The problems of the individual decision-choice actions occupy a simpler space that houses quality, quantity and time characteristics. The individuals enter the social decision-choice space with their preferences as they interact with society and nature to solve the problems of needs and wants. These individual preferences are diverse and conflicting with one another that present decision-choice problems for the social collectivity. The problems of collective decision-choice actions, therefore, are defined in a more complex quality-quantity-time space over the social set-up which is composed of

K.K. Dompere, *Social Goal-Objective Formation, Democracy and National Interest*,
Studies in Systems, Decision and Control 4,
DOI: 10.1007/978-3-319-05173-4_1, © Springer International Publishing Switzerland 2014

economic, political and legal structures that are defined by differential quality-quantity-time characteristics in a cross-sectional and time-series of social stages. The complexity of dealing with quantity, quality and time with their interactions is compounded by the problems of reconciling differences of the individual preferences in the collective decision-choice actions over the three structures of economics, politics and law with institutional configurations that must interact for social stability and progress.

It may be kept in mind that the understanding of the individual decision-choice actions is complex. Such a complexity of the understanding is amplified when it is mapped into the space of collective decision-choice actions under different regimes of political life, economic life and legal life, which together aggregate into social life. What brings complexities into the individual and collective decision-choice actions are the individuals pursuing their satisfaction without regard to the collective, and the collective pursuing the collective satisfaction and socio-political stability that define the acceptable boundaries of decision-choice behavior, as well as placing limitations on the individual decision-choice actions in the duality. Every society has individual and collective preference. The individual preferences relate to individual decision-choice sovereignty while the collective preferences relate to collective decision-choice sovereignty. The collective decision-choice sovereignty is derived by a construct from the individual decision-choice sovereignties. The collective sovereignty, when it is formed, places limitations on the practice of individual sovereignties by imposing limitations on the decision-choice sets in the political, economic and legal structures. Both sovereignties exist in a continual conflict zone in the political, economic and legal structures. Given the individual and collective preference, the exercise of sovereignties demands information and information-processing capacities in the conflict zones of decision-choice actions

In this respect, by tracing and understanding the factors that create conflicts and affect human actions in the individual-community duality, the political economists may be able to unearth static and dynamic social behavioral laws that will help to understand and explain sociopolitical conditions of societies. Alternatively stated, the social goals and objectives, the national interest and social vision, irrespective of the mode of creation, define the institutional boundaries of individual decision-choice actions. The social decision-choice space is made-up of goods and services that are legally producible, tradable and consumable in all the three structures. The legal structure places boundaries on the decision-choice space and the decision-choice elements. The legal structure is constructed under the guidance of the social vision, national interest, social goals and objectives that constitute the social characteristics of the society. The legal structure is constructed from the political structure which is created from within its internal rules. The analysis of the individual and collective decisions in the social decision space must keep in mind the social limitations on the available decision-choice elements. The understanding of the individual simpler decision-choice problems provides us with some insight to deal with the complex collective decision-choice situations that reside in the individual-community duality in continuum. The problem of the individual and collective decision-choice actions must thus be referenced to the social vision, national interest, social goals and objectives that the society is attempting to accomplish through the individual and collective socio-economic decision-choice activities.

The understanding of the problems of the formation of the social goals, objectives, national interest and social vision is the fundamental cognition of the political economy and the evolving social transformational dynamics. It is also the foundational boundaries for the study of the individual and collective decision-choice actions in a legally established social decision-choice space that contains individual and collective decision-choice bundles. The decision-choice space is not as free as is usually presented in economic theory, the theory of social choice, the theory of socio-political choices and the theory of collective action. The general decision-choice space from which an individual and the collective may select their bundles from is a variable that is sensitive to Lego-political conditions. Given the political and legal structure and their institutions, the decision-choice bundles are constrained by economic conditions. The individual and collective decision-choice problems are politico-legally restricted to be consistent with the social goals, objectives, national interest and social vision.

Let us reflect on the problem of goal-objective formation as a collective decision-choice action. The study and analysis of the micro-macro decision-choice actions may begin by assuming the existence of the social goals, objectives, national interest and social vision that are protected by the political and legal structures. In this respect, the problems of the political economy of the individual and collective decision-choice actions are forced into a straightjacket of efficient allocations within the politico-legal conditions where the national interest, social vision and the social goals and objectives are not clearly stated. Alternatively, the study and analysis of the micro-macro decision-choice actions may also begin by assuming the internal micro-macro decision-choice actions and resource limitations, and then concentrate on the analysis of how the social goals and objectives, national interest and social vision are established to define the legal boundaries of the decision-choice space that can generate the bundles available to decision-choice agents. In this respect, the decision-choice problems of the political economy involve the formation of the national interest, social vision and the social goals and objectives. The former is the tradition in the economic theory that is devoted to the study of the internal decision-choice actions within the politico-legal protective belts. The current monograph is on the latter, to examine the formation of the decision-choice to establish the protective belts through the formation of the national interest, social vision and the social goals and objectives to achieve the national interest and the social vision. The complete theory on the political economy must combine both for understanding, clarity and policy choices for any given national resource endowment.

1.2 Reflections on the Problem of Social Goal-Objective Formation

This chapter deals, in part, with a theoretical investigation into the problems of the formation of social goals and objectives of a society where the decision-choice structure is organized on the basis of general principles of democratic decision-choice process. It is devoted to stating and presenting the structure of the problem in social goal-objective formation under cost-benefit rationality. The decision-choice problem of the formation of social goals, objectives and national interest is presented as a collective decision-choice one. It is argued that every society faces two sets of decision-choice items in the social decision space. These two sets are composed of the set whose elements are individual-specific and another set whose elements are collective in nature. By explicating and distinguishing the two sets, we concentrate on the structural problem of the analysis of the set of collective decision-choice items. The understanding of the nature of the problem, and its relevance to the applications of cost-benefit analysis in the resolution of individual conflicts in collective decision-choice space, is important to the understanding of efficient utilization of democratic rules in the social formation, management and governance.

We shall consider a general decision-choice space $\mathbb{D}$ of a nation, and let $\mathbb{I}$ be the set of decision-choice items that an individual can act on in the social set up. Let $\mathbb{C}$ be the collection of decision-choice items that may be acted upon by the social collective. The space of national decision is such that, $\mathbb{D} = \mathbb{I} \cup \mathbb{C}$ with $\mathbb{I} \cap \mathbb{C} \neq \emptyset$. The structure of the decision space is such that there are decision-choice items that are strictly private, and are some that are strictly collective and there are some decision-choice elements that can be acted upon by both the individual and the collective. The set of individual decision-choice specific items may be specified as a relative complement (set difference) $\mathbb{I}^{c}$ and that of collective-specific may also be specified as relative complement $\mathbb{C}^{c}$ between $\mathbb{C}$ and $\mathbb{I}$.

$$\mathbb{I}^{c} = \left(\mathbb{I} \backslash \mathbb{C}\right) = \left\{x \mid x \in \mathbb{I} \ \& \ x \notin \mathbb{C}\right\} \tag{1.1}$$

$$\mathbb{C}^{c} = \left(\mathbb{C} \backslash \mathbb{I}\right) = \left\{x \mid x \in \mathbb{C} \ \& \ x \notin \mathbb{I}\right\} \tag{1.2}$$

$$(\mathbf{I} \backslash \mathbf{C}) \cap \mathbf{I} = \emptyset \text{ and } (\mathbf{C} \backslash \mathbf{I}) \cap \mathbf{I} = \emptyset \tag{1.3}$$

The set, $\mathbf{D}$ is said to be legally established and socially protected and $(x \in \mathbf{D})$ is said to be socially and legally admissible choice element. The element $(y \notin \mathbf{D})$ is such that $(y \in \mathbf{D}')$ and hence y is an illegal decision-choice element which cannot be acted upon by the decision-choice agents. The structure is presented in terms of a Venn diagram Figure 1.1. It may be added that the size of $\mathbf{I}^c$ and $\mathbf{C}^c$ will depend on the evolving legal structure that is created from the dynamics of the political structure. The decision-choice activities in the set $\mathbf{I}^c$ will be called the *individual decision-choice problem* while those in $\mathbf{C}^c$ will be called *collective decision-choice problem* in the political economy. All these are subject to cost-benefit balances in arriving at any decision-choice action. Figure 1.1 may be translated into a pyramidal logical representation of Figure 1.2 that will allow us to associate the elements with the analytical tools of duality, polarity, opposites, continuum, costs and benefits.

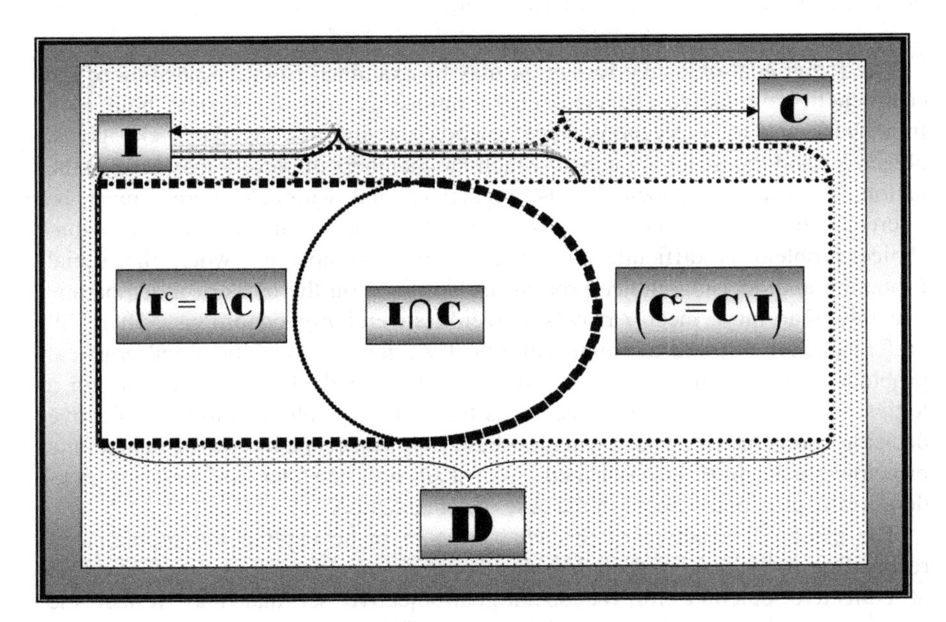

Fig. 1.1 The cognitive Geometry of National Decision-Choice Space Showing the three Relational Set of Individual and the Collective

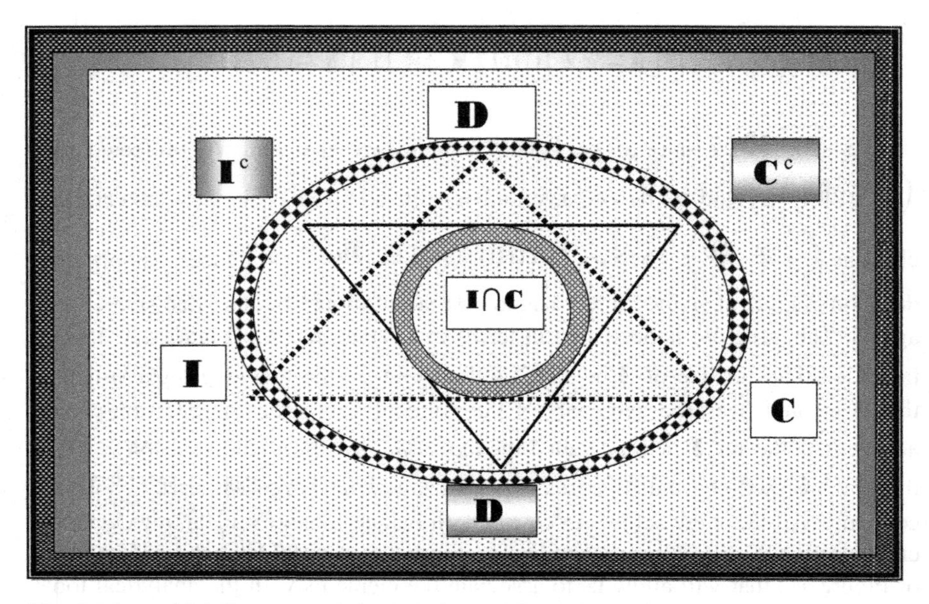

Fig. 1.2 Pyramidal Geometry of the Relations of Sets of Decision-choice Elements of the Private and Public Sector Decision-choice Systems

The social goal-objective formation may be examined from the position of the cost-benefit rationality. The cost-benefit rationality may be viewed both as a framework and as a decision-choice toolbox for constructing the social goals, objectives and the national interest, which the members of society may hold in the nation. As such, the social goals, objectives and national interest must be referenced to the cost-benefit analysis of social decisions. The collective decision-choice problem is difficult, but acquires extra complexity when the social formation and decision-choice process are designed on the basis of a democratic structure that must meet Arrow's citizen's sovereignty conditions, where all preferences are allowed to count with equal weight. The analytical and practical problem is to examine the nature of social goals and objectives and design a logical process that will bring into focus the nature of solutions to the collective decision problems of social formation based on an acceptable set of democratic principles that must be indicated. We shall not consider the case of non-democratic principles except in the case of social action.

The essential objective of this chapter is to make explicit the problems of reconciling individual conflicts in collective decision-choice space so as to arrive at collective outcome for the social-goal-objective set that may reflect the preferences of the members of the society. This social goal-objective set in addition to national interest and social vision will define the protective boundaries of decision-choice elements. The individual conflicts of preferences in the collective decision space are, reveal themselves as conflicts in the distribution of costs and benefits in the social set as decision-choice actions are undertaking to change socio-economic states. The analytical force behind the reconciliation of the individual conflicts in the collective decision space is the cost-benefit rationality. It will turn out that the reconciliation on the basis of citizens' sovereignty that every citizen's preference counts cannot be based on a majoritarian democratic decision-choice system [R7.14, pp. 139-140]. The resolution on the basis of citizens' sovereignty where every citizen's preference counts will require the introduction of weighted preferences, where such weights may be constructed from a number of social characteristics such as education, income, social awareness, degree of political participation, degree of community involvement, preference intensities and others. The corresponding socio-political institution for the collective decision-choice actions will be called *weighted-consensus democratic decision-choice system.*

Let us discuss the dimensions of social goals and objectives of a nation in relation to cost-benefit analysis and rationality. These dimensions may be related to national aspirations and interests. As such, we shall examine how national or collective goals may be formed to reflect democratic ideals in the decision-choice space in terms of participatory democracy. The goal of this discussion translates into advancing a *theory of social goal-objective formation* based on defined democratic principles of social decision making on the basic framework of cost-benefit balances and corresponding rationality. The essential objective of this theoretical framework is to investigate how cost-benefit analysis may be developed to support an increasing chance of realizing the collective or social goals without violence, and also to examine how the *fuzzy paradigm* may help to deal with ambiguities and complexities in the collective decision-choice space, where democratic participation is assumed under a given information structure [R7.15][R7.16]. The theory, therefore, may be seen as a theory of the political economy to establish the boundaries of the legally allowable decision-choice space before we can deal with the economically feasible regions of individual and collective decision-choice actions. Within the legal space of decision-choice actions, we have production of goods and services, allocation of resources and efforts, and the distribution of goods and services. Given the national resource endowment, the legally allowable decision-choice space fixes the potentially maximum n-dimensional commodity set, where such dimensionality is sensitive to the conditions of the political and legal structures viewed in terms of power distributions. The aggregate of the n-dimensional commodity set is the potentially maximum real income that is politico-legally producible. The corresponding maximum welfare of the society must take account of the phenomenon of the income distribution.

1.3 Social Goals, Objectives, National Interest and Social Vision

The collective or social goals and objectives generally drive the types of possible potential cost and benefit outcomes, and the manner in which they are distributed over the members of the society. A question arises as to whether the cost-benefit analysis could be conceptually and computationally expanded in a general framework to reduce possible conflicts among individual decision-choice rationalities, the national goals and objectives (that is, *the social goal-objective set*) on the basis of potential costs and benefits to a society, or an organization in support of national interest. In other words, can the decision-choice rationality be applied in designing social goals and objectives of a nation on the basis of individual preferences and principles of either consensus or majoritarian democracy?

Whatever meaning one attaches to the concepts of benefits and costs, or whatever measures one uses to express quantitatively defined benefit and cost concepts, the cost-benefit analysis is undertaken to rank and select those elements of the decision-choice set that optimize the general welfare positions of the society through project selections. The elements must support the attainment of collective goals and objectives of the society to the extent to which the goal-objective set reflects the collective will and interest of the society as judged by the preferences of the members. The welfare position of any nation depends on the set of actualized national goals and objectives that are used in support of its national interest and social vision. Conceptually, there is a family of social goal-objective sets that is available for selection by any nation. To each goal-objective set that may be selected from the family and for each element, there would be a corresponding cost-benefit configuration that would generate and support a particular social welfare position that satisfies the social interest of the nation. The individual goal-objective set, vision and the corresponding cost-benefit configuration must be defined within the politico-legal decision-choice space. This cost-benefit configuration will vary as the elements in the social goal-objective set alter to reflect a new set in support of a new national interest that must reflect the social collective interest, social vision and the corresponding individual preferences. The politico-legal decision-choice space is dynamic that evolves in quantity and quality to reflect new social conditions of the social set up and corresponding politico-legal order.

For example, the social goal of building an empire that is globally dominating will generate a cost-benefit configuration completely different from a corresponding cost-benefit configuration associated with a social goal of creating the best mentally and physically healthy population of a nation. Again, the social goal of either poverty reduction, or the pursuance of distributive justice will generate cost and benefit characteristics that may be completely different from the national goal of constructing and developing a war machine for conquering other nations in support of imperial aspirations. The differences may be seen in terms of resource allocation composition, and production-consumption compositions. Different sets of national goals and objectives will generate different projects for implementation, different allocations, and the projects will generate different outputs, and different cost-benefit configurations that will determine the potential social welfare of its members. The point here is simply, national goals, objectives and interests along with social vision affect projects that would be included in the set of project alternatives (*project-choice set*) for domestic and non-domestic implementations.

The social project-choice set is induced by the goal-objective set. The social project-choice set generates the potential social cost-benefit sets and the distributional impacts over the members of the society. The cost-benefit configurations, their distributional impacts, and their actual payment distributions together, define the welfare configuration of the social organization at any given point of time. The decision-choice elements and the production-consumption activities within the established politico-legal decision-choice space are carried on to support the attainment of the elements in the social goal-objective set, national interest and the social vision. As either the national interest or the social vision alters, so also do the elements in supporting the social goal-objective set and the corresponding project-choice set with the associated cost-benefit configurations that establish a new social welfare position of the social setup. The social goal-objective set defines the protective belts within which the social project-choice set evolves and takes shape. The goal-objective set is span by the national interest and the social vision.

As discussed, the implemented elements from the social goal-objective set define not only the welfare position of a nation, but also the path of her national history to any point in time. The conceived or determined social goal-objective set, not yet implemented, in addition to the held national interest, will have preponderating effects on the potential direction of the future history of the nation, as well as exert pressures on potential costs and benefits that will determine the cost-benefit output and the national welfare. The distribution of the total social cost-benefit output will define the social welfare path of the society and its individual members along the path of the evolving social history. It is on the basis of the social cost-benefit distribution, that *distributive justice* acquires economic, legal and political importance through the social budgetary decision-choice processes. It is also on the basis of the social cost-benefit distribution that social inequalities are manifested in the social set up.

Since social goals and objectives affect current costs and benefits and will affect the potential cost and benefit characteristics as well as their budgetary distributive impacts over the population; and since such potential distributive impacts will affect the welfare of individual members on the trajectory of the national history, it makes logical sense to ask a question as to how national goals, objectives and interest as well as social vision are formed or should be formed. Alternatively, the question may be stated as to whether the social vision, national goals, objectives and interest are dictatorially imposed on the members of the society or are formed in accordance with acceptable democratic rules with full citizen participation, and if they are not should they? It is possible for the social goal-objective set and the national interest to be formed in support of a particular class interest instead of the national interest. An important question must also be kept in mind in thinking about the problem of social goal-objective formation. The important question involves the relationship between the social goal-objective set and either national interest or social vision. Does the diameter of the national interest or social vision spin the circumference of the social goal-objective set, or the circumference of the social goal-objective set determine the diameter of either the national interest or the social vision? We have advanced the notion that the diameter of the set of the social vision, the national interest and the social goal-objective configuration spins the space of the resource-allocation and consumption-production decisions in the social set-up. In fact, it determines the production distribution, the consumption-product set and income distribution in any social set-up. It also determines the commodity composition composed of input composition and output composition for domestic and international trading systems.

The questions and answers surrounding the formation process of national goals and objectives, and the defined national interest and social vision are extremely important in social cost-benefit analysis of social projects for the selection of the elements of the social project set that must relate to the social goal-object set, in terms of the development of the social decision-choice space, and the possible path of national progress. The same goal-objective formation is also important to the stability of the social system on its dynamic path, because of the potential and actual cost-benefit shifting that will affect the nature of both the production-consumption activities and income distribution. The questions raised and the answers provided also relate to the understanding of the formation of the domestic policy set, and the paths of national diplomatic and non-diplomatic behaviors of nations in the international socio-economic space, relative to peace, conflict and war in the international political and power spaces. The answers to the questions of formation of goal-objective set will help to shape the direction of cost-benefit analysis of social decisions on laws, projects and policies.

For example, the objectives, goals and interests of occupying force of an imperial power and the occupied, may be diametrically opposed to one another, irrespective of the claimed benevolence and loving intentions of the imperial or occupying force. Similarly, the goal-objective set and the interest of donor countries may completely contradict the objectives of the social progress of the receiving country, in terms of choice of projects and decisions that affect the national history of the receiving country. In fact, this is critical in examining the validities of advice provided by external consultants from international agencies, as well as the whole system of socio-economic aid to countries, no matter how are the strengths of the degree of their expertness, and the generosity of the donors. Both socio-economic aid and technical consultancies must be seen in terms of the general cost-benefit space in relation to the decision to give (willingness to give), the decision to accept (willingness to accept) and at what information structure. They work within political structures with their preferences and mandates of the international institutions that they work for and within the framework set by the social goal-objective set, national interest and social vision of the countries they belong to and how these countries influence the international institutions in the global space.

Because of the existence of actual and potential conflicts due to inconsistencies in the individual and group preferences surrounding national goals, objectives and interests, and social vision, the formation of the social goal-objective set must be constructed from a broad spectrum of democratic participation in the process of setting up a hierarchy of preferences over national goals, objectives and interest around which strategic and tactical choices are to be made. The inconsistencies between the individual goal-objective set and the social goal-objective set generate the decision-choice conflict zones that require continual resolutions in the political economy for transformation and social stability. The importance of democratic participation by the individuals in setting the national goals and objectives, and hence to determine the ruling national interests and social vision, rests on the fundamental idea that individual goals and objectives and the associated costs and benefits are internalized in accordance with subjective desires and feelings that influence individual preferences. This fundamental idea is complemented by the notion that social goals and objectives are primarily externalized by the individuals in the *conflict zone* of individual and social preferences.

What is good for the collective may not be good for the individual as seen in terms of individual and collective cost-benefit balances. The assessments of social benefits and costs in the conflict zone of preferences must be in relation to the goals and objectives of the decision agents, whether these goals and objectives are social or individual or both. It is here, that one may examine the problems and difficulties encountered by Arrow when he tried to reconcile the conflicting individual preferences in the collective decision-making under democratic decision-choice process with equal weights leading to the Arrow's impossibility theorem [R2.5] [R2.125] [R7.16]. The fact is that, the only democratic decision-making process with equal weights on individual preferences, that is not dictatorial, is unanimity. A simple majority rule is a degree of dictatorship that is greater than half. The cut-off of the degree of dictatorship is the *degree of democratic tolerance* that must be specified within the democratic rules of the collective decision-making process.

Generally, we may speak of a social goal-objective set that is formed under some defined collective selection rules. Let such a set be $\Omega = \{\omega_i \mid i \in \mathbb{J}\} \subset \mathbf{C}$, where $\mathbb{J}$ is an index set for all possible social goals and objectives. It may immediately be observed that all these goals and objectives may not be within reach either for the individual or for society due to resource and technological constraints. Hence, we may speak of a feasible set of social goals and objectives, $\mathbb{A}$, that is contained in the general goal-objective set $\mathbb{A} \subset \Omega$. The feasible objective set, therefore, summarizes the social goals and objectives of the decision maker, or the beneficiary of the decision, that are within reach. When a decision is under consideration, the relevant objectives may be both economic and noneconomic, each of which has a potential benefit to either the decision agent or the beneficiary of the decision. The potential benefit is seen in terms of a set of *benefit characteristics* $\mathbb{B}$, which is associated with each element in the goal-objective set. Such a benefit characteristic set is supported by a corresponding *cost characteristic set* $\mathbb{C}$. Thus in the decision-choice space, the total characteristic set is $\mathbb{G}$, where $\mathbb{G} = \mathbb{C} \bigcup \mathbb{B}$. It may be pointed out that the benefit characteristics, just like cost characteristics of decision-choice action, are purely potential when the decision-choice action is under consideration and that they must be actualized through the practice of decision-choice and implementation actions. Since our main concern is on the social goal-objective formation under the principles of cost-benefit rationality in examining the potential social choice or programs to support the social goal-objective set as generated by the national interest and social vision, we shall examine the nature and structure of social goals and objectives, and their interrelationships, given a defined national interest, social vision and the specified nature of the society in question. Let us turn our attention to the structure and form of social goals and objectives.

1.4 A Morphology of Social Objectives and Goals

There are two important controversial decisions that encompass nation building, its development, progress and social stability. One of such controversial areas of analysis in social decision-choice space is the definition of collective national interest, social vision and the relevant set of social goals and objectives in support of the collective national interest and the social vision. The controversy is found in the process through which the national interest and social vision are established and the *admissible sets* of social objectives are constructed from the general objectives and goals of the individual members of the society in accord with their will, desires and beliefs that may be aggregated to form a collective relation to define the national aspirations and goals. Since such wills, desires and beliefs of the members of the social community may conflict with one another regarding what is good or relevant to the society, how does one reconcile the individual conflicts in the preferences of goals and objectives in the process of arriving at the admissible set of social objectives? Some of the methods of reconciliation are by consensus, voting, imposition or negotiation by an agreed upon method of selection, inclusion and some designed decision rules. It is here that the subject matter of political economy enters into the nature of decisions in understanding the structure and form of democratic social organizations of decision-choice and implementation systems.

Nonetheless, the set of objectives and goals that a society may follow is usually derived from the fundamental *ethical postulates* of the relational structure within the individual-collective duality, which underlies the socio-economic organization of production and distribution. Additionally, the derivation of the goal-objective set is influenced and supported by the collective aspiration or the social aspiration and vision from within the social set-up. Such ethical postulates either define the social ideology or are embedded in the ideology of the social organism. The ethical postulates, therefore, form part of the protective belt of the individual and collective decision-choice activities. They define the nature of the democratic norms of the evolving social decision-choice behavior, the degrees of freedom accorded to the individual in the collective decision-choice process, in addition to the practice of the principle of justice that is socially tolerated. In other words, the ethical postulates define freedom and justice in the individual decision-choice space constrained by the collective preferences and the degree of enforcement and adherence to the established democratic rules within the circumference of the national interest and social vision. The ethical postulates also impose limitations on the admissible set of social objectives in the nation and, in fact, on the admissible set of goals and objectives that an individual or a collective may even follow in the individual decision-choice space [R2.45], [R2.105], [R2.125]. As such, any serious analysis of formation of social goals, objectives and national interest must view them in the context of a particular social organization and its state of development as well as the dominant social ideology that has taken hold of the society in terms of the relationship between the individual and the community.

Different social organisms, therefore, will have different admissible social objectives even given identical preferences and national interest. It is these ethical postulates and the corresponding dominant ideology that define national differences of the decision-choice spaces, the psychology of choice, the quality of national life and the paths of social history. The social goals and objectives are abstracted from the economic and the political structures of the society in accordance with the ethical postulates that give rise to the general decision-choice framework. The legal structure is used to constrain behavior toward the fulfillment of the social goals, objectives and national interest, within the confines of the social ideology given the distribution of individual preferences and the national interest. Nonetheless, it is possible to generate a shopping list of social objectives and goals that are fairly general to most currently existing socio-economic systems (see also [R2.65] [R2.117]).

1.4.1 A Shopping List of Social Objectives and Their Mutual Relations

The social goals and objectives that nations may follow may be listed under two headings of economic and non-economic elements as they are shown in Tables 1.1 and 1.2. The economic objectives, which constitute the material life of nations, may come to determine some of the non-economic objectives and vice versa. Their mutual dependence and determination will depend on the internal and external national interest and the explicit or implicit social vision. From Table 1.1, the economic objectives (e), (f) and (g) relate to social welfare improvements while (e), (h) and (i) relate to national economic power. From Table 1.2, the non-economic objectives (a) and (b) relate to sovereignty concerns, (c, d, e, f,) relate to gross national happiness, (g, h, I, and j) relate to national political power while (d, k and i) relate to social stability of the political economy.

Table 1.1 Economic Objectives

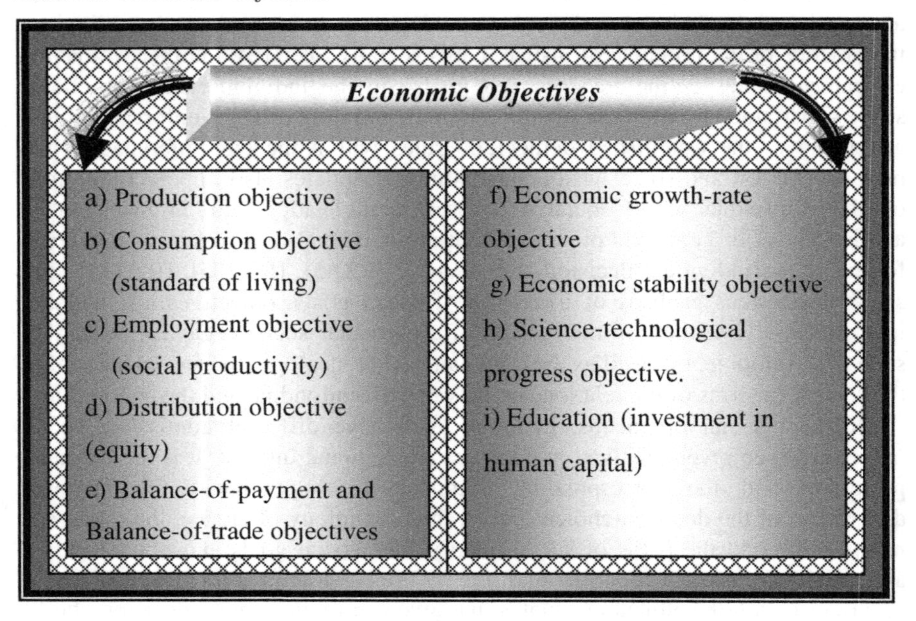

Table 1.2 Non-economic Objectives

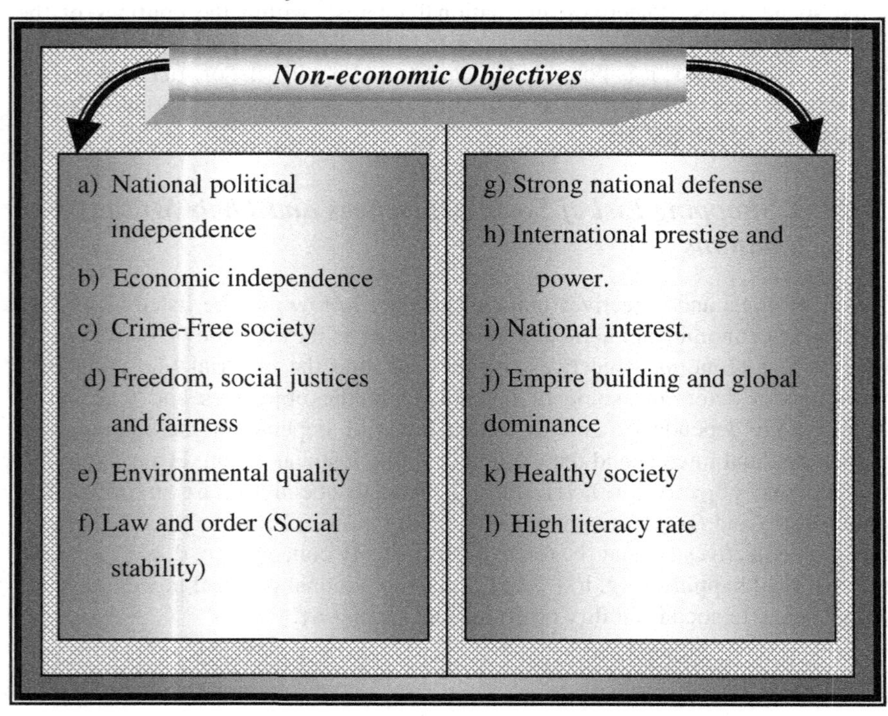

The set of social goals and objectives may include many others that meet the ideological requirements of the social organism. Among such extra objectives are individual freedom, collective freedom, national self-reliance, an altruistic society and many others qualitative characteristics. Given the set of all possible social objectives and the ideology of the social organism, it is possible to examine and define among the social goals and objectives, sets of compatible and incompatible relations in terms of their mutual support. The set of the possible relations are given in Table 1.3.

Table 1.3 Mutual Relations on the Goal-Objective Sets

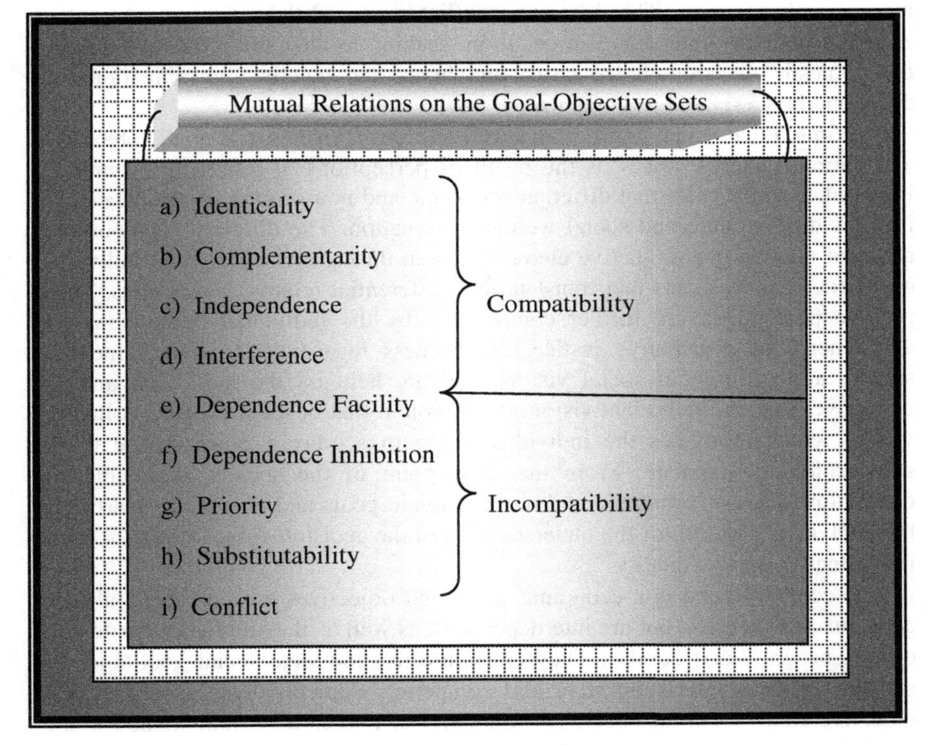

We shall consider the structure and form of each of these relations that may be established over the admissible set of social objectives given the ruling national interest. The driving force behind the relational analysis is simply to ascertain the interdependence of benefit and cost characteristics in the process of estimating benefits and costs that may be consequent of a potential social choice. Furthermore, the relational analysis allows us to introduce a game-theoretic and rent-seeking analysis of opposing preferences into the formation of the social objective set and how such a formation relates to the conflict zone of spaces of individual and social preferences, as well as at the conflict zone of space of individual national interests relative to domestic progress and international relations.

Two conflict zones will become clear in the analysis and synthesis of national goal-objective formation. They are the *domestic conflict zone* of the competing individual preferences over national interest, social vision and the supporting goal-objective set; and the *international conflict zone* in the competing inter-country interests and their supporting social goal-objective sets with potential international political instabilities and wars. These conflicts are generated by potential and actual cost-benefit configurations associated with the goal-objective sets in support of competing national interests which are wrapped in ideological belts for income and resources. The domestic conflict zone and the international conflict zone are mutually interactive in negation, seeking to alter the protective belts of the domestic decision-choice space and hence the legally allowable domestic decision-choice set and its contents.

The individual-social conflict zone of preferences over social goals, objectives and national interest arises as the result of perceptions of relatively differential importance of the roles that different economic and non-economic objectives play in individual welfare and social welfare in a nation. The differential importance attached to each goal-objective element is seen through its differential individual cost-benefit assessments that translate into differential relative perceptions. These relative perceptions are further complicated by the individual perceptions and assessments of distributive justice and fairness regarding the management of society and differential social visions that are held by the individual and the collective. It is the individual vision, in relation to that of the prevailing collective vision, which motivates the individual drive to acquire a power space in the socioeconomic organism. From the view point of the society as a unit, the distinction between economic and non-economic goals and objectives is artificial but analytically useful for the understanding of the decision-choice behavior in the individual-collective duality.

The economic and non-economic goals and objectives are not only linked in concepts and practice, but are interdependent, as well as they mutually create each other's progress and demise. They constitute the complete social goal-objective set that contains inter-supportive and competing elements. The economic goal-objective elements in the social-goal objective set, when achieved, create room to expand the non-economic goal-objective elements, which then make it possible to also improve on the economic goal-objective elements. The improvements of the social welfare position of the society and its socioeconomic development depend on the creative resolutions of the individual-collective decision-choice conflicts over the elements of the goal-objective set and the corresponding cost-benefit distribution regime. There are two dualities that create the conflict zone in the democratic decision-choice space. They are individual-collective duality in relational preferences and cost-benefit duality in the relational elements in the social goal-objective set. The relational structure may be presented in a pyramidal logic of conflicts of opposites and dualities as in Figure 1.3.

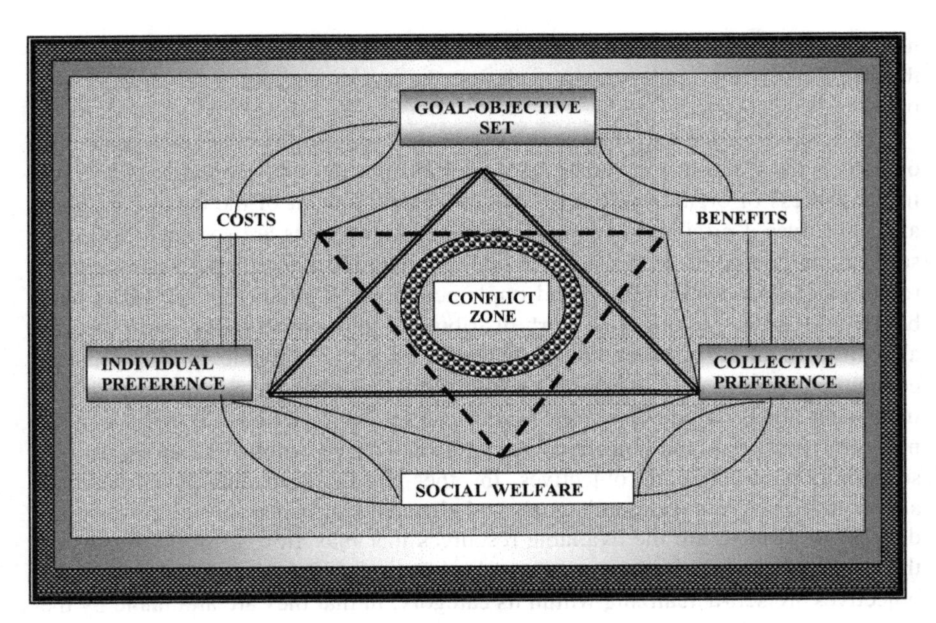

Fig. 1.3 Pyramidal Geometry of the Relation Structures of Dualities and the Conflict Zone of Democratic Decision-Choice Space of private and public Sectors as a System

In terms of social transformations, developments and analysis of social categories, the set of the economic goals and objectives constitutes the *primary category* of the social unit from which all social goals and objectives are transformed. In order words, the economic goal-objective set constitutes the collateral for the non-economic goal-objective set. The set of non-economic goals and objectives constitutes the derived category. When the basic needs of society are accomplished, then the non-economic goals and objectives, in some hierarchical order, may then be designed relative to the available resource constraints. The other objectives, over and above the basic economic objectives of the societal needs, become the vehicle for the construction, implementation and attainment of non-economic objectives. The implementation of non-economic goals and objectives requires resources. The resources must flow from the results of implementation and attainment of economic objectives of production. The non-economic objectives and goals are unattainable without the achievement of some specific economic objectives. However, the attainment of some economic objectives may be enhanced by the attainment of some non-economic objectives due to relational interdependencies. In other words, the economic and non-economic objectives exist as duality and mutually constraining. The results of the attainment of the economic objectives define the *attainment possibility surface* for, as well as the operational meaningfulness of the non-economic objective. The non-economic objectives exist as indirect economic objective in the essence of enhancing the space of economic objectives. Both the economic and non-economic goals and objectives define the *welfare possibility surface*, while the results of their implementation define the *social welfare level* at any time point.

The acknowledgement of the primacy of the category of economic goals and objectives does not underestimate the importance of the non-economic objectives in the process of organizing the society for the attainment of the economic goals and objectives. Certain non-economic goals and objectives are so much linked to some economic goals and objectives in some interdependent structures that they require simultaneous definition and implementation. A number of examples may be given. Fairness in effort or work distribution and compensation and resource allocation require simultaneous implementation with economic objectives. The economic objective of an increasing productivity and a non-economic objective of increasing levels of education and health cannot be separated. Furthermore, the non-economic goals and objectives cannot be actualized without the attainment of some important economic objectives. In other words, the means to the creation and attainment of non-economic goals and objectives, irrespective of how they are defined or formed, are the available resources that may flow from the results of the accomplished economic goals and objectives. The economic goals and objectives are self-actualizing within its category, in that they are attainable by the use of the available economic resources that may be the result of actualized economic goals or objectives from the internal structure of the social set-up. In the social complexity, the elements in the non-economic goal-objective set acquire economic characteristics during implementation.

The understanding of the relationship between economic and non-economic goals and objectives may be viewed from the fundamental notion that, in the bare essentials, there are three basic building structures of any society no matter what its level of development or size may be, measured in a specific way. The three building institutional blocks of the social organism are the economic, political and legal building structures. The economic goals and objectives are abstracted from the economic structure while the non-economic goals and objectives are constructed from the political and legal structures that respectively hold the decision-making power and rules of individual and collective socio-economic behaviors. The economic goals and objectives are the fundamental pillars, but their constructs and implementations are shaped and governed from the politico-legal structures. As presented, the economic and non-economic goals and objectives constitute a unified system in promoting and defining the path of national development and welfare where the past is crisp, the current is in the making and the future is fuzzy with configurations of cost-benefit distributions. Let us examine the structure of possible relationships from Table 1.3 and suggest a logical path that one may use in the construction of the admissible social goal-objective set. The relational structure of the institutions of the democratic decision-choice system, as it relates to relational dualities in Figure 1.3, is presented in Figure 1.4.

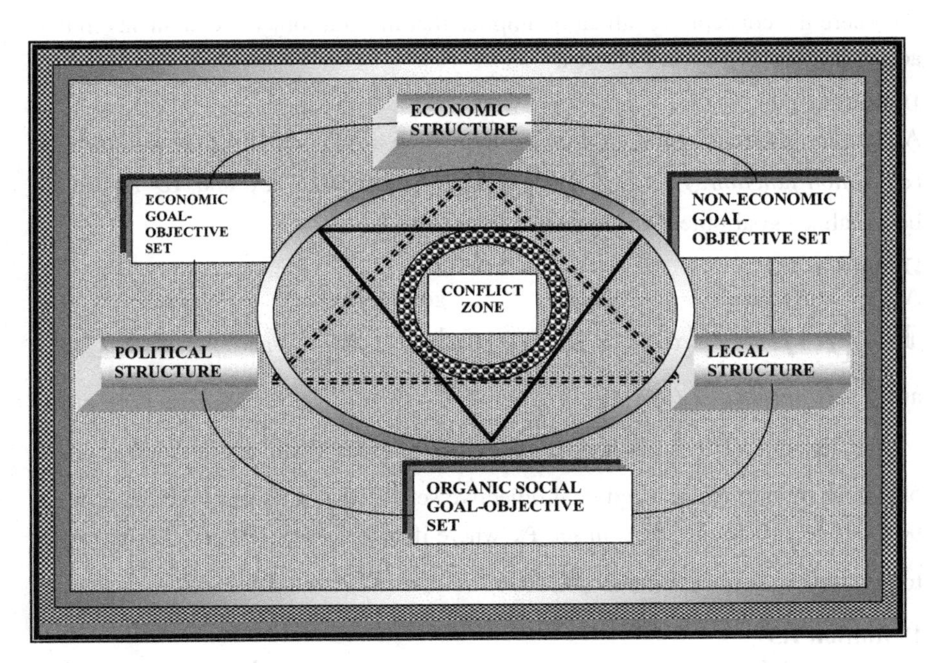

Fig. 1.4 Pyramidal Geometry of the Institutional Structures of Foundations and the Conflict Zone of Democratic Decision-Choice Space of private and public Sectors as a System

1.4.2 An Internal Relational Structure of the Goal-Objective Set

We shall consider the structures and forms of relations that may be established over either the admissible set or the set of all social goal-objective elements by compelling us to examine the benefit and cost characteristics. The examination is done in the process of identifying and measuring the benefit and cost characteristics that may be associated with any goal-objective element. The examination must further be extended to the problem of how the cost-benefit configurations tend to influence and shape the individual and collective preferences. The concept of cost-benefit configuration encompasses cost elements, benefit elements and the cost-benefit distribution among the applicable members in the implementation of the decision-choice actions. We shall begin our examination by considering the set of all possible social objectives, $\Omega = \{\omega_i \mid i \in \mathbb{J}\}$ where $\mathbb{J}$ is an index set of Ω, and a set $\mathbb{A} \subset \Omega$ is an admissible set that each nation may follow in accordance to its foundational structures of politics, law and economics. Here, the set $\mathbb{A} \subset \Omega$ is politico-legally established and constitutes a limitation on the individual and collective decision-choice action within it. Let $\mathbb{R} \subseteq \mathbb{A}$ be defined as a realized objective set where the concept of realization implies that the goal-objective elements in the admissible set $\mathbb{A}$ has been chosen.

Definition 1.1
An admissible transformation process defined on a set $\mathbb{A} \subset \Omega$ is said to be a *realization function*, $\mathcal{F}$, if $\mathcal{F}(\omega) \in \mathbb{R}$ whenever $\omega \in \mathbb{A}$ and ω is said to be a realized goal-objective element.

Definition 1.2
A realized social goal-objective set is $\mathbb{R} = \{\mu \mid \mu = \mathcal{F}(\omega) \Rightarrow \mu = \omega \text{ and } \omega \in \mathbb{A}\}$ and where $\mathbb{R} \subseteq \mathbb{A}$ and $\mathcal{F} : \mathbb{A} \to \mathbb{R}, \forall \omega \in \mathbb{A}$.

The function, $\mathcal{F}(\cdot)$, is a decision-choice mapping from the attainable goal-objective set into the realized goal-objective set $\mathbb{R}$. It is a choice function. Let us now define a relation, $\mathcal{R}$ in set $\mathbb{A}$ where if two elements $\omega_1, \omega_2 \in \mathbb{A}$ relate to each other, then we write $\omega_1 \mathcal{R} \omega_2$.

Definition 1.3
A relation $\mathcal{R}$ in $\mathbb{A}$ is said to be an *identicality relation*, $\mathbf{I}$, if $\omega_1, \omega_2 \in \mathbb{A}$, then there exist a realization function $\mathcal{F} : \mathbb{A} \to \mathbb{R}$ such that $\omega_1 \mathbf{I} \omega_2 \Leftrightarrow$

$\mu = \mathcal{F}(\omega_1) = \mathcal{F}(\omega_2)$, $\mu \in \mathbb{R}$. The goal-objective elements ω_1 and ω_2 are thus said to be identical in supporting the national interest even though $\omega_1 \neq \omega_2$.

Note:
It may be noted that two goal-objective elements are identical if they can be used to produce or accomplish the same social result, and hence only one is materially relevant In this respect, the minimum-cost goal-objective element may be selected for inclusion into the social goal-objective set. This helps to avoid the problem of duplication expense.

Definition 1.4
A relation $\mathcal{R}$ in $\mathbb{A}$ is said to be a *complementarity relation*, $\mathcal{C}$, if there exists a realization function $\mathcal{F}$ such that if $\omega_1, \omega_2 \in \mathbb{A}$ then

$\omega_1 \mathcal{C} \omega_2 \Leftrightarrow \mathcal{F}_{\omega_1\omega_2}(\omega_1\omega_2) > 0$ where $\mathcal{F}_{\omega_1\omega_2} = \frac{\partial^2 \mathcal{F}}{\partial \omega_1 \omega_2}$ and ω_1 and ω_2 are said to be complementary objectives in the sense that they are mutually augmenting and inter-supportive in accomplishing the national interest and social vision.

Note: A complementarity relation may be either benefit-enhancing or cost-reducing. If $\mathcal{F}_{\omega_1\omega_2}(\cdot) > 0$, they are said to be benefit-enhancing, and if $\mathcal{F}_{\omega_1\omega_2}(\cdot) < 0$, they are said to be cost-producing and hence their selection and implementation may be jointly undertaken since their cost-benefit configurations are mutually enforcing.

Definition 1.5
A relation $\mathcal{R}$ in $\mathbb{A}$ is said to be an *independence relation*, $\mathcal{J}$, if there exist a realization function $\mathcal{F}: \mathbb{A} \to \mathbb{R}$ such that if $\omega_1, \omega_2 \in \mathbb{A}$, then $\omega_1 \mathcal{J} \omega_2 \Leftrightarrow \mathcal{F}_{\omega_1\omega_2}(\omega_1, \omega_2) = 0$ and ω_1 and ω_2 are said to be independent objectives. They are, however, said to be dependent if $\mathcal{F}_{\omega_1\omega_2}(\bullet) \neq 0$; facility dependent if $\mathcal{F}_{\omega_1\omega_2}(\bullet) > 0$; and inhibition dependent if $\mathcal{F}_{\omega_1\omega_2}(\bullet) < 0$.

Note:
The decision-choice implication for an independent relation of goal-objective elements is that their benefit-cost configurations are independently generated, and hence the objectives are not substitutable in the support of the national interest and social vision.

Definition 1.6
A relation $\mathcal{R}$ in A is said to be an *indifference relation*, $\mathcal{I}$, if there exists a realization function $\mathcal{F}: \mathbb{A} \to \mathbb{R}$ and a ranking function, $\mathcal{U}$, defined on $\mathbb{R}$, such that if $\omega_1, \omega_2 \in \mathbb{A}$, then $\omega_1 \mathcal{I} \omega_2 \Leftrightarrow \mu_1 = \mathcal{F}(\omega_1), \mu_2 = \mathcal{F}(\omega_2)$ with $\mathcal{U}(\mu_1) = \mathcal{U}(\mu_2)$, and hence ω_1 and ω_2 are said to be *indifferent objectives* since any one of them can be used to achieve the same support for the national interest.

Note:
There is a difference between *identicality* and *indifference* relations. The indifference relation is equipped with realization and ranking functions, where the ranking function is the one that establishes the equality(that is, a double mapping) while the identicality relation is equipped with only a realization function for equality (a single mapping). The indifference relation yields the same utility in implementation while the identicality relation is of similarity in selection. All the relations are established in reference to the national interest and the social vision.

Definition 1.7

A relation $\mathcal{R}$ in $\mathbb{A}$ is said to be a *priority relation*, $\mathcal{P}$, if there exists a realization function $\mathcal{F}:\mathbb{A}\to\mathbb{R}$ and a ranking function, U, defined on $\mathbb{R}$, such that if $\omega_1,\omega_2\in\mathbb{A}$ then

$$\omega_i\mathcal{P}\omega_j \Leftrightarrow \mu_i=\mathcal{F}(\omega_i),\ \mu_j=\mathcal{F}(\omega_j)\ \text{with}\ U(\mu_i)>U(\mu_j)$$

and hence ω_i is a priority goal-objective element over ω_j, $\forall i,j\in\mathbb{J}, i\neq j$.

Note:

The implication of the priority relation is that, the elements in the realization set can be completely ranked in terms of the effectiveness in achieving the ruling national interest and social vision. It, thus, indicates the order in which programs must be constructed and resources must be assigned for the program-implementation.

Definition 1.8

A relation $\mathcal{R}$ in $\mathbb{A}$ is said to be a *substitutability relation*, $\mathfrak{S}$, if there exists a realization function $\mathcal{F}:\mathbb{A}\to\mathbb{R}$ such that if $\omega_i,\omega_j\in\mathbb{A};\ \forall i,j\in\mathbb{J}$ then

$\omega_i\mathfrak{S}\omega_j \Leftrightarrow \mathcal{F}(\omega_i)=\mu_i\in\mathbb{R}$ and $\mathcal{F}(\omega_j)=\mu_i\in\mathbb{R}$. The two objectives are then said to be substitutable for accomplishing the social vision and national interest in the social system.

Note:

If two goal-objective elements in the attainable set meet the *identicality* relation then they are substitutable in supporting the attainment of the national interest. In other words, if ${}_{\omega_i}\mathcal{I}_{\omega_j} \Leftrightarrow \omega_i\mathfrak{S}\omega_j, \forall\,\omega_i,\omega_j\in\mathbb{A},\ i\neq j$, then only one is needed to support the national interest agenda, since one is irrelevant.

Definition 1.9

A relation $\mathcal{R}$ in $\mathbb{A}$ is said to be a *conflict relation*, $\mathcal{K}$, if there is a realization function $\mathcal{F}:\mathbb{A}\to\mathbb{R}$ such that if $\omega_i,\omega_j\in\mathbb{A}$, $\forall i,j\in\mathbb{J}$, then

$\omega_i\mathcal{K}\omega_j \Leftrightarrow \mathcal{F}(\omega_i)=\mu_i \Rightarrow \nexists\mu_j\in\mathbb{R}$ such that $\mu_j=\mathcal{F}(\omega_j)$. The two objectives ω_i and ω_j, $i\neq j$ are then said to be *conflicting*.

Note:
Two goal-objective elements are conflicting in the decision-choice space if their mutual selection makes both of them difficult to support the national interest. They are incompatible in selection and implementation in the sense that they mutually enhance their costs and reduce their benefit contributions.

Definition 1.10

A relation $\mathcal{R}$ in $\mathbb{A}$ is said to be a *compatibility relation*, K, if there exists a realization function $\mathcal{F}:\mathbb{A}\rightarrow\mathbb{R}$ and a ranking function $\mathcal{U}$ defined on $\mathbb{R}$, such that for any $\omega_i,\omega_j\in\mathbb{A}$, at least one of the following conditions is satisfied:

1. $\omega_1 \mathcal{I} \omega_2 \Leftrightarrow \mu = \mathcal{F}(\omega_i) = \mathcal{F}(\omega_j)\ \mu\in\mathbb{R}\ ,\ i\neq j\in\mathbb{J}$ (identicality)
2. $\omega_i \mathcal{C} \omega_j \Leftrightarrow \mathcal{F}_{\omega_i\omega_j}(\bullet) > 0$ (complementarity)
3. $\omega_1 \mathcal{J} \omega_2 \Leftrightarrow \mathcal{F}_{\omega_i\omega_j}(\bullet) = 0$ (Independence)
4. $\left\{\begin{array}{l}\omega_i \mathcal{I} \omega_j \Leftrightarrow \mu_i = \mathcal{F}(\omega_i),\ \mu_j = \mathcal{F}(\omega_j)\\ \text{and}\ \mathcal{U}(\mu_i) = \mathcal{U}(\mu_j), \mu\in\mathbb{R}, i\neq j\in\mathbb{J}\end{array}\right\}$ (Indifference)

Definition 1.11

A relation, $\mathcal{R}$, in $\mathbb{A}$ is said to be an *incompatible relation*, $\not\mathsf{K}$, if there exists a realization function $\mathcal{F}:\mathbb{A}\rightarrow\mathbb{R}$ and a ranking function, $\mathcal{U}$ defined on $\mathbb{R}$ such that at least one of the following conditions holds, for $\omega_i,\omega_j\in\mathbb{A}$.

1. $\left\{\begin{array}{l}\omega_i \mathcal{P}_{\omega_j} \Leftrightarrow \mu_i = \mathcal{F}(\omega_i),\ \mu_j = \mathcal{F}(\omega_j)\\ \text{and}\ \mathcal{U}(\mu_i) \gtrless \mathcal{U}(\mu_j), \mu\in\mathbb{R};\ i,j\in\mathbb{J}\end{array}\right\}$ (Priority)
2. $\left\{\begin{array}{l}\omega_i \mathfrak{S} \omega_j \Leftrightarrow \mathcal{F}(\omega_i) = \mu_i\ \text{or}\ \mathcal{F}(\omega_j) = \mu_i\\ \text{but not both with}\ \mu\in\mathbb{R}\ \text{and}\ \forall i,j\in\mathbb{J}\end{array}\right\}$ (Non-substitutability)
3. $\left\{\begin{array}{l}\omega_i \mathcal{K} \omega_j \Leftrightarrow \mathcal{F}(\omega_i) = \mu_i \Rightarrow \nexists \mu_j\in\mathbb{R}\\ \ni \mu_j = \mathcal{F}(\omega_j),\ \forall i,j\in\mathbb{J}\end{array}\right\}$ (Conflict)

These relations define *conflict-resolution duality* in the goal-objective space for the support of the collective national interest and social vision as they have been agreed upon or established by the national political leadership. By such definitions, we observe that social goal-objective elements are compatible if they do not place limitations on their mutual realization, otherwise they are said to be incompatible. We may, thus, speak of the sets of compatible and incompatible social goal-objective elements in terms of the logic of opposites, dualities and continuum in reference to the acts of eliminations and transformations of the elements in the goal-objective set. An example of compatible social goal-objective elements is economic growth and increased employment. An example of incompatible social goal-objective elements may be high output growth with fossil-fuel based energy and excellent environmental quality or high rate of investment and high rate of consumption at the same time. All these relations are based on real cost-benefit imputations where all externalities must be accounted for. The most important thing to note is that an economic goal-objective element may be in conflict with some non-economic goal-objective elements, and that the space of the international political economy is also made up not only of differences in national goal-objective sets, but antagonistic national interests in such a way that the goal of global peace may not be attained in the international resource-commodity space.

We shall return to the problem of the relationships among national interests, resource-commodity needs and global peace-war space. We must keep in mind two types of conflicts. One type is the domestic conflicts produced by the domestic institutional arrangements for individual and collective decisions in the domestic economic, legal and political structures. The other type is the international conflicts produced by sovereignty institutional arrangements for the production and maintenance of the sovereignty decision-choice actions of countries based on the national interests and social visions of countries relative to the distribution of the global resource-commodity needs. The former involves the domestic political economy while the latter involves the international political economy. The two are inseparably linked as well as connected to the goal-objective formation for individual countries.

The analysis, selection and coordination of social goal-objective elements are less problematic if the elements of the set of objectives are compactable with one another. Theoretical and practical difficulties emerge, when one is confronted with incompatible social goal-objective elements for choice. When one is confronted with incompatible objectives, one attempts to resolve it. The resolution is usually done by imposing *priority* or *trade-off* relations on the elements in the set (in a sense, we set up a hierarchy of objectives) in such a way as to reflect the subjective preference of the society, government or its agent.

In the social practice of decision-choice actions, conflicts in goals and objectives are the rule in the cost-benefit analysis of social choices or projects toward the fulfillment of national interest and social vision. The nature and difficulty of conflict resolutions in the space of the social goals and objectives will depend on the specific problem in question. The social benefits, like the social costs, are always assessed in relation to the social goal-objective elements. Such goal-objective elements may be quantifiable, non-quantifiable or mixed as they relate to national aspirations, or implicitly defined interests. If they are all quantifiable, they may appear as a single value or multiple values. In some cases, it may be possible to construct a surrogate single-valued objective out of multi-objectives against which net benefits to the decision agent (or the group on behalf of which the decision is made) are assessed. The problem is, given all goals and objective that a nation may pursue, and given the differential individual preferences in a given social set-up, how do we construct an acceptable social goal-objective set that is sensitive to all members without violence and with some acceptable degree of fairness? Let us keep in mind, a lack of fairness generates an unjust social environment that creates conditions of domestic unrest, violence and social instability within the domestic institutional arrangements. This statement also holds at the international level. Global peace is unattainable when other nations violate the sovereign rights of other nations to create an unjust social environment.

For fairness and justice in the social set-up, the problem of the construction of the admissible set of the social goals and objectives may be viewed as a collective decision-choice problem. The solution may be obtained under a democratic decision-choice system of behavior where the nature of democratic decision-choice process is indicated, and by some agreement and acceptable rules of the social collectivity. Generally, the conditions of the collective decision-choice problem must be resolved before the cost- benefit analysis of alternatives may be undertaken relative to any goal-objective element. In all cases, we seek a socially admissible goal-objective set of the form:

$$\mathbb{A} = \left\{ \omega_{ij} \mid i \in \mathbb{I}, j \in \mathbb{J},\ \omega \in \Omega, \left(\sum_{i=1}^{i \le \#\mathbb{I}} 1 \Big/ \#\mathbb{I} \right) \in (0.5, 1] \right\},$$

where $\mathbb{J}$ is an index set of Ω, $\mathbb{I}$, an index set of the social collectivity, and a set $\mathbb{A} \subset \Omega$ is an admissible set. The construction of $\mathbb{A} \subset \Omega$ is the problem. This problem belongs to the class of general collective, group or social decision-choice problems. The condition $\left(\sum_{i=1}^{i \le \#\mathbb{I}} 1 \Big/ \#\mathbb{I} \right) \in (0.5, 1]$ is a decision weight which implies that at least half of the eligible decision makers agree to the goal-objective element in terms of a majoritarian principle supported by the principle of non-defective information, and where there is an individual freedom to exercise one's right to participate on the basis of preferences. It may be noted that this was also the problem of Arrow that we have previously referenced. The individual votes either in the electorate or in the decision-making core may or may not be weighted in accordance with the conditions of the decision-choice game in the collective-action space. In the solution of the conflict in the collective decision-choice space of the political economy with market institutional arrangements, in terms of what goods, how many units and in what quality are to be produced, the voting system is weighted by total units of money committed to a commodity. In this sense, the market system functions on the principle of weighted democracy where monetary weights are important determinant of democratic outcomes. This is taken to hold in economic, legal and political markets. This monetary weighted democracy is the implicit core of market solutions and privatization where the privileged money holders dominate the decision-choice environment. This monetary weighted democracy creates some kind of social system where criminal of political and corporate elites are rewarded with bonuses and journalists and reporters of the crimes are rewarded with jail-terms. In this way transparency in social decision becomes casualty of corruption.

1.5 National Interests and the Goal-Objective Formation

Throughout the discussions in this chapter, it is always indicated that the goal-objective set is formed to support a national interest or a system of national interests and social vision. The system of national interests may be divided into domestic and foreign interests. The set of domestic interests, given the political-economic system and the nature of governmental structure, consists of those elements that go to secure the internal aspirations of the nature of society which are conceived and desired. The corresponding foreign interest of the same society may be an acquisition of natural resources, broadly defined, from other lands in order to relax the domestic resource constraint and enhance domestic productive capacity. The foreign interest may also be viewed in terms of developing a national power to resist and repeal external aggression and also protect domestic resources for the benefit of the welfare of the domestic population. The point of emphasis is that the development of national power or national interest is always translated into positive and negative resource-commodity relations. The national power, directed to resource acquisition, reflects itself as imperial, colonial aspirations and military aggression in the global space. The national power developed to protect domestic resource and domestic population reflects itself as a resistant force, anti-imperialism and anti-neocolonialism.

It is analytically useful to conceptualize the national interest space $\mathbb{N}$ as a union of spaces of domestic and foreign interests. Let $\mathbb{A}_{\mathrm{D}} = \{\mathbb{d}_1, \mathbb{d}_2, \cdots, \mathbb{d}_i, \cdots, \mathbb{d}_m\}$ and $\mathbb{A}_{\mathrm{F}} = \{\mathbb{f}_1, \mathbb{f}_2, \cdots, \mathbb{f}_i, \cdots, \mathbb{f}_n\}$ be spaces of domestic and foreign interests respectively, where $\mathbb{N} = \mathbb{A}_{\mathrm{D}} \bigcup \mathbb{A}_{\mathrm{F}}$ and $\mathbb{A}_{\mathrm{D}} \bigcap \mathbb{A}_{\mathrm{F}}$ may not be empty. From these two spaces we may construct the set of national interests $\mathbb{N}$. Both the domestic and foreign interests may vary over nations and over time. The possible domestic and foreign interests may be listed as in Table 1.4.

Table 1.4 National Interest Space

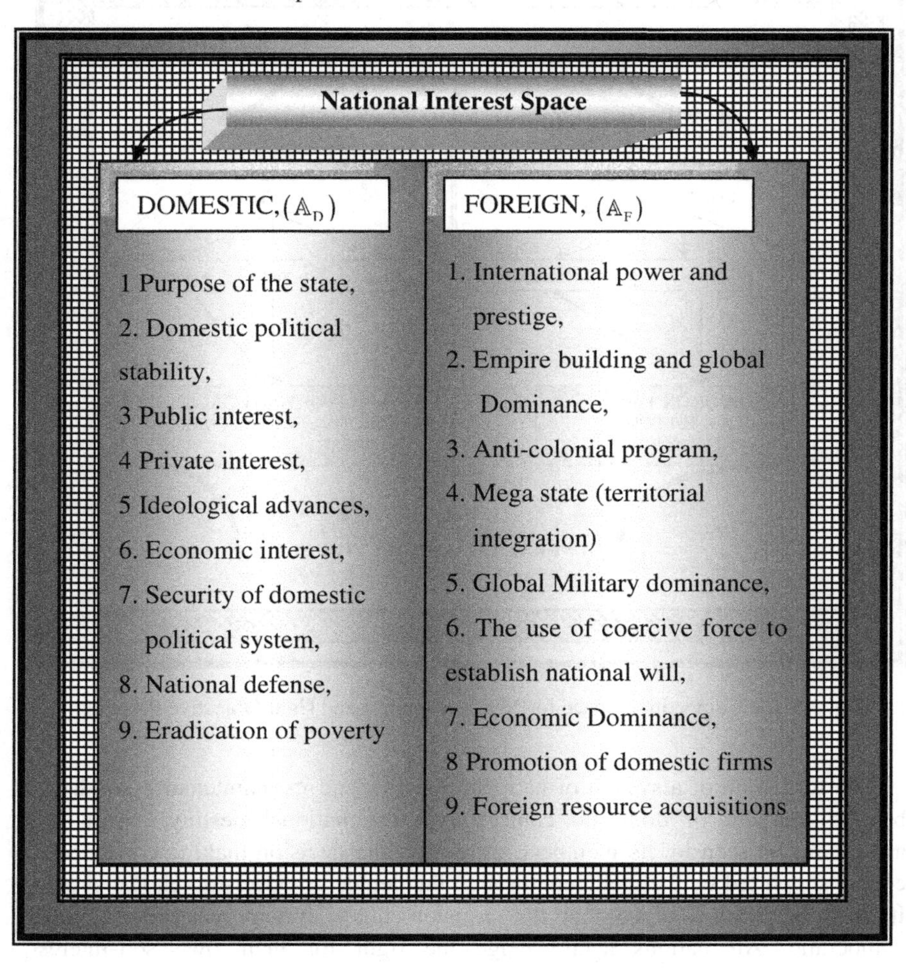

The domestic interests and the foreign interests may be inter-supportive or competitive to each other in terms of mutually negating. In some cases, trade-offs may be required to create stability between the set of national domestic interests and the set of national foreign interests. Both domestic and foreign interests may be subdivided depending on the structure of the domestic population, economic, political and legal structures, in addition to domestic commitments and foreign commitments. The general format leading to the establishment of national interests, may be viewed in terms of an integrated structure as in Figure 1.5 that shows a separation of the national interest space into domestic and foreign.

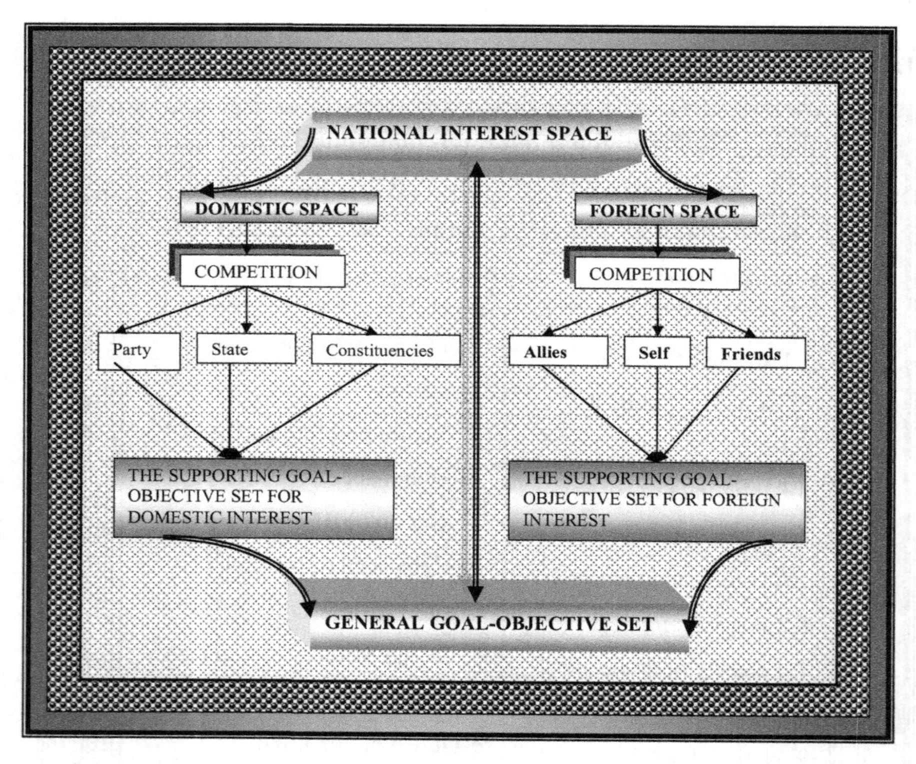

Fig. 1.5 The Competitive Space for National Interest and Goal-Objective Formation

The formation of a system of national interests and its maintenance are driven by the freedom of domestic control over the national destiny, vision and aspirations as seen by its members, especially the decision-making core, and of course supported by the members through the democratic decision-choice system. It must be acknowledged that the decision-choice system may also be non-democratic. All nations acknowledge the right for their own self-interest, self-determination, and self-vision, and yet some nations pursue an external policy of domination and tendencies toward imperial aspirations, and the establishment of direct or indirect occupation through a system of colonialism, or neocolonialism of military bases. This is the creation of domestic-foreign duality at the national level, which generates domestic internal conflicts through its effect on income distribution and resource allocation. The same domestic-foreign duality at the international level contains opposing forces, antagonistic relations and international tensions that are generated by the principle that freedom without law is chaotic, which is supported by another principle that freedom is not given. Freedom of nations, secured in the sovereign boundaries, is obtained through struggle and sacrifice. It is rested on pure national collective soul in order to achieve national mission and national interest, just as imperial tendencies rest on national collective soul, in order to achieve dominance, empire building and vision of invisibility.

The theory of national goal-objective formation may be developed with assumed institutions of democratic and non-democratic structures as part of the framework of the theory of democratic collective decision-choice system. The point of emphasis of the study of the problem of social goals-objective formation is that, it allows us to examine the problem of collective decision-choice and multi-objective decision-choice problems in a non-violent political space. It is essential to understand the social constraints imposed on the individual and group decision-choice actions in the three structures of economics, politics and law as they relate to the social goal-objective formation. The social goal-objective set, national interest and social vision define the parametric space of decision-choice actions in the social space where production, marketing and consumption of certain goods are illegal while others are legal. The parametric space, thus, spins the socially allowable decision-choice elements in addition to the structure of the institutional configuration through which social policies are implemented.

Finally, it is analytically helpful to relate the problem of the social goal-objective formation to the existing micro-macroeconomic decision-choice theories. The political economy is viewed as two subsystems of the superstructure collective decision-choice actions and the set of decision-choice substructures which are protected by the superstructure. The collective decision-choice superstructure houses the internal individual and collective decision-choice substructures and their contents. The decision-choice superstructure is made up of the social goal-objective set, national interest and social vision that constitute the institutionally organic constraint on the internal decision-choice behavior. The contents of the theory on the political economy, therefore, are viewed as two interdependent theoretical decision-choice subsystems. One is the theory of the behavior of the decision-choice superstructure and the other is the theoretical system of the internal decision-choice subsystems that are specific to individual and group decision-choice actions to accomplish the outcomes that are established by the decision-choice superstructure. The relational structure of these decision-choice theories for the understanding the behavior of the political economy is shown as a cognitive geometry in Figure 1.6. Given the outer decision-choice superstructure, an example of the decision-choice system to accomplish the elements in the social goal-objective set is geometrically structured in Figure 1.7 for a capitalist political economy. The decision-choice subsystem essentially constitutes the preoccupation of the modern economic theories and econometric analysis on the best resource allocation, commodity production and income distribution as revealed at the micro-level by neoclassical microeconomic theory and at the macro-level by the neoclassical and neo-Keynesian macroeconomic theoretical systems. Our current effort in directed toward the decision-choice superstructure, its formation and behavior under a democratic social formation that points to the setting of the outer protective belts composed of national interest, social vision and social goal-objective set.

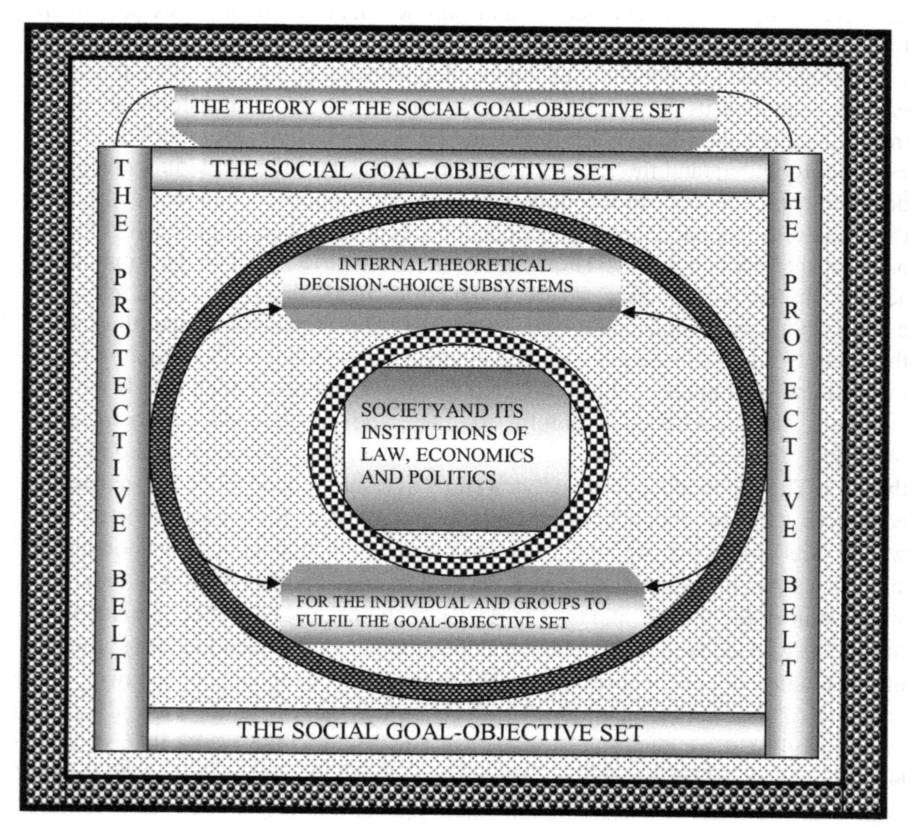

Fig. 1.6 A Complete Theoretical Structure of the Decision-Choice Systems of the Political economy

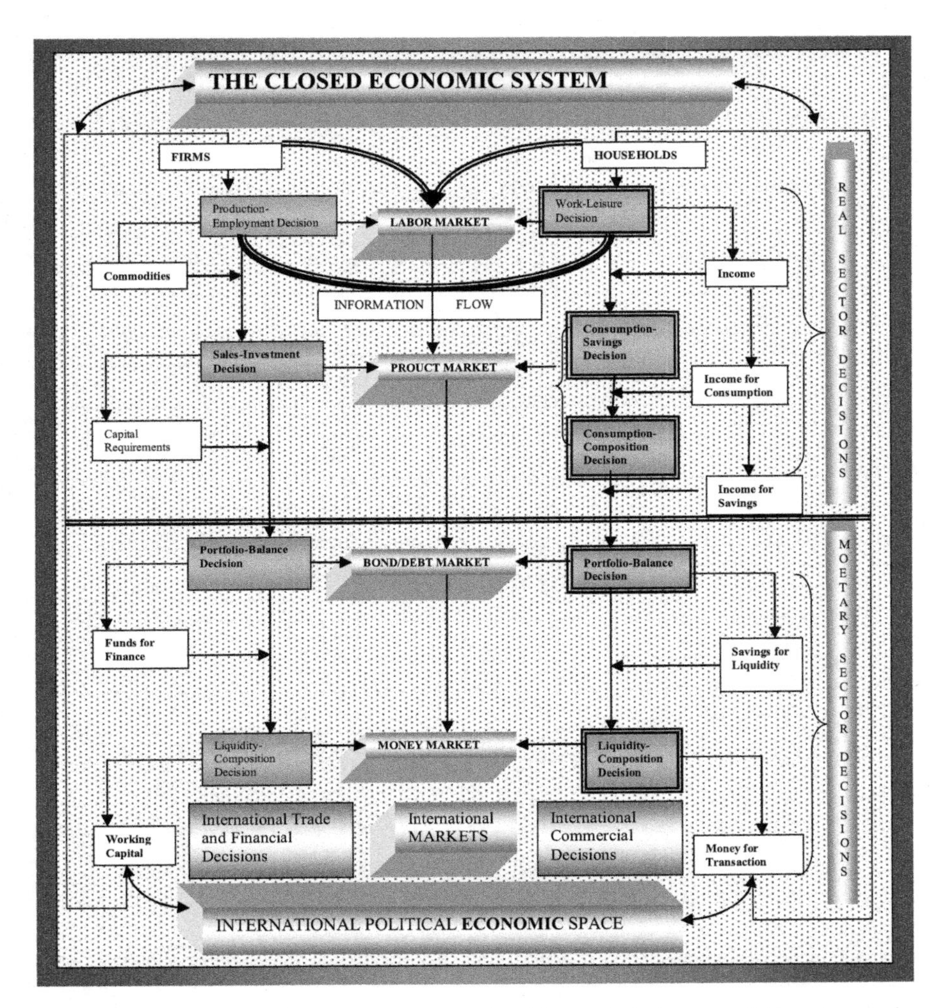

Fig. 1.7 An Example of an Organization of a Theoretical System of Decision-choice Behaviors to implement actions to Accomplish the Elements in the Social Goal-Objective Set under capitalist market mechanism.

Figure 1.7 shows the epistemic geometry for the development of the theoretical structure of microeconomic and macroeconomic theories with real-sector decision-choice structure and financial-sector decision-choice structure. The theories may be viewed from explanatory or prescriptive standpoint of knowledge production. The inputs of the theoretical constructs may be established by empirical information structure or axiomatic information structure with defined principle of information representation. The corpus of the theories is developed from the methodological principle of constructionism as a logical forward motion to process the defective and deceptive information structures and verified by methodological reductionism as a logical backward motion.

Chapter 2
A Theory of the Social Goal-Objective Formation under Democracy and Approximate Reasoning

The structure of the problem of social goal-objective formation, as a collective decision-choice one, was presented in Chapter 1 of this monograph. Such a decision-choice action may be carried on in many epistemic combinations in the social setup. The social decision-choice action involves the complete participation of the general public with all its members. The solution to the problem involves decision-choice action that requires information and information-processing capacity of the individuals and the collective in the various decision-choice calculi. The quality of the decision-choice system depends on the quality of the information structure, given the information processing capacity of the individual and the collective. The individual and collective decision-choice activities, in the political economy, involve the processing of *defective and deceptive information structures* that require the methods and logic of approximate reasoning to which we turn our attention. The nature of the defective information structure and the deceptive information structure will be discussed alongside the logic of the approximate reasoning, and its need in politico-economic thoughts in the decision-choice process.

2.1 An Introduction and the Nature of Approximate Reasoning

This Chapter is about the theory of social goal-objective formation under cost-benefit rationality, approximate reasoning and a democratic decision-choice system in support of an implicitly or explicitly defined social vision and national interest or system of national interests. It may also be viewed as a framework for determining an optimal social goal-objective set in support of the national interest (or a system of national interests). It is about the theory of social decision and choice in the socioeconomic decision-choice space, where social goals and objectives are considered as public goods that affect all members of the society. In other words, we are interested in the elements in the collective decision-choice space as a subspace of the national decision-choice space, where all members have preferences according to their cost-benefit imputations. The analytical work

K.K. Dompere, *Social Goal-Objective Formation, Democracy and National Interest*,
Studies in Systems, Decision and Control 4,
DOI: 10.1007/978-3-319-05173-4_2, © Springer International Publishing Switzerland 2014

advances a theory and sub-theory based on the logic of fuzzy decisions, mathematics and approximate reasoning as an essential part of the fuzzy paradigm. The decision-choice logic of the fuzzy paradigm is made up of the principles of opposites, dualities, continuum and contradictions, where resolutions tend to establish an optimal democratic decision-choice system in relation to the social goal-objective set. The problem, then, is to develop a framework to establish the optimal social goal-objective set on the basis of the democratic decision-choice system, given the national interest and social vision. The optimality is defined in terms of a reconciliation of conflicting individual preferences in the collective decision-choice space in solving the problem of what goal-objective set, among many, will best support the national interest and social vision of the country.

By posing the social goal-objective formation as a collective decision-choice problem in a fuzzy decision space regarding each objective, a logical opportunity is opened to us to employ the analytical toolbox of relevance-irrelevance duality where each opposite resides in the decision-choice continuum in relation to each element. The decision-choice continuum brings into the analytical work the concept of the penumbral region of decision-choice actions of the collective. By the use of the conflict in the relevance-irrelevance duality of each decision-choice element an optimal degree of *social relevance* is abstracted as a solution weight attached to each goal-objective element. The methods of fuzzy aggregations of *degree of relevance* and degree of *irrelevance* are reconciled to obtain a distribution of fuzzy optimal degrees of relevance as weights attached to each objective. The distribution of the optimal degrees of social relevance of the goal-objective elements is used to induce a social ranking on the elements of the social goal-objective set. From the optimal ranking the admissible social goal-objective set is constructed by the method of fix-level or an α-level cut of fuzzy decomposition algorithm to establish the optimal social goal-objective set. The use of the toolbox of fuzzy paradigm allows us to avoid the problem of the Arrow's paradox. It has been argued that the emergence of the Arrow's paradox is attributed to the formulation of the problem in an inappropriate mathematical space and the use of exact methods of reasoning which is inconsistent with qualitative disposition and subjective phenomenon that is governed by principles of continuum and penumbral regions of choice [R7.14] [R7.15].

The introduction of degrees of social relevance and irrelevance allows the introduction of the concept of duality into the fuzzy computable system, while the method of approximate reasoning or fuzzy logic allows the integration of subjectivity in judgment which is required in social decision-choice actions in the fuzzy continuum [R7.15] [R15.14]. Let us keep in mind that in the fuzzy paradigm, every word is a set with associated distribution of degrees of belonging to the meaning of the word's valuation. This is the *set condition* of belonging in fuzzy logical reasoning. This set condition holds for the concept of relevance and irrelevance in the valuation of the elements in the social goal-objective set. As the degree of irrelevance of any given decision-choice element decreases, the degree of its relevance increases. If the degree of irrelevance is one, then the degree of relevance is zero for a particular goal-objective element and vice versa. Every goal-objective element exists in relevant-irrelevant duality in continuum, relative to the national interest and social vision.

The nature of the problem of the social goal-objective formation involves the problem of how to reconcile two different but interdependent conflicts in social decision-choice space. One conflict involves the individual-collective duality in the social decision-making over many alternative social goals and objectives that the society may hold. Here, the individual conflicting preferences must be reconciled to create the social preferences that will produce stable social ordering. The problem lies simply in the decision-choice conflicts where some individuals may consider a goal-objective element as irrelevant, while others may consider the same goal objective element as relevant in order to achieve the national interest and social vision. The other conflict involves many relationships that the social goals and objectives may have with each other. Here, the conflicts in social goals and objectives must be reconciled to obtain optimal inter-supportive elements in the optimal social goal-objective set in support of national interest or a system of national interests no matter what they are. All these discussions bring us to the problem of collective decision-choice actions in both multi-criteria and multi-attribute decision-choice situations in order to construct an optimal admissible set of social goals and objectives. We shall deal with this problem through the logic of fuzzy decision-choice rationality and approximate reasoning for analysis and synthesis under specified information structures. For other approaches to multi-criteria and multi-attribute decision-choice problems see [R2.5][R2.66] [R2.128] [R2.132][R6.75], for the usefulness of this approach to cost-benefit rationality see [R3.1.23] [R3.1.24]and for discussions on the theory of fuzzy decisions see [R6] [R6.44] [R6.75] [R7.15]. The establishment of the national interest and social vision under the democratic decision-choice system may be treated likewise under given information structures. They may even be seen as sequential decision-choice actions, where the national interest and social vision are first established under the principle of democratic collective decision-choice system. The social goal-objective set, given the national interest and social vision, is established likewise.

The approach is to accept the existence of the space of goals and objectives that a society may follow in support of its national interest and social vision. The problem of social decision-choice action on the national interest or social vision is assumed to have been solved and hence taken as given and a constraint on the collective decision-choice actions. The individual preferences over these social goals and objectives, as accomplishing the national interest and social vision, are viewed and communicated in two sets of degrees of *relevance* and *irrelevance* as a dualistic problem in the decision-choice action of any of the social elements and any of the legal members in the social setup. The identity of each element is established by a set of characteristics that separates into collectively exhaustive and mutually non-exclusive subsets of relevance and irrelevance. These characteristics are assessed in terms of degrees of relevance or irrelevance in supporting the national interest and social vision. The degrees of relevance and irrelevance are viewed in terms of benefit-cost distribution as a problem of duality. The problem is not different from that of the price of an output or an interest rate on a coupon. A seller may consider the price of an output to be low since it is a benefit. The same seller, if turned into as a buyer, will consider the price to be too high since it is a cost. This is the *benefit-cost duality* of price in the sense that every price is both a benefit and a cost in the political economy.

The price cost-benefit duality gives rise to opposites of willingness to sell on one hand and willingness to buy on the other hand. It is here, that the concept of reservation price arises as an analytical tool. The same benefit-cost duality applies to interest rates when a person, as an investor, wants higher interest payment on his or her investment because it is a benefit, but as a borrower, the same person wants a lower interest rate because it is a cost even for the same person. The point of importance, in these discussions, is that every decision-choice element resides in a cost-benefit duality in the sense that every cost has a benefit support and every benefit has a cost support in the decision-choice space of the political economy. The distribution of the degrees of the inter-supportiveness of the cost-benefit configuration is defined in the fuzzy continuum, which gives rise to penumbral regions of decision-choice actions. Every goal-objective element may be seen by the decision-choice participants in terms of cost and benefit characteristics which are inseparably connected in a way that mutually defines their existence and identity within the duality. The cost-benefit characteristic sets define conditions of *relationality* in the goal-objective set that provides us with the quality and identity of any decision-choice element in supporting the national interest and social vision.

Just as the irrelevant-relevant duality exists in opposites and in continuum, so also the cost-benefit duality exists in opposites and in continuum. They are set in give-and-take relationships which mutually define their negations and transformations in the decision-choice space. The principle of continuum translates in the decision-choice space into the statement that every cost of a decision has a benefit support and every benefit has a cost support. The optimal decision is seen through approximations where the *bounded rationality* and the *classical rationality* are captured through the *fuzzy rationality* by the methods of fuzzy decision-choice rationality [R7.14]. The fuzzy decision-choice process allows us to formulate the problems as fuzzy set intersections. Fuzzy optimization allows us to abstract optimal solutions. The method of approximate reasoning allows the analysis and synthesis within the penumbral regions of decision-choice actions to overcome the problem of vagueness, ambiguities and subjectivity in the collective decision-choice space. Alternatively stated, the decision-choice problem is to define a fuzzy residual in the collective decision-choice space and then minimize it (see further discussions in [R7.0] [R7.30][R15.13], [R15.14]). This approach points to one important idea that at the core of constitutional social formation and governance is a proposition for the resolution of conflicts between the individual and the collective (individual-collective duality) on one hand, and between private property and its uses, and collective property and its uses. This proposition gives rise to the principle of property rights (public-private duality) in the decision-choice system of the social set up.

Under the fundamental principle of individualism, more emphasis is given to private-property rights in production and distribution to implement the elements in the social goal-objective set for accomplishing the national interest and the social vision within the continuum of the private-public duality. The application of the fundamental principle of individualism and private property rights calls into action individual self-interest based on individual net benefit realization. Under the fundamental principle of collectivity, however, more emphasis is given to the public-property rights in production and distribution to implement the elements in the social goal-objective set for accomplishing the national interest and the social vision within the continuum of the private-public duality. The application of the fundamental principle of collectivity and public property rights calls into action social or collective-self interest based on social net benefit realization. These two frameworks require different regulatory and legal regimes.

2.2 Constructing the Admissible Social Goal-Objective Set by the Method of Fuzzy Restriction: The Problem

An important difficulty emerges in the process of constructing the admissible set of social objectives and goals of a nation in a democratic decision-choice system. One may view the construction in two ways of an explanatory or prescriptive process. This difficulty is generated by the relationship between individual and social perceptions and valuations regarding social relevance of the elements that will support the national interest or a system of national interest. Some members of the society place differential weights on domestic interests and foreign interests, which together constitute the national interest, as a multi-criteria decision-choice problem given that there is sufficient non-deceptive *social information structure* defining the environment of their decision-choice activities. In this analytical framework, there is a postulate of information structure on the basis of which knowledge is abstracted, and decision-choice actions are acted upon either by the individual or by the collective. The postulate simply states that all information structures in support of decisions are *defective* in the sense that they contain the characteristics of vagueness and quantity limitations. The vagueness limitation represents the qualitative-limitative character of the information structure. The quantitative limitation represents the size-limitation of the information structure.

In democratic decision-choice systems of political economies the *defective information structures* may be complicated by a second postulate of *deceptive social information structures*. The deceptive information structure is made up of *disinformation* and *misinformation* components. The disinformation and misinformation characteristics enhance the vagueness and the spectrum of the penumbral regions of the social decision-choice activities. This principle of information structure used in this analysis is opposed to either the *principle of perfect information structure*, or the *principle of exact probabilistic information structure* used in economic and other decision theories. The perfect information structure is exact and full, while the probabilistic information structure is exact and incomplete. These two information structures are the foundations for the classical paradigm of thought in explanatory and prescriptive science and knowledge. The defective and deceptive information structures are defined in the fuzzy spaces and are handled with fuzzy logic, mathematics and soft computing. In general, the information structures available for processing as inputs into the decision-choice actions either by the individual or by the collective may be represented in terms a cognitive geometry as in Figure 2.1.

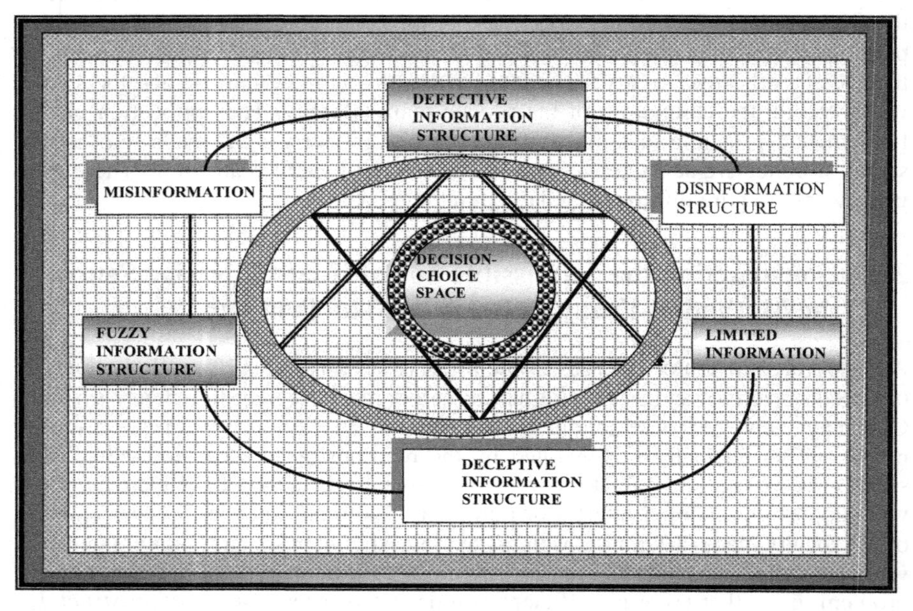

Fig. 2.1 Cognitive Geometry of the Information Structures for Epistemic Processing as Input into the Democratic Decision-Choice Space for Goal-Objective Formation National Interest and Social Vision

Let us keep in mind that the general information structure gives rise to the general uncertainty composed of possibilistic uncertainty and probabilistic uncertainty. The general uncertainty gives rise to the general fuzzy-stochastic risk which is composed of fuzzy risk and stochastic risk. The component of the fuzzy risk is amplified by the presence of deceptive information structure whether it manifests itself as misinformation or disinformation. Our ability to conceptualize these two types of uncertainties develop analytical frameworks to capture their essences and translate them into risk computations is the basic and fundamental challenge of modern logics, the corresponding mathematics and decision-choice computing. These information sub-structures and the relationships that they have to one another, to the knowledge production and decision-choice systems generate the phenomena of complexity and synergetics. The understanding of the relational structures, therefore, will offer an increasing opportunity to understand complex systems such as society, living organisms and their internal responses and develop more accurate epistemic models of representation.

2.2.1 The Problem of Social Goal-Objective Formation under Fuzzy Rationality in a Democratic Collective Decision-Choice Systems

The problem of social goal-objective formation is solved if all individual perceptions are in complete harmony with each other and that of the society as is understood, and that the vagueness and ambiguities of fuzzy residuals of individual personal decisions and the social decisions are reduced to a minimum. Thus, even if the elements in the admissible set of social goal-objective elements are compatible with one another, we still have to face the problem where such elements in the set may be incompatible with individual assessments of socially relevant goals and objectives. The problem of constructing the admissible set of social objectives may be stated as: given individual evaluations of each element relative to its social *relevance* in promoting the attainment of the national interest, how does one construct a set of social objectives that is best (in some specific sense) relative to the possible conflicting individual evaluations of social relevance in accordance with the individual preferences? This is another way of viewing the Arrow's problem of social choice with equal weights on the individual preferences. The conditions of social relevance and goal importance are determined by perceptions that are shaped by social ideology, its understanding, acceptable cultural boundaries and the interpretations of such boundaries from the available social information structure. These perceptions are subjective, ambiguous and vague which make the use of classical logic and mathematics difficult, because the decision-choice actions operate in an environment of defective information structure instead of exact and perfect information structure [R7.16].

The decision-choice constructs and the solutions to the decision-choice problems have their own rational processes that encompass analysis and synthesis. Let us keep in mind that every decision-choice action involves the constructions of qualitative equation on one hand and quantitative equation on the other hand, and then a search for solutions to the equations. In other words, every problem may be specified in terms of an equation cast in terms of either qualitative or quantitative or both structure. Every decision-choice action is also a solution to an equation that defines the appropriate problem. Every decision-choice problem whether well-defined or ill-defined has a solution. The solution is a potential that must be found. For the task at issue, we begin the analysis from a given initial conditions of a level of social welfare of the nation. From these initial conditions, a rational path to construct the social goal-objective set will be provided. Such a path is provided in Figure 2.2. The formation of the social goal-objective set is a collective decision-choice problem that is specified with a defective information structure. There are a number of methods that can be called upon to define the problem and abstract a solution. These methods are grouped under the *classical optimization methods* under the classical rationality and the *fuzzy optimization methods* under fuzzy rationality [R7.12] [R7.30]. The classical rationality works with exact information structure whose computational logic is connected to exact rigid determination. The fuzzy rationality works with defective information structure and deceptive information structure whose computational logic is connected to inexact flexible determination [R15.13] [R15.14].

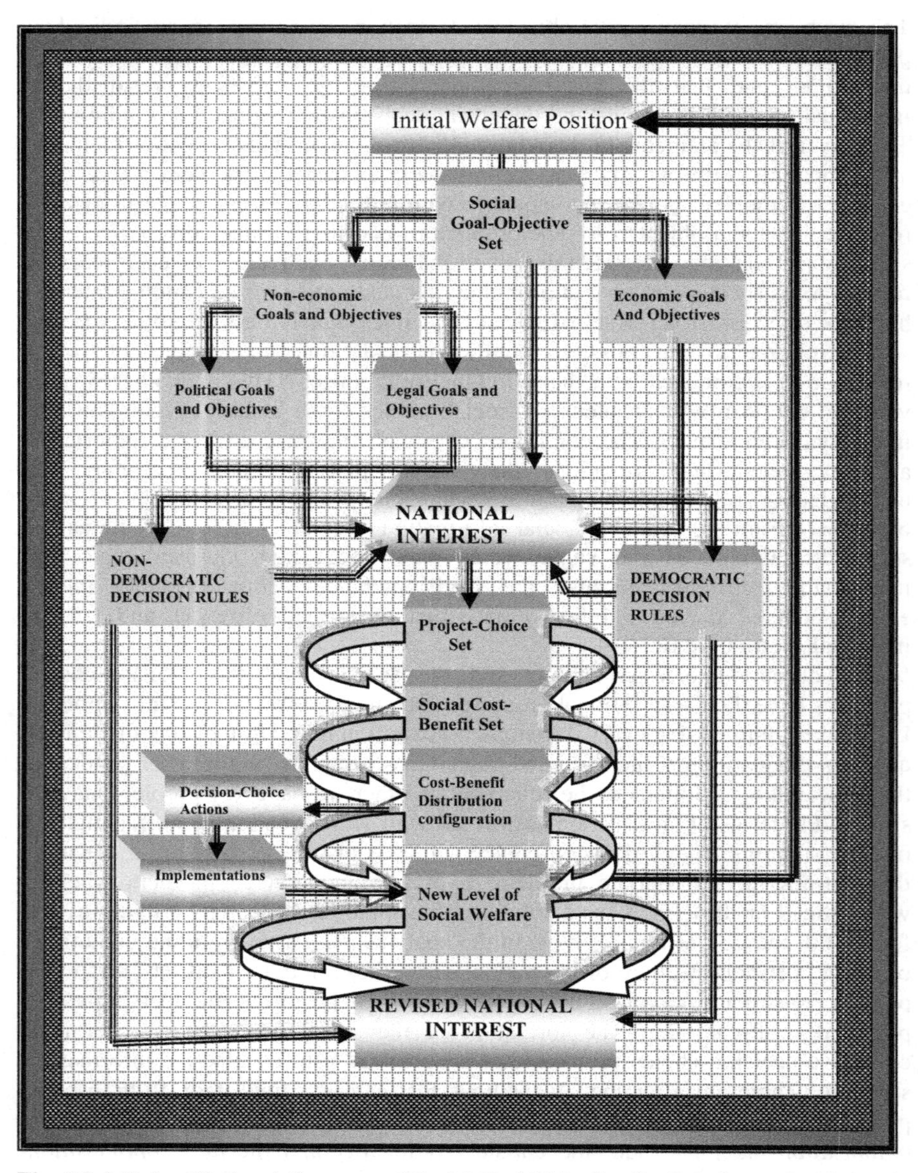

Fig. 2.2 A Path of Rational Construct of Social Goal-Objective Set Relative to the National Interest

The method to be selected will depend on how the information space is conceived and characterized by underlying assumptions. The toolbox of the classical optimization methods is based on the Aristotelian logic under an exact information structure that excludes vagueness and ambiguities which then create contradictions for the problem formulation and solution. It is incompatible with thought creation in the defective and deceptive information structures where subjectivity cannot be excluded. The toolbox of the fuzzy optimization methods accepts the environment that is characterized by all kinds of information structures and sees contradictions in terms of true-false proportions with different methods of aggregation. The aggregation methods are based on soft computing in an environment of either defective or deceptive information structure. The analytical method selected for defining the social goal-objective-formation problem begins with multi-person decision makers under a defective or non-defective information structure. The process of selecting the method of analysis is presented in Figure 2.2.

The selection process of the decision-goal objective elements is defined in a democratic decision-choice system that is sensitive to all individual preferences for the collective decision-choice action. This sensitivity may be either equal or unequal. The allowances of the sensitivities of all individual preferences make the use of classical logic and the majoritarian principle inappropriate. In this respect, the collective decision-choice actions under *majoritarian democratic decision-choice system* cannot be used. The classical logic, however, can be used to process information on all preferences, where the equal sensitivities accorded to the individual preferences are under *consensus democratic decision-choice system.* The difficulty in achieving this led to the development of the Arrow's impossibility theorem and the extensive discussions and literature on social and public choice in economics, political science, law and other related areas of thought that involves the aggregation of qualitative dispositions and subjectivities [R2.0].

The appropriate approach to the collective decision-choice problem of the goal-objective formation, where sensitivity is accorded to all the individual preferences, is the *weighted democratic decision-choice system.* The analytical process that is presented here is based on the principles of proportionality representation defined in a different structural process where social goals and objectives are viewed as collective goods that are defined in a fuzzy decision-choice space. This is the space where vagueness, ambiguity, volume limitation, misinformation and disinformation, and subjectivity of interpretations are admitted as essential elements of the information-knowledge structure. Social goals and objectives are taken to be unique in the selection and implementation. As we have stated, each goal-objective element is defined and identified by its cost-benefit duality, which also determines the element's relevant-irrelevant duality relative to the national interest and social vision. The cost-benefit duality is in relation to the goal-objective element while the relevant-irrelevant duality is in relation to the national interest and social vision. The set of characteristics of the relevant-irrelevant (relevance-irrelevance) duality space is mapped onto the set of the characteristics of cost-benefit duality space for the collective decision action. The structure is shown in Figure 2.3.

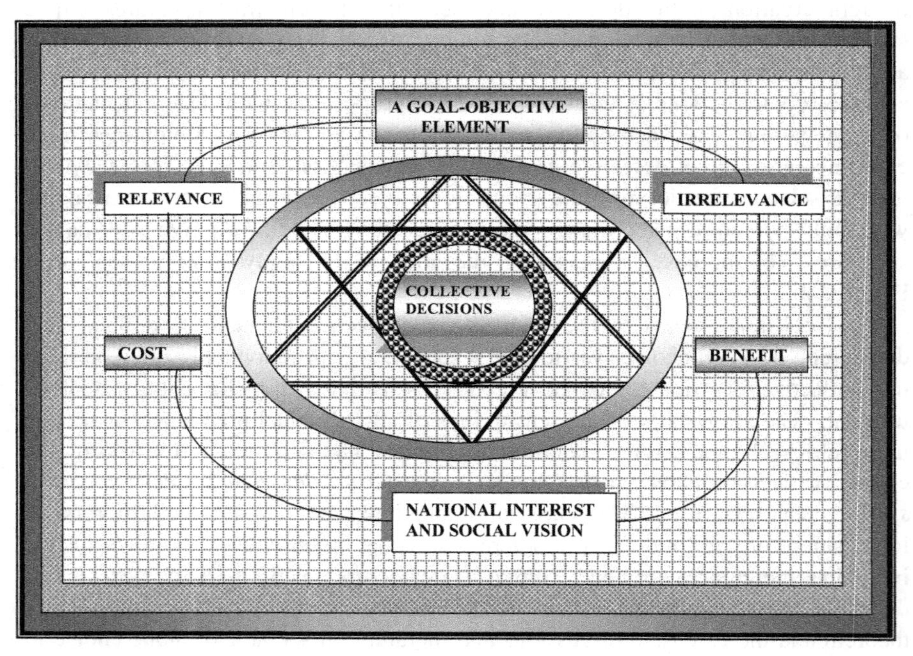

Fig. 2.3 Pyramidal Geometry of the Cost-Benefit Duality Relative to the Relevance-Irrelevance Duality of Democratic Decision-Choice Space for Goal-Objective Formation and National Interest

The costs are viewed in real terms and affect all members of society in one form or the other. The general decision-choice effects on the members, however, have a tendency for the real benefits to be restricted, and their benefit and cost distributional effects may also be distorted. The nominal costs for their implementation are tax-supported and hence obtained from the general social revenue in the support of costs of government and governance. Any member of the social group is affected by the real social costs from the selected social goal-objective set. The selected social goal-objective set may define or be defined by national interest and social vision depending on the acceptable rules of its construct. Any change in the social goal-objective set immediately affects the structure of the national interest. In a sequence of things, we may view the national interest or a system of national interest as determining the social goal-objective set, which then defines the needed set of social actions and projects that when implemented will effect and maintain the national interest and its historic path.

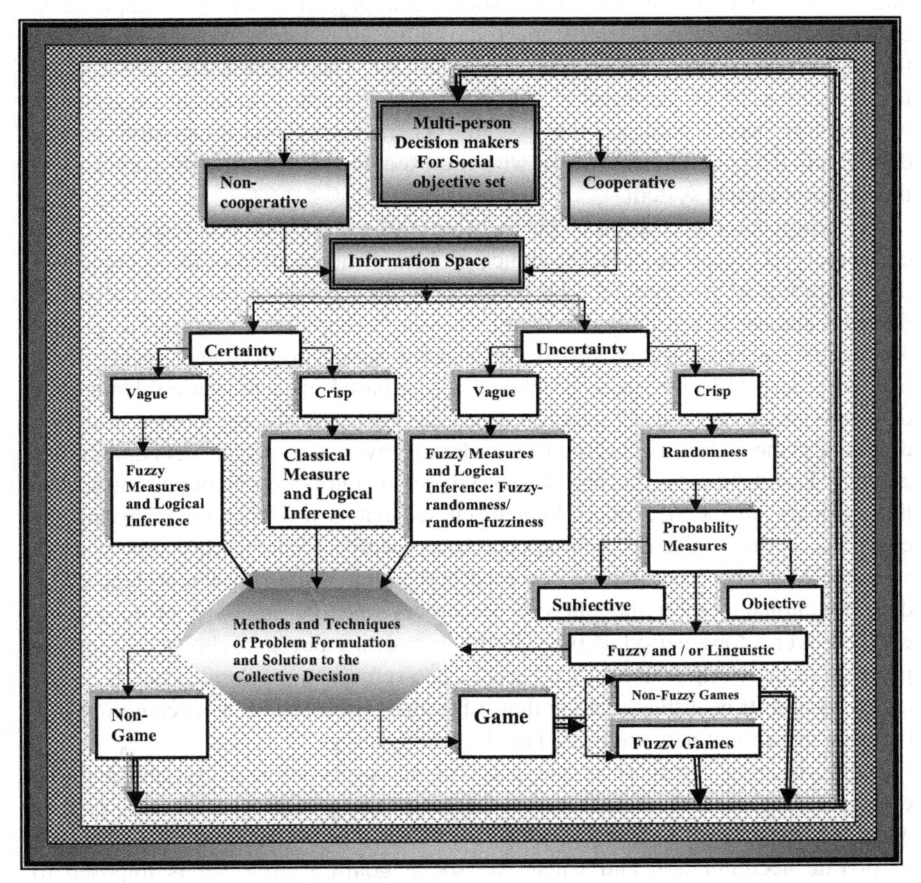

Fig. 2.4 The Structure of Method of Selection for Collective Decision in Fuzzy and Non-fuzzy Environments

In social goal-objective formations, injustices may tend to arise and do arise. The goals and objectives may be formed in such a way as to create rent-seeking opportunities for harvesting by some segments of the society depending on the nature of the social power distribution and the production-consumption enterprise system in the nation. We must keep in mind that the organization of the production-distribution-consumption system in support of the national interest and social vision may be based on either institutions of private-property rights and individual private net-benefit-incentive system, or institutions of public-property rights and collective net-benefit-incentive system or on various combinations of the two within the continuum of public-private duality or the individual-collective duality. It is not uncommon for the social goal-objective set to be imposed in either democratic or non-democratic socio-political systems, irrespective of how the supporting institutional structures are defined. The creation and existence of the institution of government is a creation and existence of actual or potential dictatorship, which violates the essential character of non-imposition in any democratic decision-choice system with the exception of the case of unanimity in the decision-choice system.

We know that the creation of the institution of government is a response to the complexity of collective decision-choice problem. The creation of the occupiers of the institution of government may be done by the partial surrender of individual decision-choice sovereignty, in faith, to the occupiers to whom a greater aggregative sovereignty is vested for general social control and social good. Similarly, any collective decision-choice action based on majoritarian democratic decision-choice system is dictatorial where the majority preference ordering dominates the minority preference ordering over the social decision-choice actions. The debate on the moral relativity of dictatorship-democratic structure, thus, centers on the size of the dictatorial set that is tolerable for resolving the conflicts in the social decision-choice space with social stability and without violence to achieve the national interest and social vision. In a non-democratic decision-choice system the class of occupiers may be imposed or may have evolved over a long period of time. In the democratic decision-choice system, the problem of the composition of the class is also subject to the democratic decision-choice action. It is at this juncture that it becomes useful to draw a separation line between theoretical democracy and applied democracy.

There is some form of rent-seeking in all socio-political systems and the corresponding production-distribution-consumption system. The extreme case is when the constructed decision-making core (elected body) becomes an instrument of private accumulation and where the social goal-objective set is imposed to abstract rent. In this case, the decision-making core (the elected and unelected officials) is placed at the service of private individuals to reap social benefits while the other members of the society serve to pay the costs in support of such private benefits within the principle of law and order of the legal structure of the political economy.

In this democratic collective decision-choice system we shall assume the existence of the principle of reasonable social justice, freedom and legal fairness in the creation of the social goal-objective set that reflects the fuzzy preferences of the members of the nations who, by the existing legal structure, have acquired the right to participate in the social decision-choice process. The problem of the democratic collective decision-choice system is to obtain an optimal social goal-objective set relative to the national interest and social vision. The problem of formation of optimal social goal-objective set in support of the defined national interest is complicated by the nature of the political economy as an institution of strategic game for determining the social cost-benefit configuration in the production-distribution-consumption system. The social cost-benefit configuration reveals itself in terms of distribution of power, income, education, health, housing, justice and punishment as it relates to what, who, where, when, and how much at the level of a domestic welfare, and at the level of international resources, state systems and sub-systems. The social cost-benefit configuration may be seen as the social incentive system. A change in the national interest and social vision may affect the social goal-objective set and the order of implementation of its elements in support of the national interest where the goal-objective set must be supported by a system of projects and social actions either managed by the private or the public or by a combination of the two.

2.2.2 Specifying the Problem of the Goal-Objective Formation in the Fuzzy Space

To tackle the problem of goal-objective formation on the principle of a weighted democratic decision-choice system that is sensitive to all individual preferences, we consider a general set of goal-objective elements, Ω, that can support different sets of national interests and social visions depending on the social set-up. The general set of the goal-objective elements, Ω, must be consistent with the ideology of the social organism in terms of the held system of values, and with generic element $\omega \in \Omega$. The set structure of the Ω will vary over different political economies depending on their respective value systems. For each element $\omega \in \Omega$, we specify the evaluation of the individual relevance-irrelevance duality in relation to the fulfillment of the national interest or a system of national interest. This duality accepts the principle of continuum and contradiction and hence rejects the Aristotelian basic principle of excluded middle in the classical laws of thought that led to the Arrow's paradox.

The Arrow's paradox simply translates to the statement that the democratic collective decision-choice system, on the principle of consensus, to generate social welfare ranking for social decision-choice action is impossible. The problem, as it has been explained elsewhere, is simply that there is no aggregation technique available to us in the classical paradigm for the summability of individual preferences in the collective democratic decision-choice space [R7.14] [R7.15] [R7.16]. It is the presence of the *excluded middle* and not its absence that conceptually helps to define the identity of the individual goal-objective element in terms of negative-positive characteristics. The identity is established by the structure of relevance-irrelevance characteristics. The relevance and irrelevance characteristics exist in a complex continuum with logical give-and-take relationships that help to define their mutual existence, negation and fuzzy residuals in concepts and thoughts. The separation of relevant and irrelevant goal-objective elements is a collective decision-choice problem requiring information and computational scheme.

To ascertain the information on relevance and irrelevance, we solicit from each socially qualified member of the community his or her evaluation of the *degree of relevance* of each social goal-objective element as well as information on the *degree of irrelevance* of the same social objective. In other words, two information bits are solicited from each individual. The rationale for such an approach is derived from possible over-evaluations if benefits are anticipated and under-evaluations if costs are anticipated since cost-benefit duality exists in decision-choice tension that provides the force of resolution. Relevance is related to the social benefit characteristics and irrelevance is related to the social cost characteristics in a continuum of the duality. Such information elicitation may be done at different times. This process may alternatively be conducted as statistical experiments in terms of sampling in order to reduce the computational size and the financial costs of developing the required information sets for decision-choice actions. The structure of the relevance-irrelevance duality as a continuum phenomenon may be stated as in Figure 2.5.

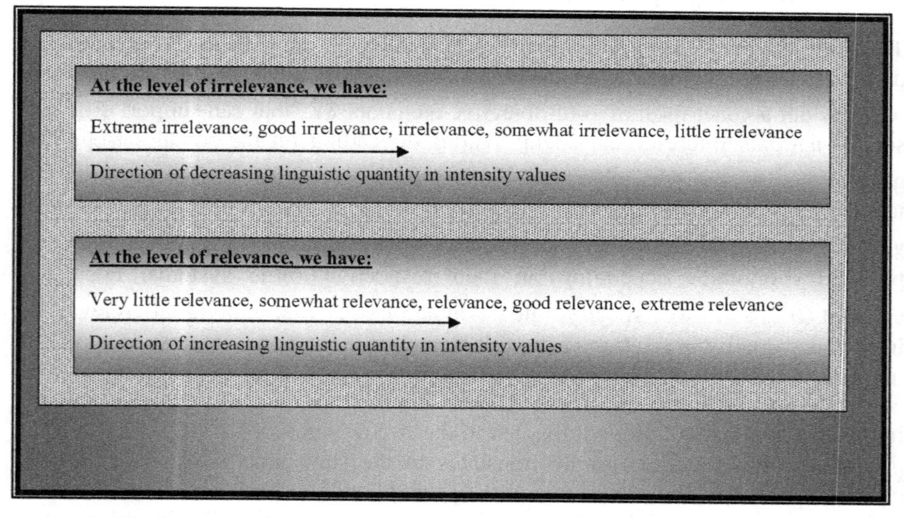

Fig. 2.5 Representation of fuzzy Variables of irrelevance and Relevance

The relevance pole is the one with a relevance characteristic set that outweighs the irrelevance characteristic set. Similarly, the irrelevant pole is one with an irrelevance characteristic set that outweighs the relevance characteristic set. The point of relevance-irrelevance indifference is when the two sets are perceived to be equal in terms of the degrees of belonging to either relevance or irrelevance relative to the national interest and social vision.

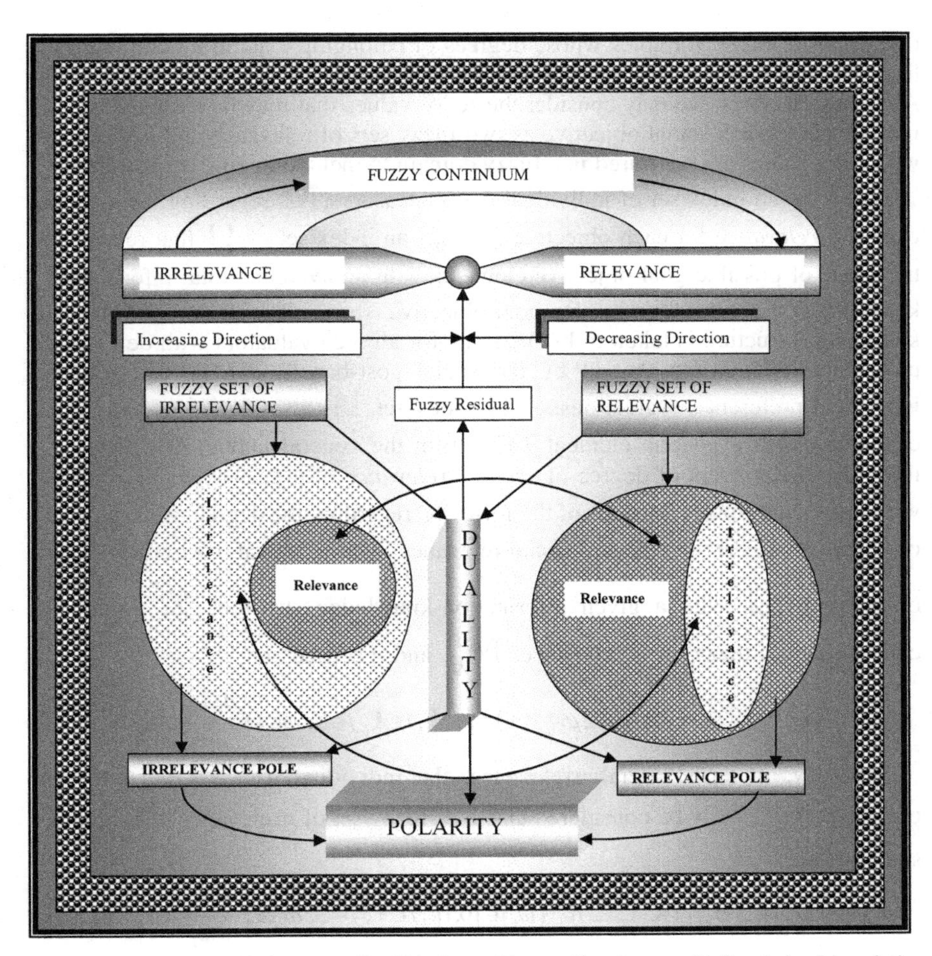

Fig. 2.6 Irrelevance-Relevance Duality in a Fuzzy Continuum Defined in Linguistic Quantities for Decision-Choice Actions

When the two sets are joined together we have a structure of continuum phenomenon supported by a system of polarities and dualities as in Figure 2.6. The structure presents a situation where there are two sets of characteristics in any of the objectives in terms of its relevance or irrelevance in supporting the national interest and social vision as they are understood by each member. It is analytically useful to view them in the decision-choice process in terms of interactions between polarities and dualities under soft computing systems. The concepts of relevance and irrelevance are linguistic variables that exist in the space of vagueness, ambiguity and subjectivity and hence are fuzzy notions, which may be represented as fuzzy variables. The relevance-irrelevance duality provides us with two opposite fuzzy variables whose degrees of belonging exist in a continuum of linguistic quantities for decision-choice actions as shown in Figure 2.6. As fuzzy variables, we may consider the set of values that may be assigned by the individuals to each social objective as two fuzzy sets of relevance and irrelevance where the values are measured in a fuzzy domain to include linguistic quantities.

Let $\mathbb{I}$ be an index set of individual fuzzy measures, $\mathbb{M}$ regarding the degrees of *social relevance* for each objective $j \in \mathbb{J}$, an index set of Ω that represents the space of possible goal-objective elements with an available total information-knowledge structure $\mathbb{K}$ in the goal-objective space. The total information-knowledge structure is taken to be the same for all the evaluators with deferential preferences which are shaped by the social cost-benefit duality that reflects individual preference evaluations. The index set $\mathbb{I}$ is also the set of individual evaluators with a generic element $i \in \mathbb{I}$ from the general public. We may also introduce the concept of degree of *social irrelevance* of a goal-objective element with individual fuzzy measures of the form $\mathbb{N}$ regarding the degree of irrelevance of objective $j \in \mathbb{J}$. In terms of social relevance, the individual, *i's assessment* of objective $j \in \mathbb{J}$ with a given information-knowledge structure $\mathbb{K}^{\mathbb{I}}$, may be considered as a fuzzy set of relevance, $\mathbb{M}_{ij}$, and specified as:

$$\mathbb{M}_{ij} = \left\{ \left(\omega_j, \mu_{\mathbb{M}_{ij}}(\omega_j) \right) \mid \mathbb{K}^{\mathbb{I}} \subset \mathbb{K}, \mu_{\mathbb{M}_{ij}}(\omega_j) \in [0,1] \text{ , } i \in \mathbb{I} \text{ , } j \in \mathbb{J}, \ \omega \in \Omega, \frac{d\mu_{\mathbb{M}_{ij}}}{d\omega_j} > 0 \right\} \quad (2.1a).$$

Similarly, in terms social irrelevance, the individual, *i's assessment* of an objective $j \in \mathbb{J}$ may be considered also as a fuzzy set of irrelevance, $\mathbb{N}_{ij}$ and be specified as:

$$\mathbb{N}_{ij} = \left\{ \left(\omega_j, \mu_{\mathbb{N}_{ij}}(\omega_j) \right) \mid \mathbb{K}^{\mathbb{I}} \subset \mathbb{K}, \mu_{\mathbb{N}_{ij}}(\omega_j) \in [0,1] \text{ , } i \in \mathbb{I} \text{ , } j \in \mathbb{J}, \ \omega \in \Omega, \frac{d\mu_{\mathbb{N}_{ij}}}{d\omega_j} < 0 \right\} \quad (2.1b).$$

The relevance and irrelevance membership characteristic functions are assumed to be differentiable functions of their inputs in the set $[0,1]$ and also defined in the same space of the characteristic set. The relevance and irrelevance membership characteristic functions are distinguished by their respective slopes. Equations (2.1a) and (2.1b) constitute a duality where $\left[1-\mu_{\mathbb{M}_{ij}}(\cdot)\right] \neq \mu_{\mathbb{N}_{ij}}(\cdot)$ and similarly $\left[1-\mu_{\mathbb{N}_{ij}}(\cdot)\right] \neq \mu_{\mathbb{M}_{ij}}(\cdot)$ since the underlining sets of information are separately elicited from the same individuals. The ideal case will be where $\left[1-\mu_{\mathbb{M}_{ij}}(\cdot)\right] = \mu_{\mathbb{N}_{ij}}(\cdot)$ and $\left[1-\mu_{\mathbb{N}_{ij}}(\cdot)\right] = \mu_{\mathbb{M}_{ij}}(\cdot)$ and hence $\mu_{\mathbb{M}_{ij}}(\cdot) + \mu_{\mathbb{N}_{ij}}(\cdot) = 1$. The ideal case will be a constraint on the fuzzy algorithmic process for fuzzy optimality.

The principles of the opposites, duality, polarity and continuum impose on the analytical structure the appropriate functional description of the fuzzy numbers for reasoning under fuzzy rationality. The appropriate fuzzy numbers are those whose membership characteristic functions are of the positive-Z, positive-S, positive-E or positive-R types for relevance or benefit. The fuzzy numbers of negative-Z, negative-S, negative-E or negative-R types are useful in specifying irrelevance or cost. The E-type represents the exponential class of functions and the R-type represents the ramp class of functions. The term $\mathbb{K}^{\mathbb{I}}$ is the social information-knowledge structure available to the public. In other words, the process of information elicitation regarding relevance and irrelevance on the basis of information-knowledge structure, avoids the possibility of fuzzy complements as constraints on the original information elicitation. The process projects an asymmetry of information elicitation as induced by the cost-benefit rationality. This conceptual approach in the information elicitation is induced by behavioral differences in individual preference evaluations of the same item if it is seen as a benefit rather than if it is seen as a cost and vice versa. Examples of the classes of these fuzzy numbers for characterizing relevance and irrelevance of the goal-objective elements are provided below for quick reference, from which we shall attend to the specification of the degrees of relevance and irrelevance of the decision making core.

2.2.3 *Relevant Fuzzy Numbers for Fuzzy Reasoning in Opposites, Duality, Polarity, Relational Continuum and Unity*

Consider the reference set of non-negative Ω of the goal-objective set. Two classes of fuzzy sets, with corresponding membership functions describing the identity of any goal-objective element $\omega \in \Omega$, may be examined: a) "x is irrelevant", b) "x is relevant" where $x \in \mathrm{R}^{+}$ is an evaluated value from an evaluator. For example $x \in [0,10]$ or $x \in [0,100]$ where zero linguistically means completely irrelevant on the scale and 10 or 100 means completely relevant on the scale of valuation. The system can also be constructed for the membership characteristic functions. Let $\mu_{\bullet}(x)$ be a generic membership function defined over $x \in \mathrm{R}^{+}$ such that $\mu(x) \in [0,1]$. The following membership functions may then be specified. We shall first deal with membership functions that express the concept of "x is irrelevant" as an evaluation for $\omega \in \Omega$, and $x \in \mathrm{R}^{+}$. The relevance-irrelevance duality may be replaced by any duality. In the case of valuation of irrelevance we have:

$\lim_{x\to 0} \mu_{\bullet}(x)=1 \Rightarrow$ complete irrelevance and $\lim_{x\to\infty} \mu_{\bullet}(x)=0 \Rightarrow$ complete relevance.

2.2.3.1 Fuzzy Numbers for "x is irrelevant"

Let us examine a class of fuzzy numbers that relate to the statement of "x is irrelevant" as an evaluation of $\omega \in \Omega$. These statements are considered as either linguistic or fuzzy variables.

2.2.3.1.1 An Exponential Fuzzy Number for "x Is Irrelevant" Description for $\omega \in \Omega$

A) $$\mu(\mathrm{x}) = \mathrm{e}^{-k\mathrm{x}} \quad , \ k > 0 \tag{2.2}$$

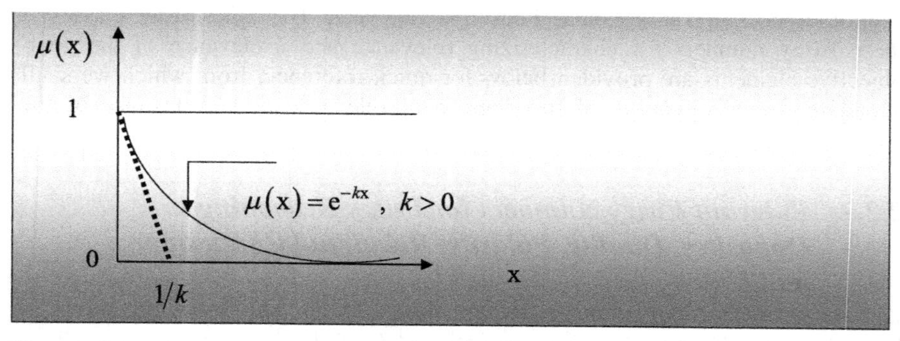

Fig. 2.7 A Graph of an E-Fuzzy Number for "x is irrelevant"

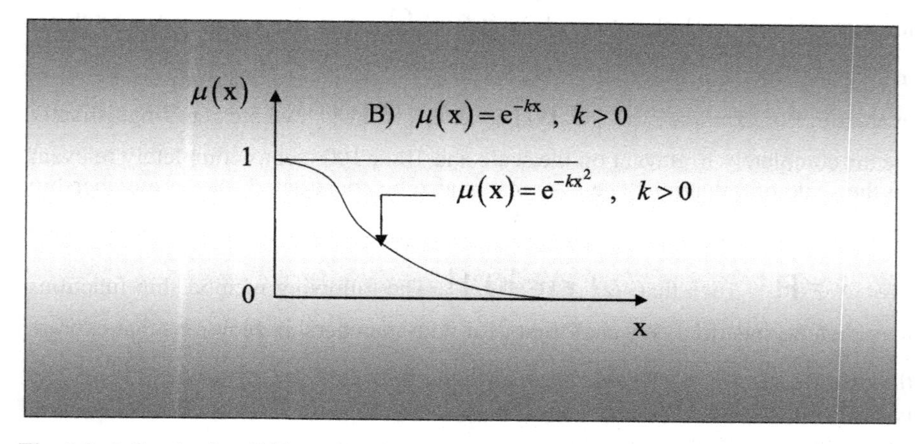

Fig. 2.8 A Graph of an E-Fuzzy Number for "x is irrelevant"

2.2.3.1.2 The Z-Fuzzy and R-Fuzzy Numbers for "x is irrelevant"

$$\mu(x)\begin{cases} =1 & , x\in[0,a_1] \\ =\dfrac{a_2-x}{a_2-a_1} & , x\in[a_1,a_2] \\ =0 & , x\in(a_2,\infty] \end{cases} \qquad \mu(x)\begin{cases} =1-ax^k, & x\in\left[0, a^{-\left(\frac{1}{k}\right)}\right] \\ =0 & , x\in\left[a^{\left(1/k\right)}, \infty\right] \end{cases}$$

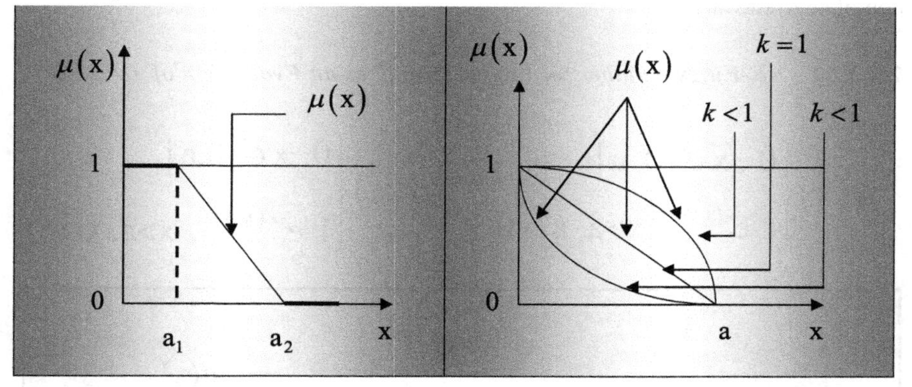

Fig. 2.9 Geometry of the Z-Fuzzy Number For "x is irrelevant"

Fig. 2.10 Geometry of the R-Fuzzy number for "x is irrelevant"

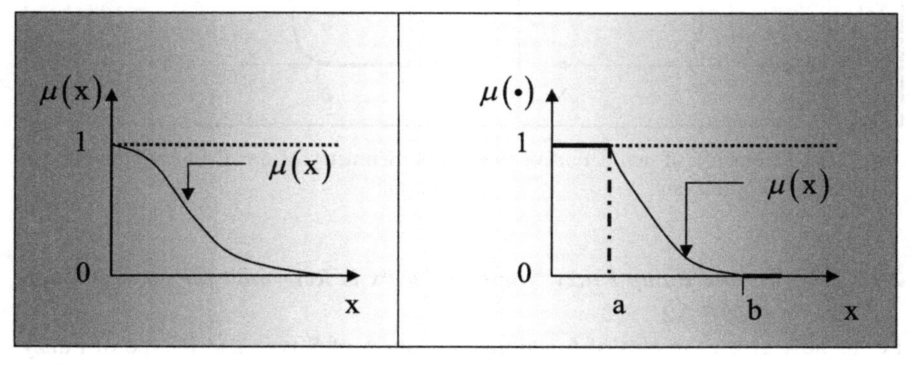

Fig. 2.11 An Inverted S-Fuzzy number for Irrelevance equation. to Figure 2.14

Fig. 2.12 Geometry of a Z-Fuzzy Number for irrelevance an equation to Figure 2.12

$$\mu(x)=\frac{1}{1+kx^2} \quad , k>1$$

$$\mu(x)\begin{cases} =1 \ , x\in[0,a) \\ =\frac{1}{2}\left\{1-\sin\frac{\pi}{b-a}\left[x-\frac{1}{2}(a+b)\right]\right\} \ , x\in[a,b] \\ =0 \ , \ x\in(b,\infty) \end{cases}$$

2.2.3.2 Fuzzy Nmbers for "x is Relevant" as an Evaluation of $\omega \in \Omega$

Let us now turn our attention to fuzzy numbers that tend to express the linguistically numerical idea of relevance, such as "a is relevant". Just as the linguistic variable "x is irrelevant", we can express relevance in varying degrees in a continuum. In the case of valuations of relevance we have:

$\lim_{x \to \infty} \mu_{\bullet}(x) = 1 \Rightarrow$ a complete relevance and $\lim_{x \to 0} \mu_{\bullet}(x) = 0 \Rightarrow$ a complete irrelevance

2.2.3.2.1 *E-Fuzzy Number for "x is relevant" as an Evaluation of $\omega \in \Omega$*

$$\mu(x)\begin{cases} 0, & x \in [0, a] \\ 1 - e^{-k(x-a)}, & x > a,\ k > 0 \end{cases} \qquad \mu(x)\begin{cases} 0, & x \in [0, a] \\ 1 - e^{-k(x-a)^2}, & x > a,\ k > 0 \end{cases}$$

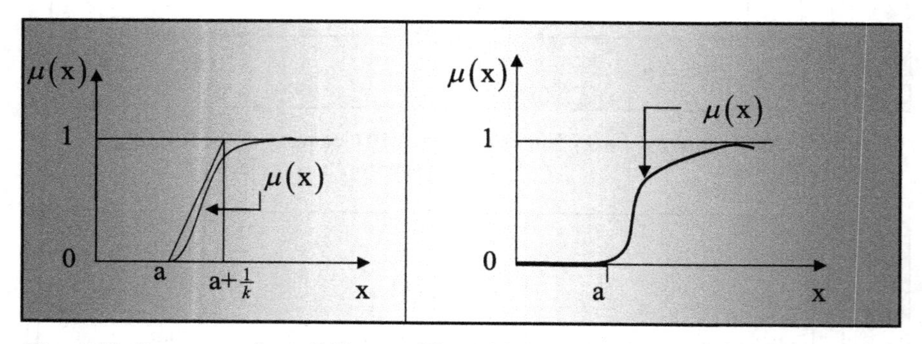

Fig. 2.13 Geometry of an E-Fuzzy Number for "x is relevant"

Fig. 2.14 Geometry of an E-Fuzzy Number for "x is relevant

2.2.3.2.2 S- *and Ramp Fuzzy Numbers for "x is Relevant" for an evaluation of $\omega \in \Omega$*

Let us now give examples of functional structures of S-type and R-type of Fuzzy numbers. Keep in mind that these fuzzy numbers can be approximated by either TFN (triangular) or TZFN (trapezoidal) fuzzy numbers depending on the nature of the opposites, polarity, duality and corresponding continuum. The functional examples, geometric examples and corresponding equations are given in Figures (2.15-2.18).

$$\mu(x)\begin{cases}0 & , x\in[0,a_1]\\ \frac{x-a_1}{a_2-a_1} & ,x\in(a_1,a_2)\\ 1 & ,x\in[a_2,\infty)\end{cases}\qquad \mu(x)\begin{cases}0\\ =a(x-a)^k, a<x<a+a^{-\left(\frac{1}{k}\right)}\\ =1, \; x\geq a+a^{-\left(\frac{1}{k}\right)}\end{cases}$$

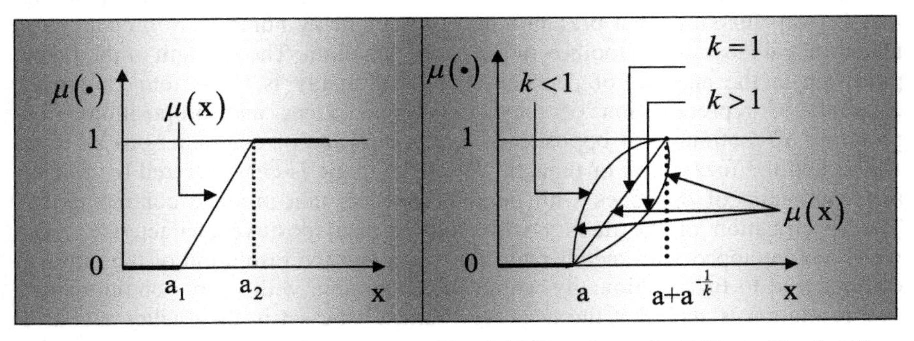

Fig. 2.15 Geometry of S-Fuzzy Number for "relevance"

Fig. 2.16 Geometry of a R-Fuzzy Number for "Relevance"

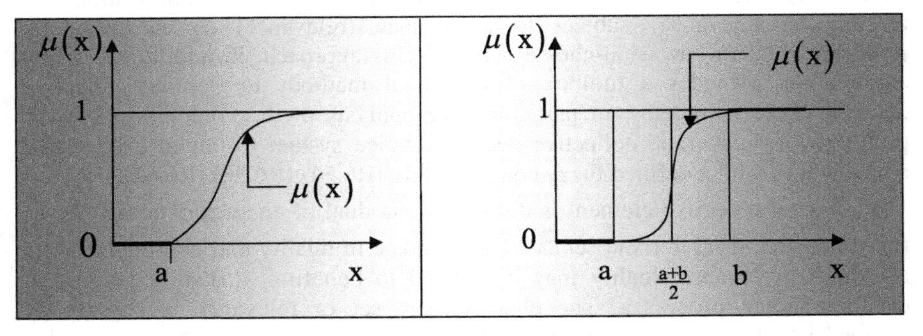

Fig. 2.17 Geometry of an S-Fuzzy Number for "relevance"

Fig. 2.18 Geometry of an S-Fuzzy Number for "relevance"

$$\mu(x)\begin{cases}=0, x\in[0,a]\\ =\frac{k(x-a)^2}{1+k(x-a)^2}, x\in(a,\infty]\end{cases}$$

Equation to Figure 2.17

$$\mu(x)\begin{cases}0 \quad x\in[0,a]\\ \frac{1}{2}\left\{1+\sin\frac{\pi}{b-a}\left[x-\frac{1}{2}(a+b)\right]\right\}, \\ \qquad x\in(a,b)\\ =1 \quad , x\in[a,\infty)\end{cases}$$

Equation to Figure 2.18

Given the appropriately relevant fuzzy numbers, the conflicts in individual cost-benefit assessments are that, higher social value is attached to an item if the item is viewed as benefit to the individual. Similarly, low evaluative social value is attached if the same item is viewed as cost to the same individual relative to the national interest and vision. Such a principle of behavior in duality helps to overcome the problem of overvaluations and under-valuations. The concepts of relevance and irrelevance are linguistic variables with qualitative characteristics whose values are subjectively defined and assessed by the corresponding membership functions that present a spectrum of fuzzy numbers in a continuum under the guidance of the toolbox of the fuzzy paradigm. The strength of the fuzzy paradigm in the analysis of decision-choice rationality is to be found in three elements of representation of inexact symbolic ideas and propositions, the principle of continuum of opposites in the acceptance of contradictions as truth values and the fuzzy laws of thought. The fuzzy logic is characterized by duality and continuum of opposites with internal conflicts that produce contradictions. The specification of the duality with negative and positive characteristic sets which are equipped with membership functions and the resolution of the internal conflict lead to fuzzy rationality within the continuum with fuzzy conditionality. The resolution is such that the negative characteristic set in the duality acts as a constraint on the positive characteristic set and vice versa. In other words, the judgment of relevance (irrelevance) is constrained by conditions of irrelevance (relevance) and cognitively formulated as maximization (minimization) of relevance (irrelevance) subject to irrelevance (relevance) as constraint on judgments and the decision-choice actions. This approach, in addition to fuzzy aggregation, provides a toolbox of analytical methods to optimal collective decision-choice problems, in particular, the analysis of the complex and fuzzy problems of democratic collective decision-choice system to obtain exact value equivalences with a defined fuzzy conditionality [R6.54] [R6.56] [R6.63].

Each goal-objective element is defined by its dual of characteristic set $\mathbb{A}$ of relevance and characteristic set $\mathbb{B}$ of irrelevance in duality and continuum. This relevance-irrelevance duality may be related to benefit-cost duality in general decision-choice processes. The characteristic set of relevance is cognitively mapped into the benefit space while the characteristic set of irrelevance is also mapped into the cost space to form a unity of creative decision-choice action in the cost-benefit space under cost-benefit rationality. In the information elicitation process, we will not accept $\mu_{\mathbb{A}'} = \left[1 - \mu_{\mathbb{A}}(\omega)\right]$ that defines the membership fuzzy complement of set $\mathbb{A}'$ as defining the data on the characteristic set of $\mathbb{B}$. The methodological approach in resolving the individual assessments in the relevance-irrelevance conflict space is to develop another characteristic data set $\mathbb{B}$ of irrelevance with a membership characteristic set $\mu_{\mathbb{B}}(\omega)$. The explanation is to be found in the classic conflict in a cost-benefit space where preferences are high for benefits and low for costs. Thus in the individual evaluative process, it is not necessary that $\mu_{\mathbb{A}'} = \left[1 - \mu_{\mathbb{A}}(\omega)\right] = \mu_{\mathbb{B}}(\omega)$

and similarly, $\mu_{\mathbb{B}'} = \left[1-\mu_{\mathbb{B}}(\omega)\right] = \mu_{\mathbb{A}}(\omega)$. In fact, $\mu_{\mathbb{A}'} \neq \mu_{\mathbb{B}}(\omega)$ and $\mu_{\mathbb{B}'} \neq \mu_{\mathbb{A}}(\omega)$. This condition ensures that in relevance we find irrelevance, and vice versa in duality and continuum, in that the degree of relevance has a degree of irrelevance as its decision support and vice versa. Complete truthful revelations will require the equality to hold as the ideal cases for all the individuals. Let us specify the fuzzy evaluative structures of relevance and irrelevance of the members of the decision-making core.

The individual fuzzy membership functions may take different forms depending on their individual-social preference ordering. From eqns. (2.1 a&b), $\mathbb{I}$ is the index set for all eligible members in the democratic collective decision-choice system. The members of the democratic collective decision-choice system for the formation of the social goal-objective set is partitioned into two that is made up of the general public which is *the principal* and the decision-making core which is *the agent.* The members of the decision-making core constitute the management of the government and the social governance. Like the index set $\mathbb{I}$, we consider an index set $\mathbb{L}$ of the social decision-making class (the decision-making core or the members of the electorate) that is constructed on the basis of some socially accepted selection rule. In this case, $\mathbb{L}$ is an index set of the decision-making core. We then elicit from the members degree of social relevance and social irrelevance associated with each the social objectives contained in Ω. The degree of social relevance (REL) or irrelevance (IIR) attached to each objective is a fuzzy number which is subjectively calculated from the perceptions and preferences of individual members in the decision-making core in terms of how they relate them to their interpretations and understanding of the national interest and social vision. The fuzzy number for relevance is a fuzzy set $\mathbb{M}_{lj}$ defined by a membership function, with a given information-knowledge structure $\mathbb{K}^{\mathbb{L}}$ available to the decision-making core and may be written as:

$$\mathbb{M}_{lj} = \left\{ \left(\omega_j, \mu_{\mathbb{M}_{lj}}(\omega_j)\right) \mid \mathbb{K}^{\mathbb{L}} \subseteq \mathbb{K}, \mu_{\mathbb{M}_{lj}}(\omega_j) \in [0,1],\ l \in \mathbb{L},\ j \in \mathbb{J},\ \omega \in \Omega, \frac{d\mu_{\mathbb{M}_{lj}}}{d\omega_j} > 0 \right\} \quad (2.2a)$$

Similarly, the fuzzy number for irrelevance (IRR) is a fuzzy set $\mathbb{N}_{lj}$ with a membership function defined as:

$$\mathbb{N}_{lj} = \left\{ \left(\omega_j, \mu_{\mathbb{N}_{lj}}(\omega_j)\right) \mid \mathbb{K}^{\mathbb{L}} \subseteq \mathbb{K},\ \mu_{\mathbb{N}_{lj}}(\omega_j) \in [0,1],\ l \in \mathbb{L},\ j \in \mathbb{J},\ \omega \in \Omega, \frac{d\mu_{\mathbb{N}_{lj}}}{d\omega_j} < 0 \right\} \quad (2.2b)$$

Equations 2.1 of (a and b) provide an individual preference public database for processing to ascertain the relevant goal-objective elements that will support the national interest and social vision. Equations 2.2 of (a and b) also provide a preference database of the members in the decision-making core for processing to ascertain the relevant goal-objective elements that will support the national interest and social vision. These equations are the *fuzzification* of the decision-choice process as the first step in the formulation-solution structure of the collective decision-choice problem in a weighted consensus democratic decision-choice system **(WCDDS).**

We may now combine these evaluations in terms of their membership values through a fuzzy aggregation for both sets to obtain fuzzy collective assessment for each social goal-objective element. These fuzzy collective assessments may be used to construct the optimal social goal-objective set relative to the national interest and social vision. The process involves two cases. One case involves a non-intra-group information sharing and hence decision-choice interactions are not allowed within the group. The second case involves a situation where intra-group information sharing through discussions and unrestricted freedom of speech are allowed, and hence decision-choice interactions are allowed to influence the individual assessment process. The first case is consistent with social decision-choice actions without political parties or specific interest groups. The second case is consistent in political party formation or formation of specific interest groups. It may be pointed out that the information-knowledge structures are such that $\mathbb{K}^{\mathbb{I}} \cap \mathbb{K}^{\mathbb{L}} \neq \emptyset$ and $\mathbb{K}^{\mathbb{I}} \subseteq \mathbb{K}^{\mathbb{L}}$ producing asymmetric information-knowledge structure in the optimal ranking process. The information asymmetry is removed if $\mathbb{K}^{\mathbb{I}} \subseteq \mathbb{K}^{\mathbb{L}}$ and $\mathbb{K}^{\mathbb{L}} \subseteq \mathbb{K}^{\mathbb{I}}$ and hence the general public and the decision-making core have the same information-knowledge structure to work with.

The term $\mathbb{K}^{\mathbb{L}}$ is the information-knowledge structure that is available to the members of the decision-making core. It might also be a situation where $\mathbb{K}^{\mathbb{I}} \subset \mathbb{K}^{\mathbb{L}} \subseteq \mathbb{K}$ which implies that the information-knowledge structure available to the public is smaller than that available to the decision-making core. This is the case where information may be classified by the agent under dubious principle of national security and public safety. Given the individual preferences, it is the presence of asymmetry of information-knowledge structures, held by the public and decision-making core, that produces differences in relevance-irrelevance assessments through the cost-benefit balances between them, that is, between the general public as the principal and the decision-making core as the agent.

2.3 Non-information Sharing and Non-Decision Interaction in Democratic Social Goal-Objective Formations

The degree of collective social evaluations of social relevance and social irrelevance for each $\omega_j \in \Omega,\ j \in \mathbb{J}$ and for the index sets $\mathbb{I}$ and $\mathbb{L}$, where there is no information-knowledge sharing, and no intra-group decision interactions, may be specified in terms of fuzzy sets. The corresponding membership functions, written for the general assessment with index set $\mathbb{I}$ and a given information-knowledge structure $\mathbb{K}^{\mathbb{I}}$ of the public or the principal may be specified as:

$$\left\{\begin{aligned}
&\mathbb{M}^{\mathbb{I}} = \left\{ \left(\omega_j, \mu_{\mathbb{M}_{\cdot j}}(\omega_j)\right) \mid \mu_{\mathbb{M}_{\cdot j}}(\omega_j) = \left[\left[\bigwedge_{i\in\mathbb{I}} \left[\mu_{\mathbb{M}_{ij}}(\omega_j)\right],\ \omega_j \in \Omega,\ j \in \mathbb{J} \right]\right]^{\mathbb{I}} \right\} \text{(REL) a} \\
&\hspace{28em} (2.3) \\
&\mathbb{N}^{\mathbb{I}} = \left\{ \left(\omega_j, \mu_{\mathbb{N}_{\cdot j}}(\omega_j)\right) \mid \mu_{\mathbb{N}_{\cdot j}}(\omega_j) = \left[\left[\bigvee_{i\in\mathbb{I}} \left[\mu_{\mathbb{N}_{ij}}(\omega_j)\right],\ \omega_j \in \Omega,\ j \in \mathbb{J} \right]\right] \right\}^{\mathbb{I}} \text{(IRR) b}
\end{aligned}\right\}$$

The index set $\mathbb{I} \cap \mathbb{L} = \varnothing$ in the sense that, the evaluative preferences of the goal-objective formation by the individuals in the public are separated from the preferences of the members of the decision-making core or the elected officials who constitute the agent. The fuzzy aggregate formation for the decision-making core (the agent) with an index set $\mathbb{L}$ and with a given information-knowledge structure $\mathbb{K}^{\mathbb{L}}$ may also be written as:

$$\left\{\begin{aligned}
&\mathbb{M}^{\mathbb{L}} = \left\{ \left(\omega_j, \mu_{\mathbb{M}_{\cdot j}}(\omega_j)\right) \mid \mu_{\mathbb{M}_{\cdot j}}(\omega_j) = \left[\left[\bigwedge_{i\in\mathbb{L}} \left[\mu_{\mathbb{M}_{ij}}(\omega_j)\right],\ \omega_j \in \Omega,\ j \in \mathbb{J} \right]\right]^{\mathbb{L}} \right\} \text{(REL) a} \\
&\hspace{28em} (2.4) \\
&\mathbb{N}^{\mathbb{L}} = \left\{ \left(\omega_j, \mu_{\mathbb{N}_{\cdot j}}(\omega_j)\right) \mid \mu_{\mathbb{N}_{\cdot j}}(\omega_j) = \left[\left[\bigvee_{i\in\mathbb{L}} \left[\mu_{\mathbb{N}_{ij}}(\omega_j)\right],\ \omega_j \in \Omega,\ j \in \mathbb{J} \right]\right]^{\mathbb{L}} \right\} \text{(IRR) b}
\end{aligned}\right\}$$

The manner in which different fuzzy information-knowledge structures are combined to induce fuzzy aggregation needs some explanation. It is equivalent to operating with the fuzzy numbers.

The fuzzy aggregation that we seek, must be consistent with the concepts and logical operations of duality and contradictions, where contradiction must be viewed in terms of commonness, and in the sense of mutual cost-benefit determination in the continuum. Social decision making is about resolution of contradictions or conflicts in costs and benefits. First, the resolution is done with individual preference ordering and secondly, given the individual cost-benefit configurations, the collective decision-choice action is resolved by reconciling the cost-benefit differences of individual preferences. An economic example will be easier to understand the relational structure of the contradiction and commonness in duality where costs and benefits characterize the same entity. For example, if we consider an assessment of a nominal price = \$10 of an item as being high and let such assessment by three individuals be $\{(10,\ 0.2),\ (10,\ 0.4),\ (10,\ 0.6)\}$ where $\{(\ 0.2),\ (0.4),\ (\ 0.6)\}$ are the individual assessments in terms of \$10.00 belonging to a set of high prices of the object, then the min (0.2) satisfies all evaluators as \$10.00 being high price. Notice that the price is fixed while the aggregation takes place over the membership functions or the fuzzy numbers. The required aggregation operator is $\min(\wedge)$. Similarly, if the information is shifted to a low price evaluation, then the max (0.6) will satisfy all of the evaluators as being low price. The required aggregation operator is $\max(\vee)$. The min-max $(\wedge - \vee)$ operators will alternate depending on the nature of the problem.

This fuzzy logical process is a search for a solution to the collective decision-choice problem in a weighted unanimity democratic decision-choice system. Equations (2.3) and (2.4) constitute a *fuzzification* of the data on the collective preferences through fuzzy aggregation which is the *second step* in the search process of the solution to the problem of constructing the social goal-objective set. Given the second step, we need to apply the toolbox of fuzzy computing to defuzzify the problem for a solution. The applications of approximate reasoning and soft computing to defuzzify the process require some logical care. In terms of what we have before us, two types of *defuzzification* process are opened to us. One way is that we may seek fuzzy optimization within groups and combine the results of optimal values to construct the final optimal goal-objective set that will give us a crisp set through a fixed-level set operator. An alternative way is that we may first utilize an inter-group fuzzy information combination and through the fuzzy optimal decision process attain a *defuzzification* to create exact (crisp) value equivalence, and a final exact goal-objective set by a fix-level set operator. Both approaches may proceed with or without fuzzy information interactions. Each of the exact-value equivalences has a fuzzy conditionality for its acceptance.

In the first approach, we take the intra-group fuzzy decision process, and first combine the fuzzy aggregates on the degree of relevance and irrelevance to obtain an intra-group fuzzy decision-choice problem for each objective. This implies that we construct two intra-group sets of decision-choice elements, and from them two fuzzy decision-choice problems are formulated. One formulation is from the general public with index set $\mathbb{I}$ and with a decision-choice problem $\Delta^{\mathbb{I}}$. The other formulation is from the decision-making core with index set $\mathbb{L}$ and with a decision-choice problem $\Delta^{\mathbb{L}}$. The formulation of the fuzzy decision-choice problem $\Delta^{\mathbb{L}}$ is the *third step* in the search process to solve the individual preference conflicts in the collective decision space under a weighted consensus democratic decision-choice system of the goal-objective formation relative to the national interest and social vision. The conflict is formulated as an objective-constraint phenomenon that may be viewed as a benefit-cost phenomenon as well as set interactions. The third step involves equations (2.5) and (2.8)

$$\Delta^{\mathbb{I}} = \left\{ \left(\omega_j, \mu_{\Delta^{\mathbb{I}}}(\omega_j) \right) \mid \mu_{\Delta^{\mathbb{I}}}(\omega_j) = \mu_{\mathbb{M}^{\mathbb{I}}{}_{\bullet j}}(\omega_j) \wedge \mu_{\mathbb{N}^{\mathbb{I}}{}_{\bullet j}}(\omega_j) \right\} \quad (2.5)$$

The right hand side of the eqn. (2.5) is such that the dot, $(\cdot)$ in $\mu_{\mathbb{M}^{\mathbb{I}}_{\bullet j}}(\omega_j)$ and $\mu_{\mathbb{N}^{\mathbb{I}}_{\bullet j}}(\omega_j)$ simply means that an aggregation has been performed over the index set $i \in \mathbb{I}$.

Given the structure of the fuzzy decision-choice problem, the solution is obtained by transforming the fuzzy decision problem into a fuzzy optimization problem as the *fourth step* and then written as:

$$\max_{\omega\in\Omega} \mu_{\Delta^{\mathbb{I}}}(\omega_j) = \bigvee_{\omega\in\Omega}\left[\mu_{\mathbb{M}^{\mathbb{I}}_{\bullet j}}(\omega_j) \wedge \mu_{\mathbb{N}^{\mathbb{I}}_{\bullet j}}(\omega_j)\right] \tag{2.6}$$

The solution to eqn. (2.6) may be obtained as the *fifth step* by transforming it into a computational fuzzy optimization problem, which will provide us with an instrument for defuzzification as stated in theorems (2.3), (2.4) and (2.5).

Theorem 2.1

The $\max_{\omega\in\Omega} \mu_{\Delta^{\mathbb{I}}}(\omega_j) = \bigvee_{\omega\in\Omega}\left[\mu_{\mathbb{M}^{\mathbb{I}}_{\bullet j}}(\omega_j) \wedge \mu_{\mathbb{N}^{\mathbb{I}}_{\bullet j}}(\omega_j)\right]$ is equivalent to

$$\left(\omega^*, \mu_{\Delta^{\mathbb{I}}}(\omega^*)\right) = \left\{ \begin{array}{l} = \max_{\omega\in\Omega} \mu_{\mathbb{M}^{\mathbb{I}}_{\bullet j}}(\omega_j) \\ \qquad \text{s.t.}\, \omega \in \left[\mu_{\mathbb{M}^{\mathbb{I}}_{\bullet j}}(\omega_j) - \mu_{\mathbb{N}^{\mathbb{I}}_{\bullet j}}(\omega_j)\right] \le 0 \end{array} \right\}$$

We can then write the optimal ranking set for the members in $\mathbb{I}$ as:

$$\mathbb{M}^{\mathbb{I}^*} = \left\{\left(\omega_j^*, \mu_{\Delta^{\mathbb{I}^*}}(\omega_j^*)\right) \mid j \in \mathbb{J}, \omega \in \Omega\right\}^{\mathbb{I}} \tag{2.7a}$$

Explicitly we have,

$$\mathbb{M}^{\mathbb{I}^*} = \left\{\left(\omega_1^*, \mu_{\Delta^{\mathbb{I}^*}}(\omega_1^*)\right), \left(\omega_2^*, \mu_{\Delta^{\mathbb{I}^*}}(\omega_2^*)\right), \cdots, \left(\omega_j^*, \mu_{\Delta^{\mathbb{I}^*}}(\omega_j^*)\right) \mid j \in \mathbb{J}, \omega \in \Omega\right\}^{\mathbb{I}} \tag{2.7b}$$

where the sequence $\mu_{\Delta^{\mathbb{I}^*}}(\omega_1^*) \ge \mu_{\Delta^{\mathbb{I}^*}}(\omega_2^*) \ge \cdots \ge \mu_{\Delta^{\mathbb{I}^*}}(\omega_j^*)$ expresses the order of degrees of relevance associated with each goal-objective subject to their degrees of irrelevance.

Similarly, we consider an intra-group fuzzy process on the construction of the social goal-objective set for the members in the decision-making core with index set $\mathbb{L}$ with a corresponding fuzzy decision-choice of the form $\Delta^{\mathbb{L}}$, where

$$\Delta^{\mathbb{L}} = \left\{\left(\omega_j, \mu_{\Delta^{\mathbb{L}}}(\omega_j)\right) \mid \mu_{\Delta^{\mathbb{L}}}(\omega_j) = \left[\mu_{\mathbb{M}^{\mathbb{L}}_{\bullet j}}(\omega_j) \wedge \mu_{\mathbb{N}^{\mathbb{L}}_{\bullet j}}(\omega_j)\right]\right\} \tag{2.8}$$

The solution to the fuzzy decision-choice problem of eqn. (2.8) is obtained by reformulating it as a constrained fuzzy mathematical programming problem where the collective degree of relevance is optimized subject to the collective degree of irrelevance.

Theorem 2.2

The $\max_{\omega\in\Omega} \mu_{\Delta^{\mathbb{L}}}(\omega_j) = \bigvee_{\omega\in\Omega}\left[\mu_{\mathbb{M}^{\mathbb{L}}_{\bullet j}}(\omega_j) \wedge \mu_{\mathbb{N}^{\mathbb{L}}_{\bullet j}}(\omega_j)\right]$ is equivalent to

$$\left(\omega^*, \mu_{\Delta^{\mathbb{L}}}(\omega^*)\right) = \left\{\begin{array}{l} = \max_{\omega\in\Omega} \mu_{\mathbb{M}^{\mathbb{L}}_{\bullet j}}(\omega_j) \\ \qquad \text{s.t.}\, \omega\in\left[\mu_{\mathbb{M}^{\mathbb{L}}_{\bullet j}}(\omega_j) - \mu_{\mathbb{N}^{\mathbb{L}}_{\bullet j}}(\omega_j)\right] \le 0 \end{array}\right\}$$

The corresponding optimal ranking sets for the decision-making core may be written as:

$$\mathbb{M}^{\mathbb{L}^*} = \left\{\left(\omega_j^*, \mu_{\Delta^{\mathbb{L}^*}}(\omega_j^*)\right) \mid j\in\mathbb{J}, \omega\in\Omega\right\}^{\mathbb{L}} \tag{2.9}$$

We may notice that $\mu_{\mathbb{M}^{\mathbb{L}^*}}(\omega_j^*) = \mu_{\Delta^{\mathbb{L}^*}}(\omega_j^*)$, and $\mu_{\mathbb{M}^{\mathbb{I}^*}}(\omega_j^*) = \mu_{\Delta^{\mathbb{I}^*}}(\omega_j^*)$, $\forall j\in\mathbb{J}$. For the proofs of theorems (2.1 and 2.2) see [R7] [R7.6] [R7.12] [R7.30] [R7.42].

Both sets $\mathbb{M}^{\mathbb{L}^*}$ and $\mathbb{M}^{\mathbb{I}^*}$ provide us with goal-objective elements that may be ranked by their corresponding membership values (fuzzy numbers) of the intra-group degrees of social relevance as attached to each goal-objective element as agreed upon by the members in each group. We may obtain the final social goal-objective set by considering the intersection of the two sets $\Delta = \mathbb{M}^{\mathbb{L}^*} \cap \mathbb{M}^{\mathbb{I}^*}$ with membership function of the form:

$$\Delta = \mathbb{M}^{\mathbb{I}^*} \cap \mathbb{M}^{\mathbb{L}^*} = \left\{\left(\omega_j^*, \mu_\Delta(\omega_j^*)\right) \mid \mu_\Delta(\omega_j^*) = \left[\mu_{\mathbb{M}^{\mathbb{I}^*}}(\omega_j^*)\right] \wedge \left[\mu_{\mathbb{M}^{\mathbb{L}^*}}(\omega_j^*)\right], j\in\mathbb{J}, \omega\in\Omega\right\} \tag{2.10}$$

where $\mathbb{M}^{\mathbb{I}^*}$ is taken as a constraint on $\mathbb{M}^{\mathbb{L}^*}$. This simply means that the sensitivity to the public on behalf of whom the decision is being made requires that the decision-making core seeks an optimal relevance of each goal-objective set subject to the optimal assessment of the public. This is an inter-group decision problem of assessments of the degree of social relevance that must be attached to each element in the social goal-objective set. Let us keep in mind the computational logic of the min. operator in producing weighted consensus in degrees of relevance and the max operator in producing degree of irrelevance in the general assessments of elements in the social goal-objective set. The analytical structure as seen in terms of principal-agent duality is such that the decision-choice behavior, in accord with the preferences of the agents, is constrained by the preferences of the principal. This is what representative democracy should entail where the elected body is accountable to the public preferences. The constrained democratic collective decision-choice problem is to reconcile the possible conflict in the principal-agent rankings. The decision-choice problem as specified in equation (2.10) may be formulated equivalently as a second fuzzy constrained optimization problem in theorem 2.3 that provides an algorithm for defuzzification.

Theorem 2.3

The fuzzy decision problem $\max_{\omega^*\in\Omega} \mu_\Delta\left(\omega_j^*\right) = \bigvee_{\omega\in\Omega}\left[\mu_{\mathbb{M}^{\mathbb{I}^*}_{\bullet j}}\left(\omega_j^*\right) \wedge \mu_{\mathbb{M}^{\mathbb{L}^*}_{\bullet j}}\left(\omega_j^*\right)\right]$

is transformable to:

$$\left(\omega^{**}, \mu_\Delta\left(\omega^{**}\right)\right) = \left\{\begin{array}{l} = \max_{\omega^*\in\Omega} \mu_{\mathbb{M}^{\mathbb{L}^*}_{\bullet j}}\left(\omega_j\right) \\ \qquad \text{s.t.}\, \omega^* \in \left[\mu_{\mathbb{M}^{\mathbb{L}^*}_{\bullet j}}\left(\omega_j^*\right) - \mu_{\mathbb{M}^{\mathbb{I}^*}_{\bullet j}}\left(\omega_j^*\right)\right] \leq 0 \end{array}\right\}$$

The solution to the fuzzy decision-choice problem appears as a goal-objective set with an optimal degree of relevance attached to each element in the goal-objective set. The new set may be written as:

$$\mathbb{M}^{\mathbb{I}^*\mathbb{L}^*} = \left\{\left(\omega_j^{**}, \mu_\Delta\left(\omega_j^{**}\right)\right) \mid j \in \mathbb{J}, \omega \in \Omega\right\}^{\mathbb{I}^*\mathbb{L}^*} \tag{2.11}$$

The proofs of the theorems (2.1, 2.2 and 2.3) follow the equivalent structures in [R3.1.23] [R6.75] [R7.30] [R7.58]. The set $\mathbb{M}^{\mathbb{I}^*\mathbb{L}^*}$ specifies the elements of the social goal-objective set with the attached optimal degrees of relevance as assessed by the general public and the decision-making core. We shall refer to this set as the primary optimal social goal-objective set. Its implementation in the organization of the production-distribution-consumption system may then proceed in accord with the elements with the highest magnitude of degree of relevance. The elements for immediate implementation may be obtained by constructing the set of $\alpha - level$ of relevance to obtain a secondary optimal social goal-objective set as:

$$\mathbb{M}_\alpha^{\mathbb{I}^*\mathbb{L}^*} = \left\{\left(\omega_j^{**}, \mu_\Delta\left(\omega_j^{**}\right)\right) \mid \mu_\Delta\left(\omega_j^{**}\right) \geq \alpha \in [0,1]\, j \in \mathbb{J}, \omega \in \Omega\right\}^{\mathbb{I}^*\mathbb{L}^*} \tag{2.12}$$

In other words, we select all those objective whose degrees of relevance is greater than α. Figure 2.18 shows the general cognitive geometry and computational scheme.

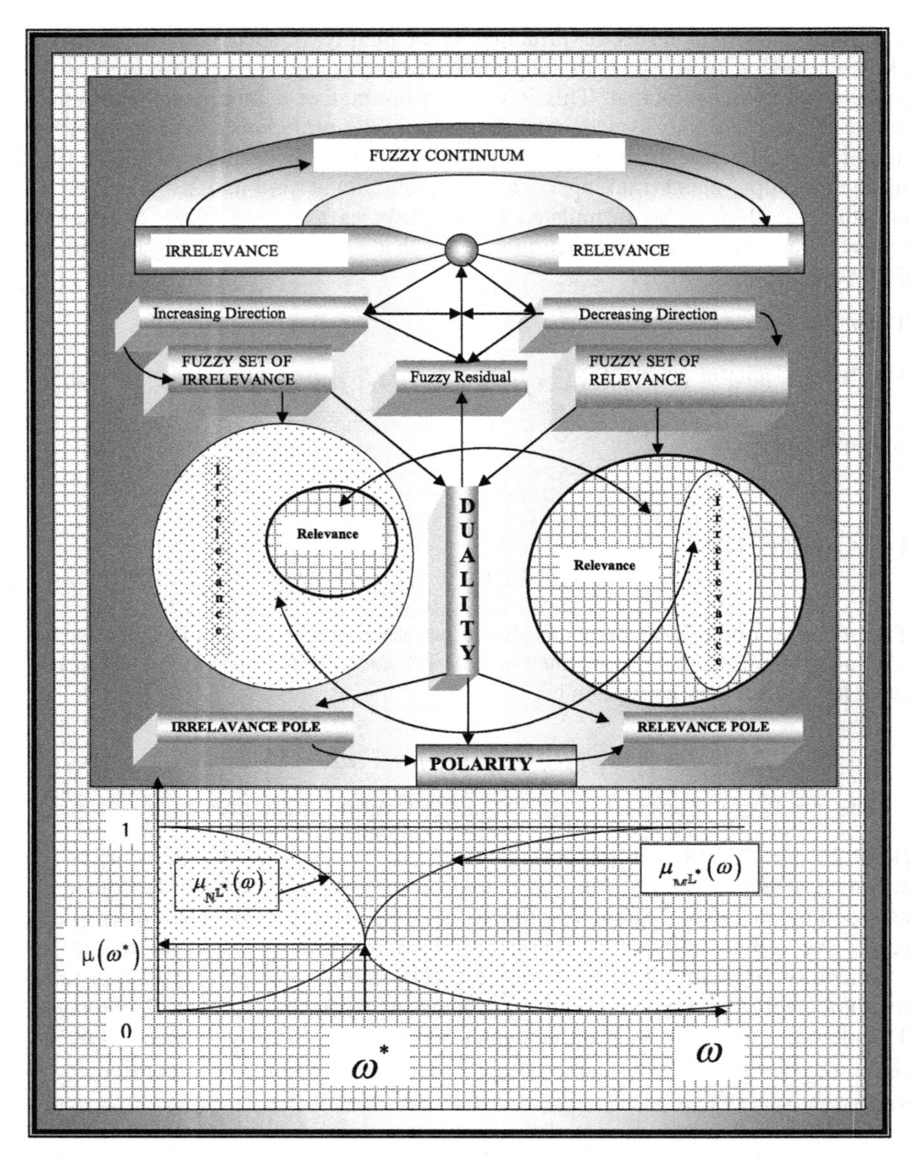

Fig. 2.19 A Cognitive Geometry and Computational Scheme for Constructing the Social Goal-Objective set on the Degrees of Relevance and Irrelevance

The solution, defined by eqns. (2.7), (2.9) and (2.11), may also be obtained by a simultaneous combination of equations (2.3) and (2.4) to define fuzzy decision set $\mathbb{M}^{\mathbb{IL}} = \left(\mathbb{M}^{\mathbb{I}} \cap \mathbb{M}^{\mathbb{L}}\right)$ that provides an aggregation of degrees of relevance in a fuzzy domain, and $\mathbb{N}^{\mathbb{IL}} = \left(\mathbb{N}^{\mathbb{I}} \cup \mathbb{N}^{\mathbb{L}}\right)$ which defines an aggregation of degrees of irrelevance, and the decision-choice problem as:

$$\max_{\omega\in\Omega} \Delta^{\mathbb{IL}} = \max_{\omega\in\Omega}\left(\mathbb{M}^{\mathbb{IL}} \cap \mathbb{N}^{\mathbb{IL}}\right) = \max_{\omega\in\Omega}\left(\mathbb{M}^{\mathbb{I}} \cap \mathbb{M}^{\mathbb{L}}\right) \cap \left(\mathbb{N}^{\mathbb{I}} \cup \mathbb{N}^{\mathbb{L}}\right) \quad (2.13)$$

Equation (2.13) may be expressed in membership equivalence as

$$\left\{\begin{array}{ll} \mu_{\mathbb{M}^{\mathbb{IL}}}(\omega) = \left\{\left[\bigwedge_{i\in\mathbb{I}}\left[\mu_{\mathbb{M}_{ij}}(\omega_j)\right]\right] \bigwedge \left[\bigwedge_{i\in\mathbb{L}}\left[\mu_{\mathbb{M}_{ij}}(\omega_j)\right]\right]\right\} & \text{(REL)} \quad \text{a} \\ \left\{ \mid \mu_{\mathbb{N}^{\mathbb{IL}}}(\omega_j) = \left[\bigvee_{i\in\mathbb{I}}\left[\mu_{\mathbb{N}_{ij}}(\omega_j)\right]\right] \bigvee \left[\bigvee_{i\in\mathbb{L}}\left[\mu_{\mathbb{N}_{ij}}(\omega_j)\right]\right]\right\} & \text{(IRR)} \quad \text{b} \end{array}\right\} \quad (2.14)$$

The fuzzy decision-choice elements from eqn. (2.14) may be written as:

$$\left\{\begin{array}{l} \mu_{\Delta}^{\mathbb{IL}}(\omega) = \mu_{\mathbb{M}^{\mathbb{IL}}}(\omega) \wedge \mu_{\mathbb{N}^{\mathbb{IL}}}(\omega_j) \\ = \left\{\left[\bigwedge_{i\in\mathbb{L}}\left[\mu_{\mathbb{M}_{ij}}(\omega_j)\right]\right] \wedge \left[\bigwedge_{i\in\mathbb{L}}\left[\mu_{\mathbb{M}_{ij}}(\omega_j)\right]\right]\right\} \wedge \left\{\left[\bigvee_{i\in\mathbb{I}}\left[\mu_{\mathbb{N}_{ij}}(\omega_j)\right]\right] \vee \left[\bigvee_{i\in\mathbb{L}}\left[\mu_{\mathbb{N}_{ij}}(\omega_j)\right]\right]\right\} \end{array}\right\} \quad (2.15)$$

The decision problem described by eqn. (2.15) uses the min-max algorithms to reconcile the conflicts in the collective assessments of degrees of relevance of the elements of the social goal-objective set by the public, and the decision-making core in order to obtain the optimal social goal-objective set that will support the national interest and social vision. The method of the decision-choice technique is to help in computing the optimal assessment value of the degree of relevance of each goal-objective element. The result is a distribution of optimal values of relevance which is then used to rank the elements in the goal-objective set. We compute a fuzzy aggregate of the preferences of both the population (principal) and the decision-making core (the agent) for relevance and irrelevance separately. The degree of relevance is then optimized subject to the degree of irrelevance as captured by their respective membership characteristic functions. In the fuzzy optimizing domain, the decision is equivalent in solving the fuzzy mathematical programming problem defined in Theorem 2.4.

Theorem 2.4

The fuzzy decision problem $\max_{\omega\in\Omega} \mu_{\Delta^{\mathbb{IL}}}(\omega_j) = \bigvee_{\omega\in\Omega}\left[\left(\mu_{\mathbb{M}^{\mathbb{I}}{}_{\bullet j}}(\omega_j)\right) \wedge \left(\mu_{\mathbb{M}^{\mathbb{L}}{}_{\bullet j}}(\omega_j)\right)\right]$ is equivalent to the fuzzy mathematical programming problem

$$\max_{\omega\in\Omega} \mu_{\Delta^{\mathbb{IL}}}(\omega) = \left\{\begin{array}{l} \max_{\omega\in\mathbb{Q}} \mu_{\mathbb{M}^{\mathbb{IL}}}(\omega) \\ \qquad \text{s.t. } \mathbb{Q} = \left\{\omega \in \Omega \mid \left[\mu_{\mathbb{M}^{\mathbb{IL}}}(\omega) - \mu_{\mathbb{N}^{\mathbb{IL}}}(\omega_j)\right] \leq 0\right\} \end{array}\right\}$$

The proof of this theorem is standard (See [R3.1.2] [R6.75] [R7.30]). The solution yields an optimal set that is ordered by the optimal degree of relevance constrained by degree of irrelevance. The optimality is defined in terms of reconciliation between the relevant-irrelevant assessment values of the general public (the principal) and the decision-making core (the agent). Let us notice that the initial constraint is transformed into a fuzzy set of residuals. The optimal solution set is of the form:

$$\mathbb{M}^{\mathbb{IL}^*} = \left\{ \left(\omega_j^*, \mu_{\Delta\mathbb{IL}}\left(\omega_j^*\right) \right) \mid j \in \mathbb{J}, \omega \in \Omega \right\}^{\mathbb{IL}} \tag{2.16}$$

The value $\mu_{\Delta\mathbb{IL}}\left(\omega_j^*\right)$, as the optimal degree of relevance attached to each objective, allows all the elements in the goal-objective set to be ranked in the order of relevance (see[R3.1.23]). The ordered goal-objective set may be reduced in size for implementation by developing an $\alpha - level$ (fixed level) set on the basis of resource availability. The fuzzy mathematical logic, in posing the social decision-choice problem in the soft-computing space, is based on the analytical duality which expresses the principles of opposites in a continuum, with conflicts within the duality due to contradictions in the continuum and commonness as a resolution in decision-choice action (see Figure 2.18). The same elements give an important interpretation of intersection of fuzzy sets where one is an objective set, and the other is a constrained set. The decision-choice problem is to optimize the membership function of the degree of relevance subject to the corresponding membership function of degree of irrelevance. Notice that the difference of the two functions expresses the functional structure of the fuzzy residual. Associated with each objective $j \in \mathbb{J}$ is an optimal degree of relevance $\alpha_j^* = \mu\left(\omega_j\right)$.

These optimal values are used to rank the elements in the goal-objective set in order of relevance or irrelevance relative to the national interest and social vision. The problem, as stated, is equivalent to optimization of benefits subject to costs in a benefit-cost space where both benefits and costs are defined, assessed and computed in a fuzzy mathematical space (see[R3.1.19].[R3.1.23] [R3.1.24]). The use of a toolbox of fuzzy paradigm in the decision-choice system rests on the notion that the information structure of the social goal-objective set is defective requiring the use of non-classical techniques for processing. The defectiveness of the information structure, as it has been discussed and illustrated in Figure 2.2.0.1, is composed of two information substructures of qualitative and quantitative dispositions. The qualitative disposition of the information structure is made up of the vagueness information sub-structure and deceptive information sub-structure. The deceptive information sub-structure is made up of disinformation sub-structure and misinformation sub-structure. The quantitative disposition of the information sub-structure is composed of a limited volume of information. All these information sub-structures exist in a complex system of interaction. In the decision-choice actions, the quantitative information may be complicated by its interactions with the qualitative disposition to define more complex mathematical space. It is here that we encounter either vague or inexact probability. The complex mathematical space to treat such a phenomenon in the problems of sciences has been referred to as a fuzzy-stochastic mathematical space with corresponding fuzzy-random variable or random-fuzzy variable [R15.13] [R15.14].

2.4 Information Sharing and Fuzzy Decision Interactions

The collective assessment values in eqns. (2.1) and (2.2) and the corresponding fuzzy decisions do not allow evaluative interaction through discussions and information sharing. The individual and collective assessments of relevance and degrees of relevance may be constrained by lack of accuracy of information. In this case, transparency is sacrificed leading to some important compromises on the principles of democratic decision-making that may be enhanced by increased information and knowledge sets through the principle of freedom of association, debating and information sharing. Some important elements of the democratic collective decision-choice system are the quality of information, honesty in information dissemination and sharing before a collective decision or assessment is made. These are especially relevant when the collective decision-choice actions are to be guided by the principles of the democratic collective decision-choice system. Information restrictions of any form lead to *informationally* suboptimal social decision-choice actions in democratic decision-making societies given the individual preferences, and that the decision-choice system is to be sensitive to the individual preferences.

An information restriction, while placing a democratic society at suboptimal levels of social decision-choice system, kills transparency in the applications of individual and collective sovereignties. It further creates conditions for corruption and rent-seeking activities, where social benefits are directed to those who are privileged to have the information, while at the same time social costs are directed to those who are deprived of the relevant information that surrounds the social decisions regarding the creation of the social goal-objective elements and the national interest which the society may pursue. In other words, the creation of an information asymmetry creates decision-choice sub-optimality, conditions of corruption and rent-seeking, where costs are socialized to the public and benefits are privatized to the private individuals who hold the socially important information. Information restriction creates information asymmetry which then distorts the general cost-benefit imputations among members in the production-distribution-consumption space. This distortion is amplified when the sector degenerates into a game of gambling, especially in the financial sector. The intensity of the distortions in the individual and collective cost-benefit imputations and the understanding of the social cost-benefit distribution will depend on the differential distribution of the information asymmetry among the members in the social set-up.

The operational nature of the information asymmetry in the society is to create illusions in individual preferences and asymmetric valuations of the costs and benefits associated with the social decision-choice actions. In this case, the effective participation in establishing the optimal goal-objective set is distorted in the favor of those who hold special information relevant to the decision-choice action and against those who are dis-informed and mis-informed. If the decision-making process admits of transparency, information and efficient dissemination without national or institutional security-classifications of information, then we must use the fuzzy product construct of the fuzzy decision space to obtain the collective assessments. Such product decision assessments appear as fuzzy sets whose membership functions for the general group $\mathbb{I}$ may be written as:

$$\left\{\begin{array}{l} \mathbb{M}^{\mathbb{I}} = \left\{\left(\omega_j, \mu_{\mathbb{M}._j}(\omega_j)\right) \mid \mu_{\mathbb{M}._j}(\omega_j) = \left[\!\left[\prod_{i\in\mathbb{I}}\left[\mu_{\mathbb{M}_{ij}}(\omega_j)\right], \omega_j \in \Omega, j \in \mathbb{J}\right]\!\right]^{\mathbb{I}}\right\} (\text{REL})\ \text{a} \\ \\ \mathbb{N}^{\mathbb{I}} = \left\{\left(\omega_j, \mu_{\mathbb{N}._j}(\omega_j)\right) \mid \mu_{\mathbb{N}._j}(\omega_j) = \left[\!\left[\prod_{i\in\mathbb{I}}\left[\mu_{\mathbb{N}_{ij}}(\omega_j)\right], \omega_j \in \Omega, j \in \mathbb{J}\right]\!\right]^{\mathbb{I}}\right\} (\text{IRR})\ \text{b} \end{array}\right\} \quad (2.17)$$

Similarly, the product decision-choice assessments for the decision-making core may be written as:

$$\left\{\begin{array}{l} \mathbb{M}^{\mathbb{L}} = \left\{\left(\omega_j, \mu_{\mathbb{M}._j}(\omega_j)\right) \mid \mu_{\mathbb{M}._j}(\omega_j) = \left[\!\left[\prod_{i\in\mathbb{L}}\left[\mu_{\mathbb{M}_{ij}}(\omega_j)\right], \omega_j \in \Omega, j \in \mathbb{J}\right]\!\right]^{\mathbb{L}}\right\} (\text{REL})\ \text{a} \\ \\ \mathbb{N}^{\mathbb{L}} = \left\{\left(\omega_j, \mu_{\mathbb{N}._j}(\omega_j)\right) \mid \mu_{\mathbb{N}._j}(\omega_j) = \left[\!\left[\prod_{i\in\mathbb{L}}\left[\mu_{\mathbb{N}_{ij}}(\omega_j)\right], \omega_j \in \Omega, j \in \mathbb{J}\right]\!\right]^{\mathbb{L}}\right\} (\text{IRR})\ \text{b} \end{array}\right\} \quad (2.18)$$

Equations (2.17) and (2.18) may be combined to develop a fuzzy product decision problem $\Delta^{\mathbb{IL}}$ with a corresponding membership function $\mu_{\Delta^{\mathbb{IL}}}(\omega)$.

The interactive decision assessment process requires us to combine group product interactions for the relevance and irrelevance to develop a distribution of the degrees of collective relevance and irrelevance (importance) between the decision-making core $\mathbb{L}$ and the individual democratic participants in the index set $\mathbb{I}$. This allows the construct of the social goal-objective set under increasing openness in democratic decision-choice systems. The decision-making core will be the Congress (The Houses of Senate and Representatives) in the United States of America or the members of the Parliament in any parliamentary system of social formations and others. The decision-choice structure, as presented, may also apply to localities, broadly defined and with some political autonomy of governance within the nation that is formed under the principle of federation. The combination of eqns. (2.17) and (2.18) allows us to develop an aggregate representation of relative relevance as:

$$\left\{\begin{array}{l} \hat{\mathbb{M}}_{\mathbb{IL}} = \left\{\left(\omega, \hat{\mu}_{\hat{\mathbb{M}}_{\mathbb{IL}}}(\omega)\right) \mid \hat{\mu}_{\hat{\mathbb{M}}_{\mathbb{IL}}}(\omega) = \left[\!\left[\prod_{i\in\mathbb{I}}\left[\mu_{\hat{\mathbb{M}}_{ij}}(\omega_j)\right]\right]\!\right] \wedge \left[\!\left[\prod_{j\in\mathbb{L}}\left[\mu_{\hat{\mathbb{M}}_{ij}}(\omega_j)\right]\right]\!\right]\right\} \text{a (REL)} \\ \hat{\mathbb{N}}_{\mathbb{IL}} = \left\{\left(\omega, \hat{\mu}_{\hat{\mathbb{N}}_{\mathbb{IL}}}(\omega)\right) \mid \hat{\mu}_{\hat{\mathbb{N}}_{\mathbb{IL}}}(\omega) = \left[\!\left[\prod_{i\in\mathbb{I}}\left[\mu_{\hat{\mathbb{N}}_{ij}}(\omega_j)\right]\right]\!\right] \wedge \left[\!\left[\prod_{j\in\mathbb{L}}\left[\mu_{\hat{\mathbb{N}}_{ij}}(\omega_j)\right]\right]\!\right]\right\} \text{b (IIR)} \end{array}\right\} \quad (2.19)$$

The information on irrelevance is used as the constraint on the optimal assessment of the degree of relevance (importance) associated with each social objective to specify the information–interactive assessments as a fuzzy decision-choice problem of the weighted consensus of the democratic collective decision-choice system in the form: $\hat{\Delta}^{\mathbb{IL}} = \hat{\mathbb{M}}_{\mathbb{IL}} \cap \hat{\mathbb{N}}_{\mathbb{IL}}$ with a membership characteristic function defined in the form:

$$\hat{\mu}_{\hat{\Delta}^{\mathbb{IL}}}(\omega) = \left[\left(\hat{\mu}_{\hat{\mathbb{M}}^{\mathbb{IL}}}(\omega)\right) \wedge \left(\hat{\mu}_{\hat{\mathbb{N}}^{\mathbb{IL}}}(\omega)\right)\right] \tag{2.20}$$

The problem is to maximize $\mu_{\hat{\Delta}^{\mathbb{IL}}}(\omega)$ for each $j \in \mathbb{J}$.

Theorem 2.5

The $\underset{\omega \in \Omega}{\text{opt}} \mu_{\hat{\Delta}^{\mathbb{IL}}}(\omega) = \underset{\omega \in \Omega}{\text{opt}} \left[\left(\hat{\mu}_{\hat{\mathbb{M}}^{\mathbb{IL}}}(\omega)\right) \wedge \left(\hat{\mu}_{\hat{\mathbb{N}}^{\mathbb{IL}}}(\omega)\right)\right]$ is equivalent to

$$\begin{cases} \underset{\omega \in \mathbb{Q}}{\text{opt}} \left(\mu_{\mathbb{M}^{\mathbb{IL}}}(\omega)\right) \\ \qquad \text{s.t. } \mathbb{Q} = \left[\hat{\mu}_{\hat{\mathbb{M}}^{\mathbb{IL}}}(\omega) - \hat{\mu}_{\hat{\mathbb{N}}^{\mathbb{IL}}}(\omega)\right] \le 0 \end{cases}$$

Let us notice in a continuous form that as the degree of relevance of a goal-objective element increases, that of irrelevance falls $\left(\frac{d\hat{\mu}_{\mathbb{N}_{\mathbb{IL}}}(\omega)}{d\omega}\right) \le 0$ and $\left(\frac{d\hat{\mu}_{\mathbb{M}_{\mathbb{IL}}}(\omega)}{d\omega}\right) \ge 0$. This is consistent with fuzzy duality, fuzzy continuum, the principles of opposites and contradictions with the implied logic. The solution of eqn. (2.20), after applying the theorem 2.5, appears in the form:

$$\hat{\mathbb{M}}^{\mathbb{IL}^*} = \left\{\left(\omega_j^*, \hat{\mu}_{\hat{\Delta}^{\mathbb{IL}}}\left(\omega_j^*\right)\right) \mid j \in \mathbb{J}, \omega \in \Omega\right\}^{\mathbb{IL}} \tag{2.21}$$

The term $\hat{\mathbb{M}}^{\mathbb{IL}^*}$ is a set of all social goal-objective elements with a distribution of optimal degrees of relevance to all the elements. The set of the optimal degrees of relevance imposes a ranking possibility, and allows us to arrange the goal-objective elements in an ascending or descending order of fuzzy magnitudes. In other words, equations (2.7a), (2.16) and (2.21) are all defined in an ordered field of relevance and irrelevance. The concepts of relevance and irrelevance as linguistic variables of the same entity are seen in an ordered continuum, where as the value of the characteristic function of the linguistic variable of irrelevance decreases, the value of the membership characteristic function of the linguistic variable of relevance increases in value and vice versa. Thus, corresponding to every value of the variable relevance, there is a value of the variable irrelevance as its conditional support in establishing the analytical duality. The measure of the individual optimal degree of belonging to relevance relative to the national interest and social vision provides qualified acceptance.

2.5 Alternative Computational Schemes: Solely on Relevance

Alternatively, we may consider the citizen's aggregate assessment and the decision-making core's aggregate assessment of the degree of relevance as attached to each objective in constructing the goal-objective set. In this frame, it is useful to consider the optimization of the decision-choice objective as the assessment of the decision-making core, where the community assessment is used as a constraint on the selection of the decision-making core (e.g. the elected officials). In other words, the elected officials or the members of the representative decision-making core compute the degree of relevance attached to any goal-objective element by taking into account the assessments of degree of relevance of the legal population. This is consistent with a political system and process where the decision-making core serves the public interest. In this way, we take eqn. (2.14a), construct the decision-choice problem as $\Delta^{\mathbb{IL}} = \left(\mathbb{M}^{\mathbb{I}} \cap \mathbb{M}^{\mathbb{L}}\right)$ for the case of non-decision interaction or non-information sharing and compute the distribution of optimal degrees of relevance by Theorem 2.6

Theorem 2.6

$$\underset{\omega_j \in \Omega}{\text{opt}}\, \mu_{\Delta^{\mathbb{IL}}}(\omega_j) = \begin{cases} \underset{\omega \in \mathbb{Q}}{\text{opt}} \left\| \bigwedge_{i \in \mathbb{L}} \left[\mu_{\mathbb{M}_{ij}}(\omega_j)\right] \right\| \\ \text{s.t. } \mathbb{Q} = \left\{ \omega_j \in \Omega \,|\, \left\| \bigwedge_{i \in \mathbb{I}} \left[\mu_{\mathbb{M}_{ij}}(\omega_j)\right] \right\| - \left\| \bigwedge_{i \in \mathbb{L}} \left[\mu_{\mathbb{M}_{ij}}(\omega_j)\right] \right\| \leq 0 \right\} \end{cases}$$

For the case of interaction and information sharing we use the eqn. (2.19a) and construct a fuzzy product decision of the form:

$$\mu_{\Delta^{\mathbb{IL}}}(\omega_j) = \left\| \prod_{l \in \mathbb{L}} \left[\mu_{\mathbb{M}_{lj}}(\omega_j)\right] \wedge \prod_{i \in \mathbb{I}} \left[\mu_{\mathbb{M}_{ij}}(\omega_j)\right] \right\|, \quad \omega_j \in \Omega \tag{2.22}$$

We then compute the optimal value of the membership function with Theorem 2.7 by finding the value $\omega_j^* \in \Omega$ that optimizes $\mu_\Delta(\omega_j)$.

Theorem 2.7

$$\underset{\omega_j\in\Omega}{\mathrm{opt}}\,\mu_{\Delta^{\mathbb{L}}}(\omega_j)=\begin{cases}\sup\limits_{\omega_j\in\mathbb{Q}}\left[\left\{\prod\limits_{l\in\mathbb{L}}\mu_{\mathbb{M}_{lj}}(\omega_j)\right\}\right]\\ \qquad\text{s.t. } \mathbb{Q}=\left\{\omega_j\in\Omega \mid \left[\prod\limits_{l\in\mathbb{L}}\left[\mu_{\mathbb{M}_{lj}}(\omega_j)\right]-\prod\limits_{i\in\mathbb{I}}\left[\mu_{\mathbb{M}_{ij}}(\omega_j)\right]\right]\le 0\right\}\end{cases}$$

The system of solutions to the problem as specified in eqns. (2.22) in Theorems (2.6 and 2.7) will appear as:

$$\hat{\mathbb{M}}^{\mathbb{L}^*}=\left\{\left(\omega_j^*,\hat{\mu}_{\hat{\Delta}^{\mathbb{L}}}(\omega_j^*)\right)\mid j\in\mathbb{J},\omega\in\Omega\right\}^{\mathbb{L}} \tag{2.23}$$

Eqn. (2.23) provides a distribution of optimal degrees of relevance of the elements in the goal-objective set relative to the national interest and social vision. The elements may be arranged as an ordered set induced by the optimal degrees of relevance as in eqn. (2.24):

$$\mathbb{G}^{\mathbb{L}^*}=\left\{\omega\in\Omega \mid \begin{pmatrix}\omega_1^*\\ \hat{\mu}_{\hat{\Delta}^{\mathbb{L}}}(\omega_1^*)\end{pmatrix}\succcurlyeq\begin{pmatrix}\omega_{1+j}^*\\ \hat{\mu}_{\hat{\Delta}^{\mathbb{L}}}(\omega_{1+j}^*)\end{pmatrix} j\in\mathbb{J}\right\}^{\mathbb{L}} \tag{2.24}$$

The symbol $\succcurlyeq$ means socially more relevant objective relative to the national interest and hence preferred as compared to others, and hence for $\forall j\in\mathbb{J}$ $\omega_1\succcurlyeq\omega_{1+j}\Rightarrow\hat{\mu}_{\hat{\Delta}^{\mathbb{L}}}(\omega_1^*)\ge\hat{\mu}_{\hat{\Delta}^{\mathbb{L}}}(\omega_{1+j}^*)$. From eqn. (2.24) we can develop a fixed-level or an $\alpha-\text{level}$ set by the method of fuzzy decomposition, where the α is determined by conditions of resource availability. In this respect, the admissible social goal-objective set for implementation in support of the national interest, $\mathbb{G}_\alpha^{\mathbb{L}}$ is.

$$\mathbb{G}_\alpha^{\mathbb{L}^*}=\left\{\omega\in\Omega \mid \begin{pmatrix}\omega_1^*\\ \hat{\mu}_{\hat{\Delta}^{\mathbb{L}}}(\omega_1^*)\end{pmatrix}\succcurlyeq\begin{pmatrix}\omega_{1+j}^*\\ \hat{\mu}_{\hat{\Delta}^{\mathbb{L}}}(\omega_{1+j}^*)\end{pmatrix}\text{ and }\hat{\mu}_{\hat{\Delta}^{\mathbb{L}}}(\omega_{1+j}^*)\ge\alpha,\forall j\in\mathbb{J}\right\}^{\mathbb{L}} \tag{2.25}$$

2.6 Weighted Preferences and Goal-Objective Formation under Democratic Principles

The fuzzy decision processes, as specified in equations (2.5), (2.6) and (2.19), (2.20) and (2.22), are such that equal weights are assigned to the evaluators irrespective of the position that one occupies in the social decision-choice system. This is a case where individual preferences are valued equally in the democratic social formation. Equality of preferences in goal-objective formation and national interest definition would require certain conditions to be met in the democratic decision-making system. The conditions are such that, among other things, the following postulates must be met:

P.1 *The postulate of information sufficiency*:
All individuals must have equal access to all relevant information on the basis of which the national interest and social vision are set as well as the nation is governed and managed.

P.2 *The postulate of information asymmetry*:
There should not be asymmetric information conditions between the legal general population (the principal) and the decision-making core (the agent). There is no deceptive information structure.

P.3 *The postulate of individual information-processing capacity*:
All individual members, involved in the collective decision-choice system, must have equal information-processing capacity and ability within reasonable limits to determine what is good for the collective.

P.4 *The postulate of equal participation*
All individuals must have equal access to, and must participate in one form or the other.

These postulates are essential for efficient functioning of the democratic collective decision-choice system of the political economy.

Even if these principles are met in the governance and management of the democratic society, the implementation of the elements obtained from the social goal-objective formation relative to the implied national interest and social vision, on the principle of equal participation, will be difficult to implement in our modern industrial social set-ups. Furthermore, income distribution, power relations and resource availability in practice are such that equal weights on preferences from the beginning are practically impossible. To get around this difficulty where participation is *a right* that confers on the individual an opportunity that may or may not be exercised, we introduce the concept of *weighted preferences* which will capture the impact of individual preferences on the goal-objective selection process. The weighting process admits of irregularities in the decision-choice process due to ignorance, choice of non-participation, possible resource asymmetry, information-knowledge asymmetry that may arise, rates of social involvement and expert judgments. In this respect, we partition the set of non-decision making core into *social classes*. The criteria for partitioning the society into these classes will depend on time, social conditions, the level of national development and politico-legal institutional framework of democratic governance and management. These classes may vary in terms of ignorance, information-processing capacity, commitment to participate and others. Members of each class are assumed to have equal weights within the class while such weights vary across the groups; in other words, there is intra-group equality and inter-group differences in preference imputations. Preferences, quantity and quality of social information-knowledge structure and cognitive limitations of processing capabilities harbor within them ambiguities, vagueness and subjectivities that tend to affect equally weighted preferences in the social decision-choice process.

Let the index set of the classes be $\mathbb{T}$ where $(\mathbb{I}_\tau \subset \mathbb{I}, \tau \in \mathbb{T})$ is the index set of the $\tau-\text{class}$, such that $\bigcup_{\tau\in\mathbb{T}} \mathbb{I}_\tau = \mathbb{I}$ and $\bigcap_{\tau\in\mathbb{T}} \mathbb{I}_\tau = 0$. The condition $\bigcap_{\tau\in\mathbb{T}} \mathbb{I}_\tau = 0$ may be relaxed to incorporate overlapping classes. We may now specify the distributive weights of preference in the goal-objective selection process as $\beta_\tau \in [0,1],\ \tau \in \mathbb{T}$. If the value

$\beta_\tau > 0$ and $\beta_{i\tau} = 0,\ i \in \mathbb{I}_\tau\ \tau \in \mathbb{T}$, then we interpret the condition as, the ith-individual in $\tau - class$ has no impact on the goal-objective decision process. On the other hand, if $\beta_\tau = 0$, then all the members in the $\tau - class$ have no impact in the goal-objective formation process. The explanation of the condition $\beta_\tau = 0$ may be found from the legal structure that defines the operations in the political sphere. The values of β_τ's may be constructed from a number of factors such as income, education and social awareness and others. With this in mind, we can specify the two assessments of degrees of relevance and irrelevance by the general population on the basis of intra-class weighted degrees of relevance and irrelevance for a given information-knowledge $\mathbb{K}^{\mathbb{I}} \subset \mathbb{K}^{\mathbb{K}} \subseteq \mathbb{K}$ as:

$$\left\{\begin{array}{ll} \mathbb{M}^{\mathbb{I}} = \left\{ \left(\omega_j, \mu_{\mathbb{M}._j}(\omega_j)\right) \mid \mu_{\mathbb{M}._j}(\omega_j) = \left[\!\left[\bigwedge_{\tau\in\mathbb{T}} \bigwedge_{i\in\mathbb{I}_\tau} \left[\beta_{i\tau}\mu_{\mathbb{M}_{ij}}(\omega_j)\right],\ \omega_j \in \Omega,\ j \in \mathbb{J} \right]\!\right] \right\}^{\mathbb{I}} & \text{(REL) a} \\ \mathbb{N}^{\mathbb{I}} = \left\{ \left(\omega_j, \mu_{\mathbb{N}._j}(\omega_j)\right) \mid \mu_{\mathbb{N}._j}(\omega_j) = \left[\!\left[\bigvee_{\tau\in\mathbb{T}} \bigvee_{i\in\mathbb{I}_\tau} \left[\beta_{i\tau}\mu_{\mathbb{N}_{ij}}(\omega_j)\right],\ \omega_j \in \Omega,\ j \in \mathbb{J} \right]\!\right] \right\}^{\mathbb{I}} & \text{(IRR) b} \end{array}\right\} \quad (2.26)$$

The ranking of the elements of the social goal-objective set, associated with the general legal public, is induced by the solutions to the optimal fuzzy decision-choice problems of each goal-objective element. This may be specified as $\Delta_{\mathbb{I}_j} = \mathbb{M}._j \cap \mathbb{N}._j$ which is $\mu_{\Delta_{\mathbb{I}_j}}(\omega_j) = \left[\mu_{\mathbb{M}._j}(\omega_j)\right] \wedge \left[\mu_{\mathbb{N}._j}(\omega_j)\right]$. The optimal ranking problem requires a fuzzy optimizing solution that can be obtained with Theorem 2.8 with the understanding that the dot $(\cdot)$ indicates that fuzzy aggregations have been performed.

Theorem 2.8

$$\underset{\omega_i\in\Omega}{\text{opt}}\,\mu_{\Delta_{\mathbb{I}_j}}(\omega_j) = \left[\!\left[\begin{array}{l} \underset{\omega_i\in\mathbb{Q}}{\text{opt}}\left[\mu_{\mathbb{M}._j}(\omega_j)\right]^{\mathbb{I}} \\ \qquad \text{s.t. } \mathbb{Q} = \left\{\omega_i \in \Omega \mid \left[\mu_{\mathbb{M}._j}(\omega_j)\right] - \left[\mu_{\mathbb{N}._j}(\omega_j)\right] \le 0\right\}^{\mathbb{I}} \end{array} \right.$$

The theorem simply states that the solution to the weighted decision-choice problem is obtained by optimizing the membership function of the weighted assessment values of relevance, subject to the net value between the assessment values of weighted relevance and irrelevance, and that this is equivalent to optimizing the fuzzy decision function. The solutions to these fuzzy optimization problems allow us to construct the set of optimal ranking $\mathbb{G}^{\mathbb{I}^*}$ in the order of decreasing degrees of relevance for the general legal population in the form:

$$\mathbb{G}^{\mathbb{I}^*} = \left\{ \omega_i \in \Omega \mid \begin{pmatrix} \omega_1^* \\ \widehat{\mu}_{\hat{\Delta}^{\mathbb{I}}}\left(\omega_1^*\right) \end{pmatrix} \succcurlyeq \begin{pmatrix} \omega_{1+j}^* \\ \widehat{\mu}_{\hat{\Delta}^{\mathbb{I}}}\left(\omega_{1+j}^*\right) \end{pmatrix}, \ \forall j \in \mathbb{J} \text{ and } \mathbb{K}^{\mathbb{I}} \right\}^{\mathbb{I}} \quad (2.27)$$

Similarly, we may construct weighted assessments of the decision-making core by assigning differential weights $w_{ik} \in [0,1]$ to their assessments. The weights depend on the party affiliation. Let us suppose that the members in the decision-making core are group under political party affiliations. Let the index set of the members in each political party be $\mathbb{L}_k$, $k \in \mathbb{P}$, such that $\mathbb{L} = \bigcup_{k \in \mathbb{P}} \mathbb{L}_k$ and $\bigcap_{k \in \mathbb{P}} \mathbb{L}_k = 0$ with $\mathbb{P}$ defining the index set of the available political parties in the decision-making core. The dot indicates after fuzzy aggregation. We may now specify a weighted fuzzy set of relevance and irrelevance as in eqn.(2.28), given the information knowledge structure $\mathbb{K}^{\mathbb{L}} \subseteq \mathbb{K}$ available to the social decision-making core.

$$\left\{ \begin{array}{ll} \mathbb{M}^{\mathbb{L}} = \left\{ \left(\omega_j, \mu_{\mathbb{M}._j}\left(\omega_j\right)\right) \mid \mu_{\mathbb{M}._j}\left(\omega_j\right) = \left[\!\left[\bigwedge_{k \in \mathbb{P}} \bigwedge_{i \in \mathbb{L}_k} \left[w_{ik} \mu_{\mathbb{M}_{ij}}\left(\omega_j\right) \right], \ \omega_j \in \Omega, \ j \in \mathbb{J} \right]\!\right]^{\mathbb{L}} \right\} & \text{(REL) a} \\ \\ \mathbb{N}^{\mathbb{L}} = \left\{ \left(\omega_j, \mu_{\mathbb{N}._j}\left(\omega_j\right)\right) \mid \mu_{\mathbb{N}._j}\left(\omega_j\right) = \left[\!\left[\bigvee_{k \in \mathbb{P}} \bigvee_{i \in \mathbb{L}_k} \left[w_{ik} \mu_{\mathbb{N}_{ij}}\left(\omega_j\right) \right], \ \omega_j \in \Omega, \ j \in \mathbb{J} \right]\!\right]^{\mathbb{L}} \right\} & \text{(IRR) b} \end{array} \right\} \quad (2.28)$$

The problem of fuzzy optimal ranking may be posed as constrained interactions between assessment values of fuzzy sets of relevance and irrelevance. This implies a formation of a fuzzy decision-choice problem of the form $\Delta^{\mathbb{L}} = \left(\mathbb{M}._j \cap \mathbb{N}._j\right)$ which also implies that $\mu_{\Delta^{\mathbb{L}}}\left(\omega_i\right) = \mu_{\mathbb{M}._j}\left(\omega_j\right) \wedge \mu_{\mathbb{N}._j}\left(\omega_j\right)$. The optimal ranking of the elements in the social goal-objective set requires an optimization of $\mu_{\Delta^{\mathbb{L}}}\left(\omega_i\right) = \mu_{\mathbb{M}._j}\left(\omega_j\right) \wedge \mu_{\mathbb{N}._j}\left(\omega_j\right)$ whose solution is obtained by using Theorem 2.9.

Theorem 2.9

$$\underset{\omega_i \in \Omega}{\text{opt}} \mu_{\Delta_{\mathbb{L}_j}}(\omega_j) = \left\| \begin{array}{l} \underset{\omega_i \in \mathbb{Q}}{\text{opt}} \left[\mu_{\mathbb{M}_{\cdot j}}(\omega_j)\right]^{\mathbb{L}} \\ \text{s.t. } \mathbb{Q} = \left\{\omega_i \in \Omega \mid \left[\mu_{\mathbb{M}_{\cdot j}}(\omega_j)\right] - \left[\mu_{\mathbb{N}_{\cdot j}}(\omega_j)\right] \le 0\right\}^{\mathbb{L}} \end{array} \right.$$

Theorem 2.9 is of the same structure as that of (2.8). The solutions to these fuzzy optimization problems allow us to construct the set of optimal ranking, $\mathbb{G}^{\mathbb{L}^*}$ for the decision-making core in the form:

$$\mathbb{G}^{\mathbb{L}^*} = \left\{ \omega_i \in \Omega \mid \begin{pmatrix} \omega_1^* \\ \hat{\mu}_{\hat{\Delta}^{\mathbb{L}}}(\omega_1^*) \end{pmatrix} \succcurlyeq \begin{pmatrix} \omega_{1+j}^* \\ \hat{\mu}_{\hat{\Delta}^{\mathbb{L}}}(\omega_{1+j}^*) \end{pmatrix}, \ \forall j \in \mathbb{J} \text{ and } \mathbb{K}^{\mathbb{L}} \right\}^{\mathbb{L}} \quad (2.29)$$

Equations (2.27) and (2.29) provide us with two optimal ranking of the elements in the social goal-objective set. One is the optimal ranking by the general legal public (the principal) and the other is the optimal ranking by the decision-making core (the agent). The two optimal rankings may not be the same and hence must be reconciled. The reconciliation may proceed with two algorithmic forms. The first one is to compute a new fuzzy set of degrees of relevance as optimally arranged by the two groups whose indices are $\mathbb{L}$ and $\mathbb{I}$ by the method of fuzzy union to obtain $\mathbb{G}^{\mathbb{IL}} = \mathbb{G}^{\mathbb{L}^*} \cup \mathbb{G}^{\mathbb{I}^*}$ which may be explicitly writing as:

$$\mathbb{G}^{\mathbb{IL}} = \left\{ \omega_j \in \Omega \mid \begin{pmatrix} \omega_j^* \\ \hat{\mu}_{\mathbb{G}^{\mathbb{IL}}}(\omega_j^*) \end{pmatrix} = \begin{pmatrix} \omega_j^* \\ \hat{\mu}_{\hat{\Delta}^{\mathbb{L}}}(\omega_j^*) \end{pmatrix} \bigvee \begin{pmatrix} \omega_j^* \\ \hat{\mu}_{\hat{\Delta}^{\mathbb{I}}}(\omega_j^*) \end{pmatrix}, \ \forall j \in \mathbb{J} \mid \mathbb{K}^{\mathbb{L}} \text{ and } \mathbb{K}^{\mathbb{I}} \right\}^{\mathbb{IL}} \quad (2.30)$$

This computation may change the ranking order of the degrees of relevance and hence reordering is required to obtain:

$$\mathbb{G}^{\mathbb{IL}^*} = \left\{ \omega_j \in \Omega \mid \begin{pmatrix} \omega_1^* \\ \hat{\mu}_{\mathbb{G}^{\mathbb{IL}}}(\omega_1^*) \end{pmatrix} \succcurlyeq \begin{pmatrix} \omega_{1+j}^* \\ \hat{\mu}_{\mathbb{G}^{\mathbb{IL}}}(\omega_{1+j}^*) \end{pmatrix}, \ \forall j \in \mathbb{J} \mid \mathbb{K}^{\mathbb{L}} \text{ and } \mathbb{K}^{\mathbb{I}} \right\}^{\mathbb{IL}^*} \quad (2.31)$$

It is possible in eqn.(2.30) that either the optimal assessment of the decision making core may come to dominate that of the public and vice versa. To avoid this situation an alternative computational scheme may be set up. In this case, we may take the fuzzy set of relevance as assessed by the decision making-core $\mathbb{L}$ and combine it with the fuzzy set of irrelevance as assessed by the general public $\mathbb{I}$. The logic is to optimize the aggregate assessments of relevance by the decision-making core subject to what the general legal public considers as irrelevant objectives relative to the national interest. This require that we combine eqns. (2.26b) and (2.28a) and write:

$$\left\{\begin{array}{l} \mathbb{M}^{\mathbb{L}} = \left\{ \left(\omega_j, \mu_{\mathbb{M}_{\cdot j}}(\omega_j)\right) \mid \mu_{\mathbb{M}_{\cdot j}}(\omega_j) = \left[\!\left[\bigwedge_{k\in\mathbb{K}} \bigwedge_{i\in\mathbb{K}_k} \left[w_{ik}\mu_{\mathbb{M}_{ij}}(\omega_j) \right], \ \omega_j \in \Omega, \ j \in \mathbb{J} \right]\!\right]^{\mathbb{L}} \right\} \text{(REL) a} \\ \\ \mathbb{N}^{\mathbb{I}} = \left\{ \left(\omega_j, \mu_{\mathbb{N}_{\cdot j}}(\omega_j)\right) \mid \mu_{\mathbb{N}_{\cdot j}}(\omega_j) = \left[\!\left[\bigvee_{\tau\in\mathbb{T}} \bigvee_{i\in\mathbb{I}_\tau} \left[w_{ik}\mu_{\mathbb{N}_{ij}}(\omega_j) \right], \ \omega_j \in \Omega, \ j \in \mathbb{J} \right]\!\right]^{\mathbb{I}} \right\} \text{(IRR) b} \end{array}\right\} \quad (2.32)$$

We then construct a fuzzy decision-choice problem of the form $\Delta^{\mathbb{IL}} = \mathbb{N}^{\mathbb{I}} \cap \mathbb{M}^{\mathbb{L}}$ and optimize its membership function $\mu_{\Delta^{\mathbb{IL}}}(\omega)$. A similar procedure may be undertaken for the weighted fuzzy product decision. In all these cases, the constructed optimal goal-objective sets such as $\mathbb{G}^{\mathbb{IL}^*}$ are contained in the general goal-objective set, Ω that is $\left(\mathbb{G}^{\mathbb{IL}^*} \subset \Omega\right)$.

The politico-legal calculus in the political economy, functioning within the principle of collective democratic decision-choice system, is that the preferences of the agent (elected members) are constrained by the preferences of the principal (the general legal public). In actual practice of the democratic decision-choice system under a representative process, the constraint of the principal's preferences on the preferences of the agent may be made ineffective through a legal process or active disinformation and misinformation, multiplicity of fear or monetary activities, or a reward-punishment process in the political space. In this respect, an actual functioning of a democratic collective decision-choice system may be reduced to an effective dictatorial decision-choice system. In this case the preferences of the decision-making core become the effective constraint on the decision-choice behavior of the principal

2.7 Implementation Decomposition Analytics by α-Level Cuts

The sets $\mathbb{G}^{\mathbb{IL}^*}$ may be very large and not all of the elements may be admissible at any moment of social and political time due to resource constraints and ideological shifts in the national and global political spaces. As such, we may construct an implementable set $\mathbb{S}$ of the optimal social goal-objective elements on the basis of either equation (2.25) or (2.31) and the use of the method of fixed-level or an $\alpha - level$ set of fuzzy decomposition theorem [R3.1.23] [R3.1.24][R6.63] [R6.55][R6.75]. First we notice that the elements in both

$\mathbb{G}^{\mathbb{IL}^*}$'s have different fuzzy values of degrees of social relevance from which we must construct the set of implementable goal-objective set. We construct an ordinary set of implementable resource-constraint goal-objective set by α-level cuts as:

$$\mathbb{G}_{\alpha}^{\mathbb{IL}^*} = \left\{ \mu_{\mathbb{G}^{\mathbb{IL}^*}} \mid \mu_{\mathbb{G}^{\mathbb{IL}^*}}\left(\omega_j^*\right) > \alpha \ , \ \forall j \in \mathbb{J} \ , \ \alpha \in [0,1] \ , \ \omega^* \in \Omega \right\}_{\mathbb{I}} \quad (2.33).$$

We may define the non-implementable set of social goal-objective elements as a complements of the constructed implementable sets respectively as:

$$\mathbb{G}_{\alpha}^{\mathbb{IL}^*} = \left\{ \mu_{\mathbb{G}^{\mathbb{IL}^*}} \mid \mu_{\mathbb{G}^{\mathbb{IL}^*}}\left(\omega_j^*\right) \leq \alpha \ , \ \forall j \in \mathbb{J} \ , \ \alpha \in [0,1] \ , \ \omega^* \in \Omega \right\}_{\mathbb{I}} \quad (2.34).$$

By the method of α-decomposition, we can partition the optimal goal-objective set for sequential implementations. Let α_1 be the lower bound of the set of implementable social goal-objective elements, then we can specify $\alpha_{1+\ell} \geq \alpha_1, \forall \ell \in \mathsf{L}$, the index set of implementable stages such that $\bigcup_{\ell \in \mathsf{L}} \mathbb{G}_{\alpha_{1+\ell}}^{\mathbb{IL}} = \mathbb{G}_{\alpha}^{\mathbb{IL}}$ where $\alpha_{1+\ell}$ is the highest degree of relevance. The α's may be constructed on the basis of budget proportions of the government's revenue flows and budgetary process.

2.8 Reflections on the Fuzzy Computing Process

There are a number of observations that must be noted. To do this, a few definitions are required to provide some clarity to the observations and reflections that we are undertaking. These definitions may also be useful in the understanding and appreciation of the analytical structure of the social goal-objective formation under a democratic collective decision-choice system. The definitions are about relative differences of cardinality of ordinary and fuzzy sets, and how they may help to specify the required α to be used in the decomposition of the optimal goal-objective set for implementation.

Definition 2.1

If $\mathbf{A}$ is an ordinary set with an index set $\mathbf{J}$ and generic elements, $x_j \in \mathbf{A}$ with a membership function $\mu_{\mathbf{A}}\left(x_j\right) = 1$, $\forall j \in \mathbf{J}$, then the *cardinality* of $\mathbf{A}$ (card. $\mathbf{A}$) is defined as

$$\text{card.}\mathbf{A} = \sum_{j \in \mathbf{J}} \mu\left(x_j\right) = \#\mathbf{J} .$$

Definition 2.2

If $\mathbb{A}$ is a fuzzy set with an index set $\mathbb{J}$ and a generic element $x_j \in \mathbb{A}$ with a membership function, $\mu_{\mathbb{A}}(x_j) \in [0,1]$, $\forall x_j \in \mathbb{A}$, then the fuzzy cardinality of $\mathbb{A}$ (Fcard $\mathbb{A}$.) is defined as

$$\text{Fcard.}\mathbb{A} = \sum_{j \in \mathbb{J}} \mu_{\mathbb{A}}(x_j) \leq \#\mathbb{J}$$

It is useful to note that the process of constructing the socially admissible goal-objective set, on the basis of a defined set of democratic rules of decision making, involves matrices and vectors of characteristics and preferences. Care must be taken when the social goal-objective set is being constructed. Furthermore, the α-level cuts in the decomposition process cannot be conducted arbitrarily. One way out to obtain α is to appeal to *expert judgments*, or the judgments of key political administrators whose business is the management of the socio-economic system, instead of simply using the budget distribution over programs. This can be done through an elicitation of judgment information regarding the fixed-level value of social relevance from a set of experts, $\mathcal{S}$, irrespective of the nature of individual objectives. To obtain the cut-off point of degree of social relevance and hence the set of implementable optimal goal-objective elements, we consider the elicited information as a fuzzy set $\mathbb{B}_i$ and define a fuzzy set of experts' (or administrators') evaluations of degree of relevance as:

$$\mathbb{B}_i = \left\{ \left(x, \mu_{\mathbb{B}_i}(x) \right) \mid x \in (0,1) \text{ and } \mu_{\mathbb{B}_i}(x) \in [0,1] \text{ , } i \in \mathcal{S} \right\} \quad (2.35)$$

We note that x is the degree of fix-level cut-off and $\mu_{\mathbb{B}_i}(x)$ is the degree of confidence that the experts associate to their assessments of taking x as the α. For the rationale in justifying the use of experts' assessments see [R15.18]. Because of problems associated with information elicitation, we construct a survey set $\mathfrak{S}$ from the public and elicit from each member in the survey an assessment of degree-cutoff of social irrelevance in terms of fuzzy set that may be written as:

$$\mathbb{B}_\ell = \left\{ \left(x, \mu_{\mathbb{B}_\ell}(x) \right) \mid x \in (0,1) \text{ and } \mu_{\mathbb{B}_\ell}(x) \in [0,1] \text{ , } \ell \in \mathfrak{S} \right\} \quad (2.36)$$

Each of the membership functions has the properties that:

$$\lim_{x \to 1} \mu_{\mathbb{B}_i}(x) \to 1 \text{ with } \tfrac{d\mu}{dx} \geq 0 \text{ , } \forall i \in \mathcal{S}$$

$$\lim_{x \to 1} \mu_{\mathbb{B}_\ell}(x) = 0 \text{ with } \tfrac{d\mu}{dx} \leq 0, \ \forall \ell \in \mathfrak{S}$$

Equations (2.35) and (2.36) may be combined to define a fuzzy decision problem Δ where the membership function may be specified as:

$$\Delta = \left\{ \begin{array}{l} \left(x, \mu_{\Delta}(x)\right) \mid \mu_{\Delta}(x) = \\ \qquad \left[\bigvee_{i \in \mathcal{S}} \mu_{\mathbb{B}_i}(x) \right] \bigwedge \left[\bigvee_{\ell \in \mathfrak{S}} \mu_{\mathbb{B}_\ell}(x) \right], \forall x \in (0,1) \end{array} \right\} \tag{2.37}$$

The optimal degree of relevance to be used in selecting social objectives that will constitute the admissible social goal-objective set for implementation is obtained by optimizing $\mu_{\Delta}(x)$ to obtain $\alpha = \mu_{\mathbb{B}_i}(x^*) = \mu_{\mathbb{B}_\ell}(x^*)$. The optimal value of α is then obtained by solving the following fuzzy mathematical programming problem.

$$\sup_{x \in (0,1)} \mu_{\Delta}(x) = \left\| \begin{array}{l} \max_{x \in \mathfrak{B}} \left[\bigvee_{i \in \mathcal{S}} \mu_{\mathbb{B}_i}(x) \right] \\ s.t. \\ \mathfrak{B} = \left\{ x \in (0,1) \mid \left\| \left[\bigvee_{\ell \in \mathfrak{S}} \mu_{\mathbb{B}_\ell}(x) \right] - \left[\bigvee_{i \in \mathcal{S}} \mu_{\mathbb{B}_i}(x) \right] \right\| \geq 0 \right\} \end{array} \right. \tag{2.38}$$

From the solution to eqn.(2.8.4) we obtain:

$$\alpha = x^* = \mu_{\mathbb{B}_i}^{-1}(x^*) = \mu_{\mathbb{B}_\ell}^{-1}(x^*) \tag{2.39}$$

The α-value, $\alpha = x^*$ becomes the support of the crisp set of implementable goal-objective set. The $\alpha = x^*$ – value obtained from the experts may be compared with the one obtained from the budget distribution proportions. The result of the comparison will reveal whether enough resources have been budgeted for the implementation of the needed elements of the goal-objective set in support of the defined national interest.

It is important to note that the membership functions for degrees of relevance and irrelevance must be carefully specified or selected to include socioeconomic parameters that will capture resource constraints of the socioeconomic system. Even if α is determined according to equations (2.37) and (2.38) of the fuzzy optimization process, a practical problem exists in the construction of $\mathbb{B}$, the information set of the experts. The practical problem involves the cost of obtaining the fuzzy information. To get around this problem, the method of statistical survey may be used to reduce the size of the relevant set of the evaluators. The construction of $\mathbb{B}$ may also be approached through the evaluation and combination of expert judgments. In the construction of the admissible set of social goals and objectives by the method of fuzzy restriction, we assumed that decision agents have legal authority (permissibility condition) to do so , as well as the institutional configuration (institutional conditions) is right and consistent with the permissibility conditions. It would be useful to examine the implied sovereignty rights of decision agents and permissibility accorded by the institutional configuration in the formation of social goals and objectives in the social space in support of the national interest and social vision. Such sovereignty rights involve political and legal rights that have been established as part of the democratic decision-choice process. The mechanism for constructing the optimal goal-objective set is through the individual and collective reconciliation assessments of degrees of relevance and irrelevance associated with each element of the possible social goal-objective set which a society may follow in support of its national interest and social vision. The legal authority and the domain of power permissibility as accorded by the institutions of decision-choice process involve the creation and protection of social goals and objectives. The creation and protection of social goals and objectives are discussed in the subsequent Chapters.

2.9 Conflicts in the Social Goal-Objective Formation, Political Platforms, Social Policies and Democracy

Let us turn our attention to the discussions on the problem and solution of social goal-objective formation and how they relate to the various political platforms of political parties, social policies, national interest and social vision in the political economy. The decision-choice items involving the social goal-objective formation must make sense with the established institutions and new institutions that must be created. They must also relate to the ideological configuration of the social system and the nature of the political economy. The ideological configuration may have one central ideological pillar called the grand ideology around which the sub-ideologies of political parties revolve to simplify and determine the individual and collective decision-choice behaviors. The ideologies of the political parties function through the institutional configuration which encapsulates the grand ideology as well as forms its protective belt to give the political economy its organizational coercion and social stability. The party ideologies and the grand ideology live in an unstable tension and unity under a mutual give-and-take process, where the party ideologies seek to alter the grand ideology and the grand ideology seeks to restrict the acceptable domain of variation of the party ideologies. Each party ideology shapes the party's preferences in establishing the goal-objective set in a manner that tends to establish the party's platform, and define the party's position in relation to social policy in the nation, social vision and the national interest.

We must face the question as to how the party's preferences are formed to determine the party's platform in relation to national interest social vision and the supporting social goal-objective set. Furthermore, how do such preferences relate to the individual preferences as well as affect individual democratic decision-choice action regarding the collective decision-choice process? What factors influence the decision-choice behavior of the political elites? Figure 2.20 represents a framework for understanding the relationships among grand ideology, party ideologies, party preferences, national interest, social vision, social goal-objective set, decision-choice actions and democratic collective decision-choice systems.

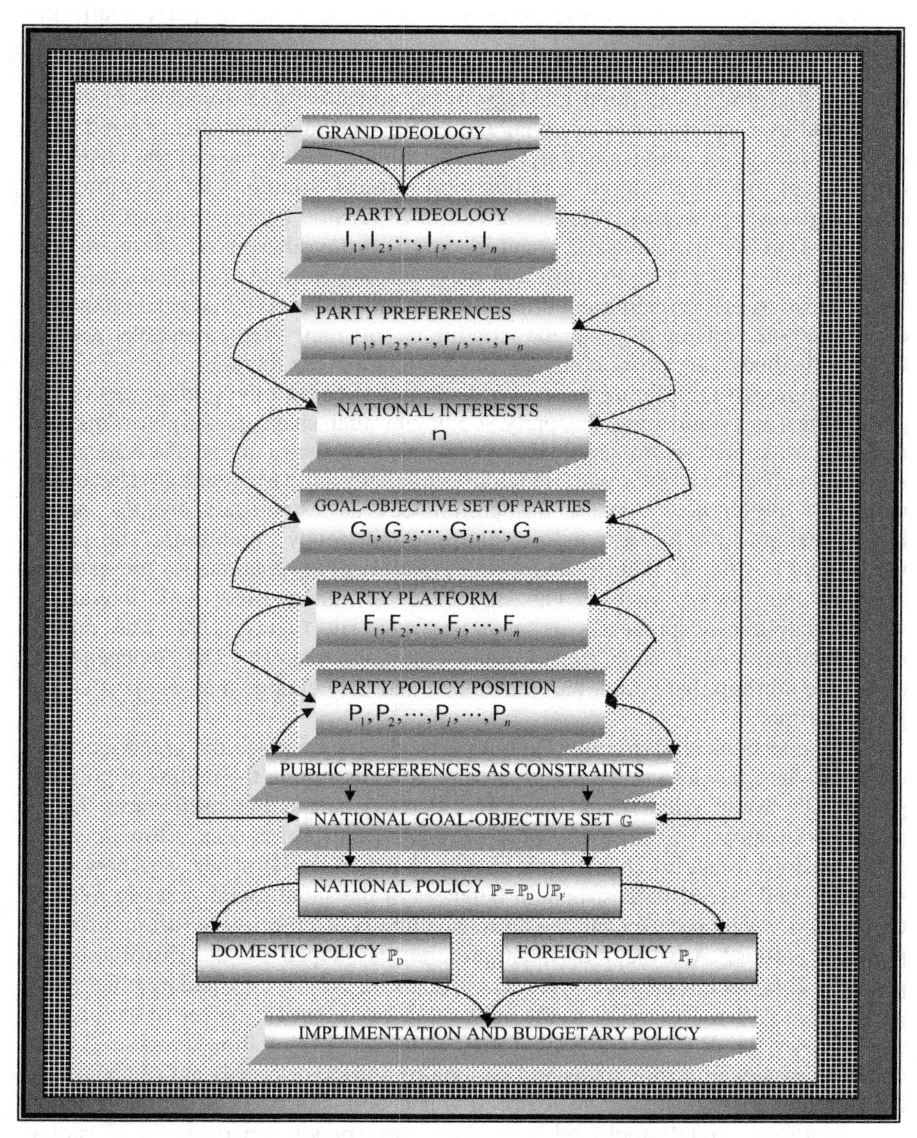

Fig. 2.20 The Process of National Goal-Objective Formation and Ideology in a Multi-Party Collective Democratic Decision –Choice System

Let us notice that in the political economy, there is the Grand National ideology, $\mathbb{G}$. In a multiparty political economy, there are also several party ideologies. Let the parties be $\mathbb{I}_1, \mathbb{I}_2, \cdots, \mathbb{I}_i, \cdots, \mathbb{I}_n$ with corresponding preferences, $\mathbb{R}_1, \mathbb{R}_2, \cdots, \mathbb{R}_i, \cdots, \mathbb{R}_n$, and party goal-objective sets $\mathbb{G}_1, \mathbb{G}_2, \cdots, \mathbb{G}_i, \cdots, \mathbb{G}_n$. Corresponding to these is a set of party platforms $\mathbb{F}_1, \mathbb{F}_2, \cdots, \mathbb{F}_i, \cdots, \mathbb{F}_n$ that supports the party positions, $\mathbb{P}_1, \mathbb{P}_2, \cdots, \mathbb{P}_i, \cdots, \mathbb{P}_n$, that must be conform to the grand ideology in the collective decision-choice system. The grand ideology is taken as given and established in an historical period. The structure of such a grand ideology is also evolving with time and the changing structure of the institutional configuration toward solutions of existing and emerging social problems. There is a continual tension between the set of party ideologies on one hand and the grand ideology on the other hand. There are also continual tensions among the different party ideologies in the political economy. These tensions are defined around the conflicts in preferences that relate to social decision-choice elements and the conditions of their implementable actions through the budgetary process and the means of achieving them.

One approach in defining the party ideologies is to take the preferences of the party elites as establishing the party ideologies that shape party platforms and preferences. It also seems reasonable to assume that the majority of each party's members subscribe to their party ideology, and that such party ideology shapes the members' preferences toward the preferences as established by the party-ideological position, corresponding platform and policy conditions. The party-ideological position is then related to the grand ideology that shapes the national interest and social vision. The sequence of interrelated and interactive sub-decision-choice processes is also presented in Figure 2.20. The parties represent decision-choice groups with common interest. The independents constitute a group that is non-party affiliated and may change their support at any political time.

Some questions tend to arise: Does a party-ideology replace the grand ideology or modify it to be close to the party-ideology after assuming the realms of political power? To what extent does the party ideology shape the preferences of its members in support of the party platform and social preferences? How does the party-platform relate to cost-benefit configuration of social states, and how does the individual assess the cost-benefit balances of the same social states in relation to the set of social policies and the corresponding goal-objective set from different political parties? Are the individual preferences shaped by the distribution of potential cost-benefit balances over social states, or are they formed by blind party preferences without reference to the individual cost-benefit balances of the general society? Are the different party-platforms constructed in relation to national cost-benefit balances of social states and national progress, or cost-benefit balances of the different party-positions, their progress and dominance? How do the different party-platform capital stocks relate to their corresponding party-political capital and national aspirations and interest? The answers to these questions are crucially important to the national spirit and its personality as well as to how they relate to the institutional configuration that encapsulate the national spirit and personality. The answers to these questions are not easy, but provide us with reasonable understanding of the political conflicts and social process under multiparty democracy in decision-choice activities to establish the social goal-objective set in support of the national interest and social vision through democratic collective decision-choice systems. These questions will be taken up in the next chapter.

The grand ideology will point to the nature of the institutional configuration that can be constructed for the production-distribution-consumption system to implement the elements in the optimal social goal-objective set. Any theory of decision-choice activities in the production-distribution-consumption system is developed by assuming the existence of the social goal-objective set, national interest and social vision on the basis of which the problems of allocation-production-consumption efficiency are defined and solved. A complete theory of the political economy will proceed in two stages of the theory of social goal-objective formation and the theory of production-distribution-consumption decision. The former shows how the social goal-objective set, national interest and social vision are established. The theoretical structure of the former, then, becomes a constraint on the latter, which is a decision theory about how the social goal-objective set is efficiently accomplished. In general, however, the theory of the social goal-objective set in the political economy is not constructed. It is assumed where the goal-objective set, the national interest and social vision are taken as given. An example of the former is the neoclassical theory of production, distribution and consumption, where the theory concentrates substantially on the various structures of the market as well as on the optimal decisions in the various markets, where the presence of completion of all forms among politico-economic decision-choice agents allows the use of game-theoretic analysis. It is in this framework that game theory offers an analytical tool for decision-choice activities. In this respect three theoretical structures of game theory that correspond to information structures come into play. They are the theory of stochastic games with exact and limited information structure, the theory of fuzzy games with vague and full information structure and the theory of fuzzy-stochastic games or stochastic-fuzzy game with vague and limited information structure. The theoretical structures of the non-stochastic game require active development, especially the fuzzy-stochastic games [R5.1] [R5.2] [R5.7] [R7.14] [R7.16].

Chapter 3
The Dilemma in the Democratic Collective Decision-Choice System: The Games

This chapter connects to previous chapters of discussions on the theory of social goal-objective formation on the basis of cost-benefit analysis as applied to the political economy under a specified set of conditions of the democratic collective decision-choice process. A number of questions were raised about the roles that different parties play in the collective decision-choice actions in the social goal-objective formation when an institution of government is introduced. The government is charged with overall management and administration of the social organization through the creation of a social decision-making core under multi-partisan conditions. The differences in party ideologies, preferences, policy positions and platforms generate conflicts in the collective decision-choice space involved in the goal-objective formation even given the same national interest and social vision. The different preferences over national interest and social vision may come to complicate the construct of the theory of democratic collective decision-choice system as applied to the social goal-objective formation as part of the overall administration of the social set-up. The formation of political parties is an organizational calculus to simplify the decision-choice process by consolidating similar preferences and thus to reduce the complexities in extreme conflicts over the individual preferences. It may be seen as a first step in the preference aggregation in the decision-choice process in creating and protecting the elements in the social goal-objective set.

The discussions in the previous chapters are extended into conceptual frameworks of cooperative and non-cooperative games under the logic of soft computing. The theoretical frameworks present the social goal-objective formation as a game in defining the relative sizes of the private-public-sector duality in the social production-distribution-consumption structure in the organizational form of the social set up. The organizational form is disaggregated into private and public forms as they relate to the provision of elements in the production-distribution-consumption space. Besides the political party systems, there are three groups of theoretical relevance given the individual preferences and decision-choice sovereignties. They are the public-sector advocates, private-sector advocates and the social decision-making core. The game is played among the

K.K. Dompere, *Social Goal-Objective Formation, Democracy and National Interest*, Studies in Systems, Decision and Control 4,
DOI: 10.1007/978-3-319-05173-4_3, © Springer International Publishing Switzerland 2014

groups of advocates of public-sector interest, private-sector interest and the social decision-making core (the elected official) under a specified set of democratic collective decision-choice conditions. The advocates may or may not belong to any of the legally established political parties. However, their public-private-sector preferences may be shaped by the prevailing party-ideologies at the period of analytical interest. It is argued that, democracy is a good-evil duality in the politico-economic game under any given legal institutional structure. Adam Smith puts it in a different way in his book *The Theory of Moral Sentiments*

> *The man of systems... seems to imagine that he can arrange the different members of a great society with as much* ***ease as the hand arranges the different pieces upon a chess-board have no other*** *principle of motion besides that which the hand impresses upon them; but that, in the great chess-board of human society, every single piece has a principle of motion of its own, altogether different from the legislature might choose to impress upon it* [pp. 380-381]

This statement expresses the conflicts of decision-choice actions in the individual-community duality in the social system composed of individual sovereignties and community sovereignty regarding decision-choice actions around what constitutes individual interest and community (public) interest. In these discussions, the concept of community is used in a broad general sense to cover a collective, society and the public.

The individuals and the community reside in dualistic conflicts and in a never-ending continuum. The political calculus for resolving these conflicts may be seen as the creation of an institution of government that must be administered by a class. In a democratic social formation, the class is constituted in accordance with the rules of the democratic collective decision-choice system where the legal individuals are allowed to exercise their sovereignties in accord with their preference. The democratic collective decision-choice system is such that parts of the individual decision-choice sovereignties are transferred voluntarily to the administering class. This process creates a mega sovereignty that is vested in the institutions of government and used by the administering class to govern. This process is used to reduce complexities in the collective decision making in the social set-up with the view of retaining the individual decision-choice sovereignty with the belief that they are participating in the social affairs that affect their lives. The process leads to the voluntary creation of principal-agent duality with an asymmetry of sovereignties and possible asymmetry of preferences under cost-benefit rationality. The presence of asymmetry of sovereignties brings a dilemma into the social practice of the democratic collective decision-choice system. In practice, a deceptive information structure is created to take advantage of the system in order to promote goal-objective elements that are inconsistent with rules of the collective decision-choice system. We turn our attention to the problem of the dilemma. Let us keep in mind that there are three decision-choice steps in the political economy.

3.1 The Basic Foundation of the Dilemma in Democracy: The Principal-Agent Problem

A democratic decision-choice system presents a dilemma in the social goal-objective formation where the distribution of potential cost-benefit configuration forms the driving force of the conflicts through the practice in the social decision-choice space. The dilemma is revealed through the nature of the accepted representative social decision-choice process where the elected members of the social decision-making core may completely pursue social goals and objectives that reflect their personal interest and preferences with ideological twists, rather than the collective preferences of the community, party members or the established national interest. The democratic decision-choice system presents a principal-agent duality with possible conflicts of interest. The conflict of interest may be induced by individual or party interests. Here, the principal is the general public and the agent is the decision-making core composed of the elected officials.

Like the structure of the principal-agent decision-choice problem, the elected officials may act as if they are not accountable to the public or the party platform

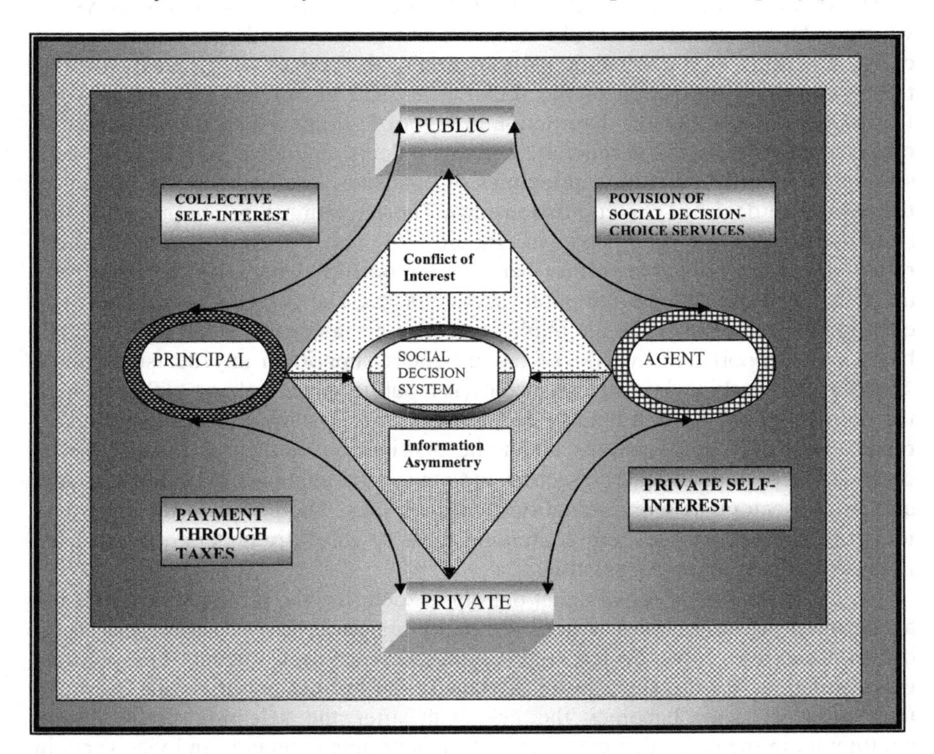

Fig. 3.1 A Cognitive Geometry of the Relationships among Principal-Agent and Private-Public Sector Dualities, and Interest Conflict Zones in Democratic Decision –Choice Systems

in directions of social decisions and choices. They may decide on the principle of personal self-interest rather than on the principle of collective self-interest. They may also use the concept of public interest to restrict freedoms and dismantle social and even natural rights under the principle of law and order. Here, the elected officials, the members of the decision-making core, are given the trust of general social decision making to improve the general welfare of the political economy through appropriate public-private sector balances. The general population constitutes the principal whose behalf are decision-choice actions, concerning freedom and justice, are undertaken by the decision-making core. The principal pays the members of the decision-making core for their services in governance and management through a complex tax system. The principal-agent duality with differential cost-benefit interests generates continual conflicts where such conflicts increase in intensity under an asymmetric information structure. The resultant direction of the cost-benefit conflicts that define differential preferences involves the interplay of politico-economic forces that require sub-coalition formations and negotiations in the collective decision-choice space, given the legal structure that helps to establish the cost-benefit configuration of socioeconomic policies.

The dilemma arises in the practice of the democratic decision-choice system where a democratic choice leads to an establishment of a dictatorial system of decision making with no/or little participation of the collective. It reveals itself in the creation and protection of the social goals and objectives which may yield forwardness of cost-benefit distribution in favor of groups and to the advantage of others. Analytically, the democratic collective decision-choice system involves a system of polarities, dualities, principles of opposites, conflicts and continuum in the pyramidal decision-choice relationships. These relationships are presented in a simplified cognitive geometry in Figure 3.1. In the last analysis, these relationships are reducible to cost-benefit relationships which, on the aggregate, are distributed over the private-public sector spaces. The costs and benefits always exist in dualities. The epistemic implication here, for a theory, is that every benefit has a cost support and every cost has a benefit support in any decision-choice situation. Similarly, every private sector has a public sector support without which the political economy will tend to become a chaotic system. Freedom and justice relate to the fairness questions about the distribution of the costs and benefits between the private and public sectors and among the members in both the private and public sectors, as seen in dualities and polarities. The power of the analytical tools of duality, opposites and continuum must be carefully understood and used in these dualistic and polar relations.

For example, the principal-agent duality is such that the principal becomes the agent when decision-choice actions are being undertaken to constitute the social decision-making core, which then constitutes a new agent by political transformation. By the same political transformation, the public as the agent, in the voting process, becomes the principal, after the selection process in a continuum, in the sense that every agent has its principal-support and vice versa in the democratic collective decision-choice process. The analytical structure also holds for cost-benefit duality. The debate on the size of government is simply on the question of: what is the efficient size of the public sector in support of the private sector and vice versa in the social formation? The choice of the private-public sector's relative combination has always generated tension in the political economy due to its relationship to the ownership of the means of production and how the ownership confer the decision-choice sovereignty rights. The decision-choice sovereignty rights in turn affect income distribution, freedom and justice. In fact, the whole debate about the appropriate private-public sector combination extends to capitalist-communist duality. Let us keep in mind that there are three decision-choice steps in the political economy leading to three decision-choice theories under either a democratic or non-democratic decision-choice system. The relational structure of the steps, problems and theories is illustrated as an epistemic geometry in Figure 3.2.

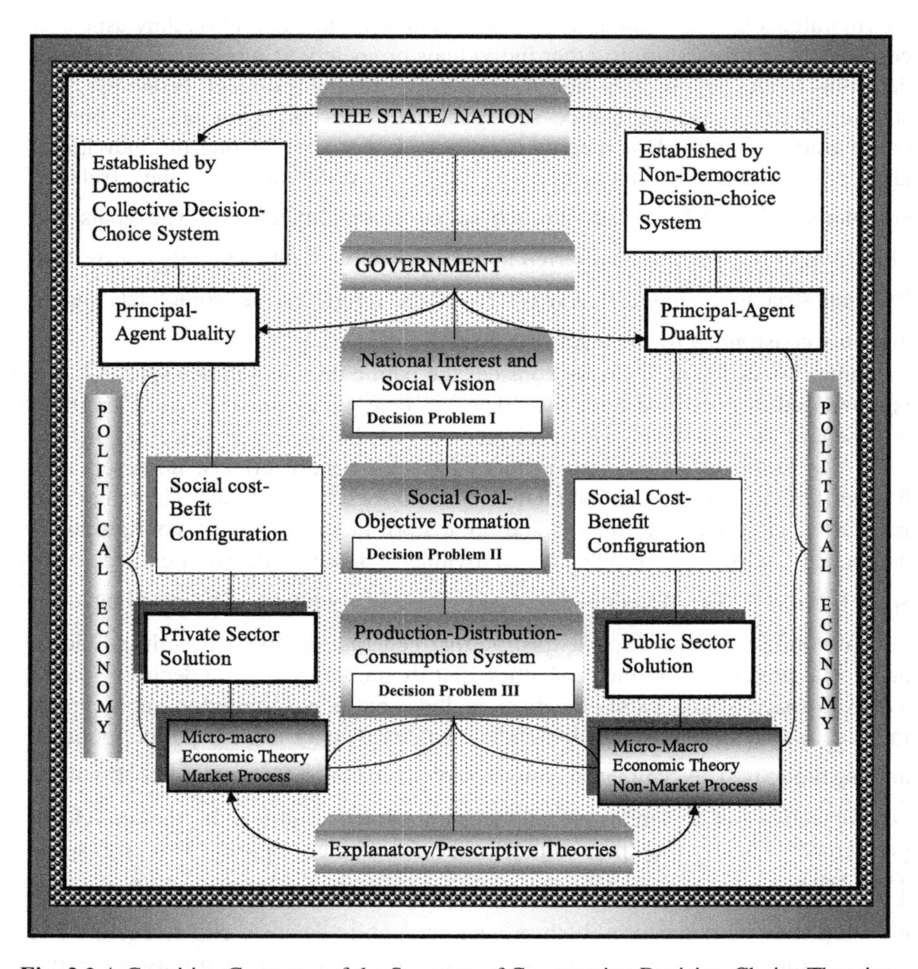

Fig. 3.2 A Cognitive Geometry of the Structure of Constructing Decision-Choice Theories on the Political Economy

The nature of the social decision-choice problem and the dilemma that it imposes on the efficiency and fairness of the democratic social organization or decision-choice system may be explained by considering the general decision-choice space, Θ composed of general goals and objectives that all decision agents, both individual and collective, may pursue. The general decision-choice space Θ, may be partitioned into two decision-choice sub-spaces of private, Ψ and public, Ω decision-choice sub-spaces where $\Theta = \Psi \bigcup \Omega$, and $\Psi \bigcap \Omega \neq \varnothing$. The condition, $\Psi \bigcap \Omega \neq \varnothing$, means that there are overlapping areas of decision-choice actions which may be simultaneously undertaken by the private and public sectors. The private decision-choice space is composed of characteristics where only personal decision-choice actions are undertaken, and where the cost-benefit configurations accrue to the individuals with or without externality. Any aspect of cost-benefit configuration to other individuals or society as a result of the private decision-choice action is a social externality in the political economy. The structure of this decision-choice sub-space with the corresponding cost-benefit configuration is individual-specific. Conflicts do not arise among the individuals and society on the elements in the set difference or the (relative complement) (Ψ / Ω) except through social externality effects. The goals and objectives in this set will be referred to as completely private and only constrained by legal structure. Similarly, (Ω / Ψ) will be referred to as completely public and only constrained by the legal structure.

Things are different when we consider the public decision-choice space Ω that is composed of goals and objectives that the society may pursue, and where their cost-benefit configurations affect all members of the society irrespective of their relative position in society and history. The cost-benefit effects, however, may have differential impacts on the members of the social collectivity. They may also generate conditions that may come to distort the cost-benefit configurations associated with a completely individual private decision-choice sub-space (Ψ / Ω). The decision-choice sub-space Ω, with its corresponding cost-benefit configurations, is social-specific that generate conflicts in both individual and social preferences in the production-distribution-consumption decision space . This sub-space of goals and objectives includes but is not limited to public goods in an economic sense of decision-choice actions. It may include public, political and legal goods. The symmetric complement, or the set difference, (Ω / Ψ) of the goal-objective set generates social decision-choice conflicts in the individual preferences. It is the pure public decision-choice space. The sub-space $(\Psi \bigcap \Omega)$ is an impure public decision-choice space since some goal-objective elements may be pursued by both private and public decision-choice agents. The understanding of the nature of the social goal-objective formation must recognize these three sub-spaces and how they affect the distribution of the individual and collective cost-benefit balances. The nature of these sub-spaces is shaped by the institutions of politics, law and economics. Our main analytical concern is on the pure (Ω / Ψ) and impure $(\Psi \cap \Omega)$ public decision-choice sub-spaces to which we shall attend. The geometry of the decision-choice space and sub-spaces is shown in Figure 3.3.

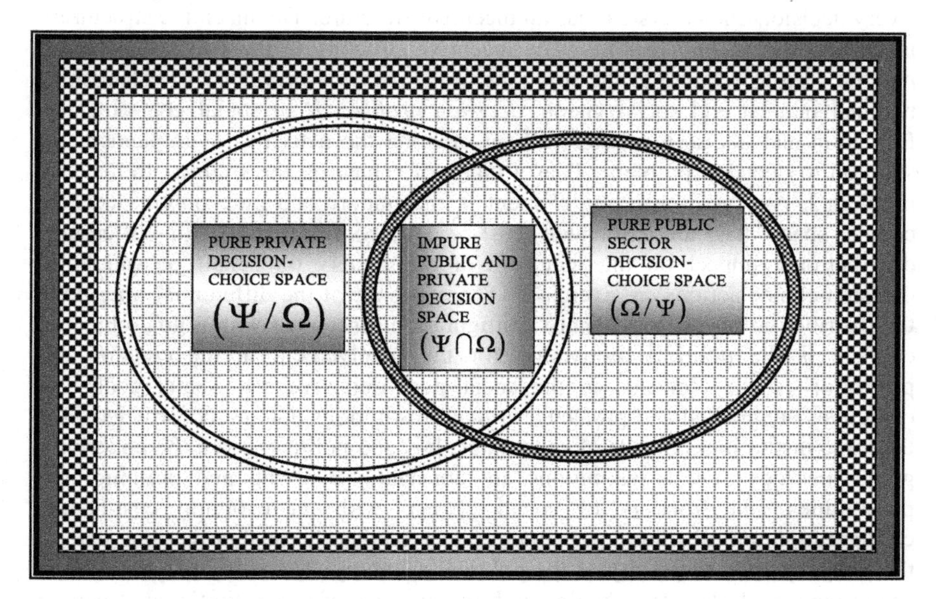

Fig. 3.3 The Partition of the General Decision-Choice Space into Public and Private Subspaces

NOTE 3.1

The decision-choice sub-spaces relevant for understanding the production-distribution-consumption game are provided below.

$$\left.\begin{array}{l}(\Psi / \Omega)=\{x \mid x \in \Psi \text{ and } x \notin \Omega\}: \text{Pure private decision-choice space} \\ (\Omega / \Psi)=\{x \mid x \in \Omega \text{ and } x \notin \Psi\}: \text{Pure public decision-choice space} \\ (\Psi \cap \Omega)=\{x \mid x \in \Psi \text{ and } x \in \Omega\}: \text{Impure public or private decision-choice space} \\ (\Psi \cup \Omega)=\{(\Psi / \Omega) \cup (\Omega / \Psi) \cup (\Psi \cap \Omega)\}: \text{Private-public decision-choice space}\end{array}\right\} \quad (3.1)$$

The above categories of the social decision choice space reflect the fundamental problem of institutional polarity, duality and continuum with private and public sector polarities with embedded individual-collective dualities. In this respect the institution of government is seen as playing a mediating role in the conflicts within the private-public-sector polarity, and individual-collective duality in the social decision-choice space in relation to strategies and counter strategies in the production-distribution-consumption activities.

3.2 The Creation and Protection of Social Goals and Objectives

Every decision-choice system has an incentive structure. The incentive structure is made up of costs and benefits. The incentive structure in the political economy may be seen as relating to the solutions to the *decision-problem I* of establishing national interest and social vision, *decision problem II* of social goal-objective formation and *decision problem III* for creating the activities to implement the solution to decision problems II as a mode of achieving the solution to decision problem I. Any goal-objective element in the impure and pure public decision-choice space has an associated cost-benefit configuration with differential distributional impact on the members of the society on behalf of which the decision-choice actions are exercised. The existence of the differential impacts of the distribution of the cost-benefit configuration associated with a goal-objective element is the political motivation for the involvement in the creation and protection of the elements in the social-goal objective set. Social goals and objectives are created and protected through institutions that are constructed from the three institutional blocks of economics, law and politics, given the underlying grand ideology. These institutions are constantly evolving and defining new possibilities and frontiers of social change with an evolving cost-benefit configuration and distributional structure to establish the qualitative-time path of the integrity and complexity of the political economy. The study of formation of social goals and objectives begins with the analysis of the institutional structures and how they are used to organize the individual preferences to effect social preferences in the decision-choice space of goals and objectives that establish a social policy spectrum of the political economy.

Differences in individual preferences create conflicts which lead to social tension within the general public in the social decision-choice space. This tension reveals itself through the perceived injustice in the distribution impacts of the cost-benefit configuration associated with the possible elements in the social-goal-objective set. The presence of social tension motivates the formation of interest groups. The presence of interest groups gives rise to power relations which then lead to the creation of politico-economic games and strategy formations in the social decision-choice actions to construct an admissible social goal-objective set. The objective of the decision-power relations is to create an organic social policy and corresponding social sub-policies relative to the national interest. Individual differences, associated with the national interest and social vision, create difference preference order of the elements in the goal-objective set. Given the national interest and social vision, different cost-benefit distribution induces different preference order of the social goal-objective set. The social decision-choice actions take place within the institutions of law, politics and economics where each decision-choice action has a cost-benefit support. The institutional arrangements of law, politics and economics are created to allow the elements of the goal-objective set to be implemented in support of the national interest and social vision.

3.2.1 Institutional Structures, the Social Goal-Objective Set and Decision-Choice Foundations

The process of creating the admissible set of social objectives and goals is abstracted from the rules and regulations imposed from the legal structure through the political organization and management. The rules and regulations affect the individual and collective behaviors in the total decision-choice space in the sense that every decision-choice action operates through the cost-benefit rationality, which is directly or indirectly related to individual and collective economic interests in the political economy. The prevailing legal structure is always the construct of the political organization of the decision-choice system. The decision-choice space is embedded in the economic structure of societies. Every social system and decision-choice actions which are made within it, are composed of three basic structures of a) the economic structure, b) the political structure and c) the legal structure. These three structures are the pillars of the political economy that contains all other sub-institutions. The economic structure defines the foundation of life and the conditions from which comfort and material processes are created to support and enhance physical, spiritual and mental dimensions of life. All fundamental private and public decisions in support of life are made directly in the economic structure. All control mechanisms are created in the politico-legal structures. Any other decision-choice action is, by reductionism, traceable to the economic structure, at the last analysis, where *costs* and *benefits* are assessed and balanced for action. In other words, the physical, mental and spiritual dimensions of life are reducible to cost-benefit configurations. It must be kept in mind that these costs and benefits associated with decision-choice actions may be measurable or non-measurable; they may be monetarily measurable or non-monetarily measurable in multi-dimensional quality-quantity relationships [R3.1.23] [R3.1.24] [R3.2.6] [R15.34].

The legal structure presents an integrated system of rules, regulations and established norms on the basis of which socially admissible decision-choice actions of both private and public sectors are made in the economic, political and legal structures, and related extensions. The legal structure imposes restrictions on the economic structure. The restrictions are seen in terms of goals and objectives that may be individually and socially pursued. Additionally, the legal restrictions define the acceptable behavior in the decision-choice space of the specified society. Similarly, the legal structure imposes restrictions on the political structure in terms of socially acceptable decision-choice actions and behavior according to the established rules and regulations by the existing legal order and the constitutional framework of governance. In the last analysis, however, all decision-choice actions are reducible to cost-benefit balances.

The political structure, on the other hand, defines the boundaries and the distribution of the decision-making power among the members regarding the controls and manipulations of activities in the economic structure, legal structure and the political structure itself. Power is conferred on the individual or group of individuals to indulge in social decision-choice activities in the political structure as prescribed and guided by the established legal rules and regulations in the legal structure. The right of personal decision-choice actions and the degree to which one can exercise this right in the pure private decision-choice space is established in the political structure acting through the legal structure in legal codes. By the nature of a complex social system, the legal rules lead to the creation and establishment of a principal-agent relationship between the population and the social decision-making core as we have earlier explained. The preferences and interests of the members of the decision-making core, acting collectively as an agent of the population, may diverge from the preferences and interests of the population, which exists as the principal, and on behalf of which social decision-choice actions are to be made. In this logical frame, a government is by organization, just a social decision-making instrument that encompasses a potential for good and evil in unity.

In other words, a government, as an organizational entity, exists as a good-evil duality under a continuum principle, where good fades gradually into evil, and evil gets gradually transformed into good, through human decision-choice actions under a defined rationality [R7.14]. The government is a totality of simultaneous existence of good and evil characteristics defined not in terms of Aristotelian logical representation. In Aristotelian laws of thought, the government is viewed in terms of dualism in such a way that it is either good or evil but not both. The set of good characteristics of government has a supporting set of evil characteristics and vice versa in the political economy. The relative evil-good combination that will prevail at any time depends on the constitution of the members in the decision-making core as democratically or non-democratically constituted and have come to inherit the government as an institution. The sustainability of the prevailing relative good-evil combination will depend on the tolerance level of the principal. Changes in the composition of the members may or may not change the good-evil balances that reside in the potential relative to the actual in the continuum. It is the potential for the change of the good-evil combination which provides the argument for periodic elections and the principle of term limits for the members in the social decision-making core. The good-evil characteristics are defined in terms of fuzzy logical representations in the good-evil duality under the principle of continuum [R7.13] R7.15].

The directions of movement within the good-evil duality, and the proportionate good-evil composition in unity, depend on the decision-choice dynamics of the set of the social goals and objectives, which may be formed from within the society and subject to its admissible politico-ideological protective belt. Additionally, they will be influenced by the manner in which these goals and objectives are determined as well as the principal-agent relative roles in the social decision-choice space. The individual and collective decision sovereignties become affected as well as shaped by the behavioral nature of the *decision-making core* on which social decision-making power has been conferred on within the political structure through the legal structure. The quality of any society will depend to a greater extent on the choice of the good-evil combination of the decision-making core within the continuum. The collective will of the decision-making core, when it is formed to assume the role of the agent, overrides the will and preferences of other members of the society (the principal) even in completely democratic social formation, where social decisions are to be constructed on the basis of decision-choice *citizens' sovereignty* through an exercise of voting. The social goals and objectives, at any moment of time, must be formed to give direction to the designs and implementations of projects and programs that must correspond to the elements in the goal-objective set and in relation to the defined *national interest* and social vision.

The questions to be answered, then, are: How are the national goals and objectives formed and how is the national interest determined under principal-agent duality as applied to a nation? What role does the principal play and what role does the agent play in the democratic decision-choice system? To what extent are the national interests, social vision, the supporting goals and objectives shaped and established by the relative positions of the principal and agent decision-choice power relations in a democratic social formation? How much good and how much evil is the principal willing to accept from the agent? To what extent can the agent control the collective decision-choice sovereignty of the principal? To what extent can the principal affect the collective decision-choice behavior of the principal? What role does an effective manipulation of the information structure play in affecting the exercise of decision-choice sovereignties? This is a classic African paradox in dualities and continuum (The *Asantrofi-Anoma* problem [R7.13] [R15.16]) which Catullus alluded to in 58 B.C in Rome [R15.29]. The *Asantrofi-Anoma* problem is a general analytical representation of decision-choice elements expressing good-evil duality that resides in the same entity. In this analytical framework, every good has evil support and every evil has a residence of good in the same element. Alternatively stated, the problem, in its simpler form, postulates that every decision-choice element exists in a cost-benefit duality such that any choice element comes with benefits and costs and that the benefits cannot be taken without the associated costs. In a general form, one may view the good as positive and the evil as negative. In this way, one models positive-negative duality in a continuum. Our task is to examine the problem and solution of goal-objective formation at the existence level of the principal-agent decision-choice structure in democratic social set-ups under a multi-partisan electoral system.

3.3 The Goal-Objective Formations as Socio-Political Games

The channels through which social goals and objectives are formed are plagued with conflicts. The conflicts reflect differential preferences within the general public, between the public and the social decision-making core and within the social decision-making core. The differential preferences are due to the perceived differential cost-benefit distributions over the elements of the social goal-objective set. The formation of the set of socially admissible goals and objectives is simply a politico-economic *game* whose reward in the final analysis is greater benefits relative to the costs of social decision-choice actions. The environment of the game is defined by deferential ideologies, party platforms and interpretations of national interests and social vision. The defined environment must be related to potential cost-benefit balances of decision-choice actions of different social states as they tend to affect the behavior of the principal-agent duality and the power configuration. We shall refer to the goal-objective formation as *social-goal-objective game* under the multi-party democratic collective decision-choice system. This social-goal-objective game is embedded in the grand game called the *political-power game* and played in the in the political structure supported by the *political market* just as economic games are played in the economic structure supported by the general economic market. The political game is, however, connected and supported by the behaviors of the structures and markets in the economic and legal frameworks The integrated relational environments of the economic and legal frameworks relative to the nature of the social decision-making power is illustrated in Figure 3.4.

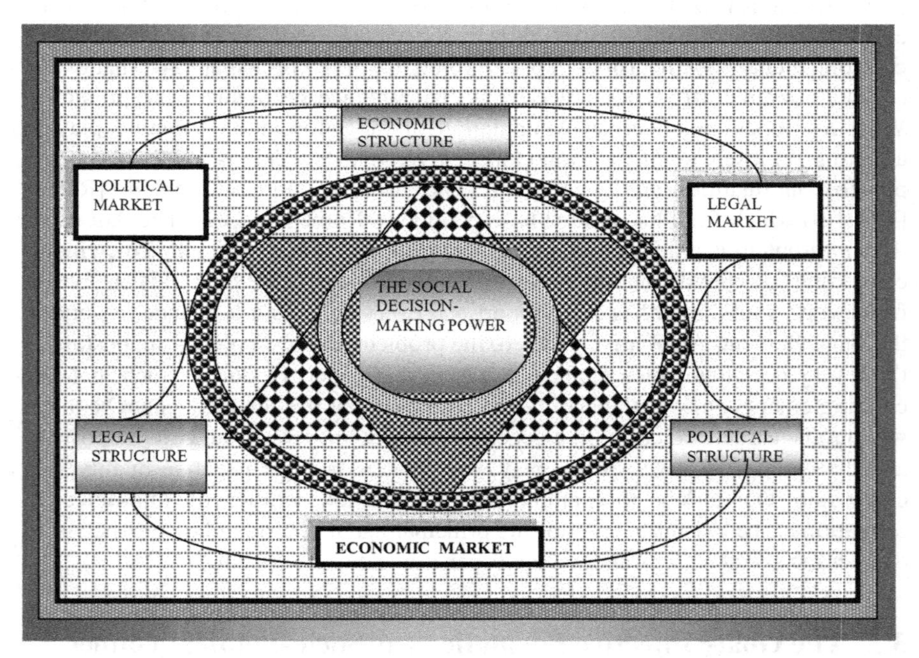

Fig. 3.4 The Structure of the Decision-choice System and the Corresponding Markets for the Socio-Political Game

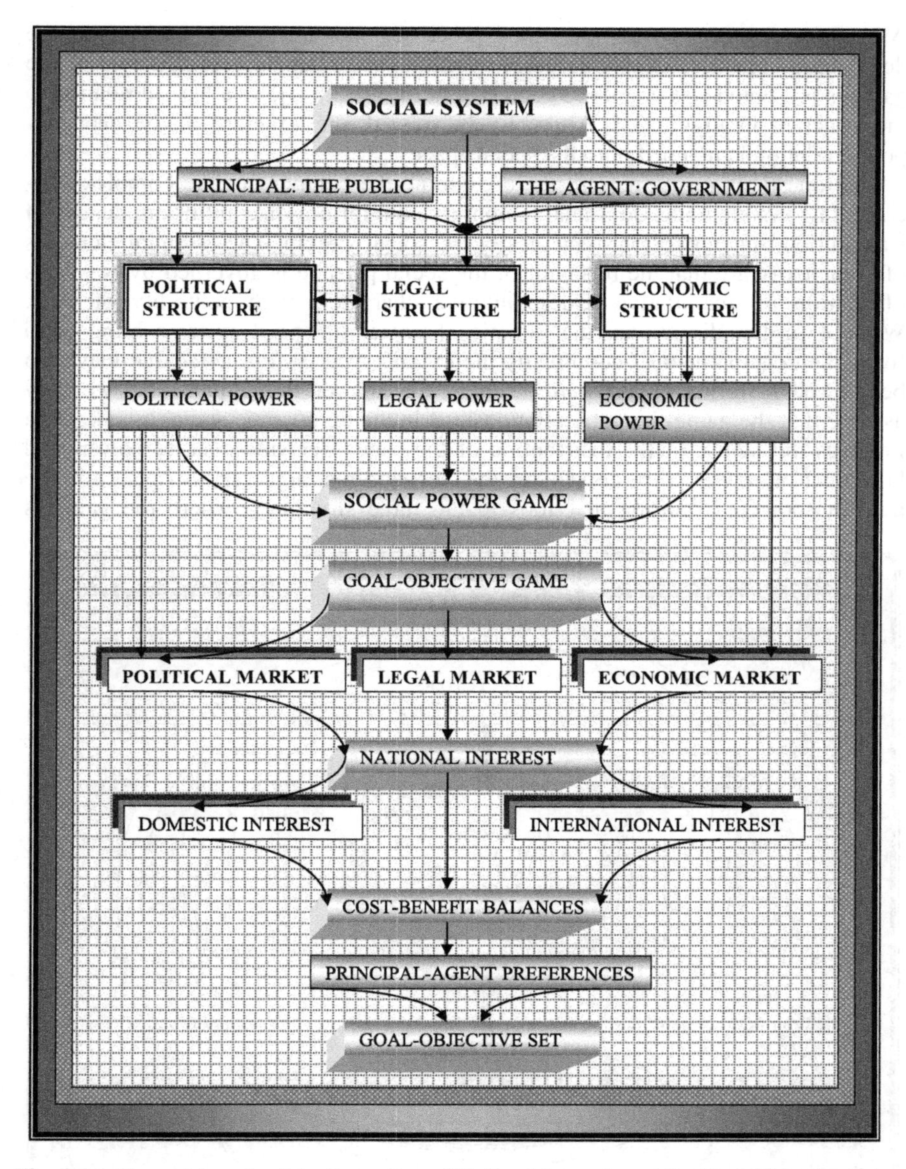

Fig. 3.5 A Geometric path of the Formation of National Interests and the Goal-Objective Set

A question thus arises as to what is the structure and the nature of the game in the political market, and whether the game is played in the same way as those in the economic market. The structure and the nature of the social goal-objective game must reveal to us the rules of the game. The rules of the game are determined at the level of the political structure and coded as legal rules in the legal structure which alternatively affects the behavior of the economic structure.

All of these appear as an integrated processes into an organic process in the democratic collective decision-choice system. The organizational structure of the game is presented as a cognitive geometry in Figure 3.5 and in a logical relation to the structure in Figure 3.4. Alternatively viewed, the structure of the decision-power relation is mapped onto the social space.

The questions raised require us to examine the behavior of the complex self-correcting system under continual temporary equilibrium-disequilibrium duality of substitution-transformation dynamics of the social system. The essential elements of complexity are that the behavior of the social particle is governed by two laws of motion of quantitative and qualitative characters with neutrality of time. Without further discussion, the organic character of social systems brings into focus the development of complexity theory and *synergetic science* that may also be viewed in terms of *energetic science*. To answer these questions, we may assume that the game-theoretic relationships in the democratic social organization involves the nature of the distribution of the social decision-making power, and how such a power is bestowed on the decision-making core as agent for the general society, regarding activities in the three structures, where the game is

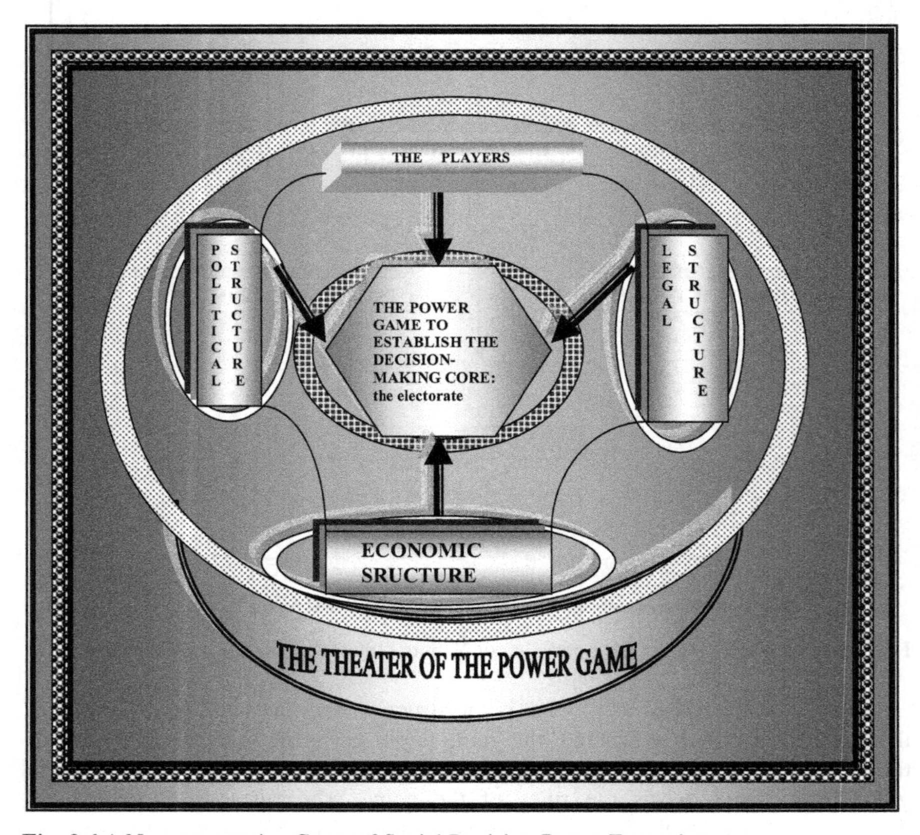

Fig. 3.6 A Non-cooperative Game of Social Decision-Power Formation

played by the players without coalition in the three sectors as shown geometrically in Figure 3.6. The first step in the game is to establish the social decision-making core in a democratic society, where the process in the decision-core formation allows the preferences of the citizens to be expressed in support of the competing members for an entry into the decision-making core in a democratic decision-choice system.

In such a society, the establishment of the decision-making core is through an electoral process which is established by the legal order through the *political structure*. The electoral process acquires an intense interest as we critically examine the process of the formation of the social goals, objectives, national interest and social vision from a game-theoretic viewpoint. It may be noted that the prize of winning the political-power game is the right, bestowed on the members by the implied *legal structure*, to make social decisions which may alter the fundamental relationships among the three structures in favor of the winners defined in terms of their interests, or *national interest* , where such interests may be established by personal preferences or party ideology. The reward of the game to the public is to create a political agent, and the reward for the candidates is to become a member of the agents and hence be paid for the social services rendered. Furthermore, it must be understood that every society, irrespective of how it is organized, has legal rules of establishing the decision-making core (the agent) for determining the social goals, objectives, national interest and social vision. The selection of the members into the core is done by voting to allow the citizens to express their preferences over who should make social decisions in a democratic collective decision-choice system. The whole process is about the settlement of who will control the social decision-choice power to help to define the national interest, social vision and to approve the elements in the social goal-objective set in order to shape the nation's progress and the collective welfare of its members.

The voting process to establish the decision-making core requires an important decision-choice interpretation of the implications of casting an individual vote for any political candidate who seeks to enter the class of the decision-making core. The implication is abstracted from the legal system of the social decision-making under conflict of individual preferences in the collective decision-choice space without physical violence and destruction. When one exercises one's vote, one ultimately surrenders one's legal and social rights and a preference ordering to make social decisions regarding selections of social goals, objectives, national interest, social vision, and their management where social governance is bestowed on the social decision-making core. The voting right and individual preference ordering are surrendered not to the person that one votes for, but to the candidate that actually wins the political game in the voting space. In this respect, the right to vote is also the right to surrender one's right of making social decisions to a member of the decision-making core under the established rules of the game. The right to vote is also a right to willingly surrender one's decision-choice sovereignty over national affairs. It is also a right to create dictatorial class and to be governed by this class in accordance with the established rules of the democratic collective decision-choice system. A right to vote is a right to create a constraint on individual freedom to exercise individual sovereignty in the social decision-choice space. The act of voting initially empowers the public as the temporally agent to establish the decision-making core who holds the temporally position as the principal. After the voting process, the same act of voting disenfranchises the public who then becomes the temporally principal that must be subservient to the decision-making core that is now empowered to act on the behalf of the public. The voting process performs dualistic action. It enfranchises and disenfranchises the public. It also empowers and dis-empowers the social decision-making core.

After the voting decision, the Lego-political rules lead to the creation and establishment of the *principal-agent duality* with the resulting principal-agent problem. At the voting space, the public assumes the conditions of the agent and the members seeking to enter the decision-making core constitute the principal on behalf of which the voting decision-choice actions are exercised. At the level of social decision and the management of the political economy, the roles are reversed where the public is transformed into the principal and the decision-making core is transformed into the agent (the government). The public and the decision-making core constitute categories that may form a duality in the sense of opposite forces in tension. The role reversal is a categorial transformation through qualitative laws of motion. The freedom to exercise one's sovereignty in the democratic collective decision-choice system leads to the freedom to willingly commit oneself to servitude without force and without violence. Here comes an important dilemma, where the freedom to vote is also a process of willingly taking away one's freedom and right to participate in a democratic social decision-choice system that involves important affairs of the social set-up. This dilemma is not paradoxical in the knowledge construct. It is explainable through the conceptual system of dualities, principles of opposites, continuum, tension and categorial conversions. Opposites live in unity and in mutual negation through conversions of categories with forces of internal conflicts. It is easy to conceptualize the dilemma if the opposites are viewed in terms of cost-benefit duality where every cost has a benefit support and every benefit has a cost support. As it has been pointed out in the previous discussions, there is no benefit that is free of cost and vice versa. This is a classic *Asantofi-Anoma problem* that we have also pointed out in the earlier discussions. The game that we have discussed is based on the social conditions of free voting without violence to establish the decision-making core even though political parties may or may not be present.

When this decision-making core is established by the voting process a different game is ensured to intensify the dilemma and analytical complexities of the democratic decision-choice system. The game simply involves the social goal-objective setting that will guide the path of the national interest and history through the processes of production-consumption duality and income distribution in line with the established social vision. Once the members of the public have disfranchised themselves through the electoral decision-choice process, a second game is then played among the members of decision-making core, $\mathbb{D}$, the class of the private-sector advocates, $\mathbb{P}$, and the class of public-sector advocates, Π. Here, a game with all forms of possible political coalitions is established as is shown in Figure 3.7. The reason for coalition formations is seen in terms of power distribution and power concentration to affect the direction of decisions. Coalition formation is equivalent to military alliance to concentrate the power of decision-choice sovereignty to change the center of gravity of preferences.

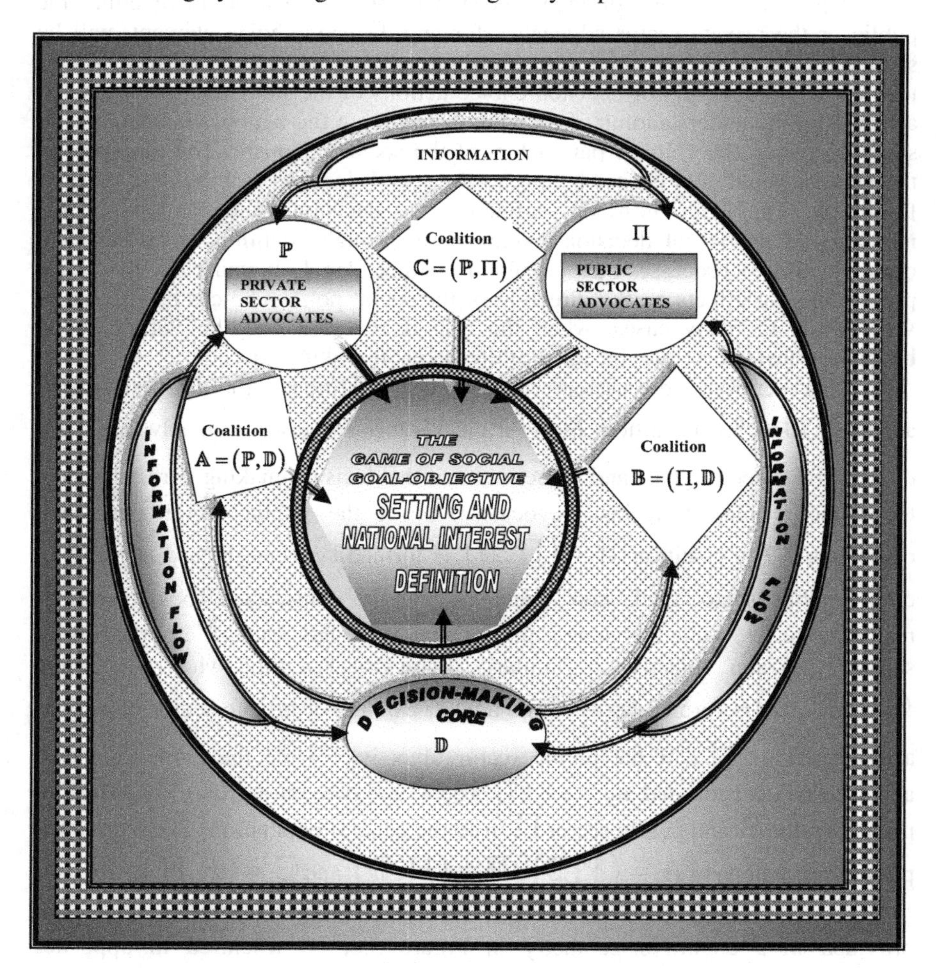

Fig. 3.7 The Environment of the Game of Social Goal-Objective Setting with Varying Coalitions

There are three possible mega-coalition formations that can emerge, given a democratic collective decision-choice system from which the three major social interest blocks, political-party affiliations and constituencies shape the structure of the game. We shall assume that the members of each group are defined by common interests, and hence have similar social preferences or preference ordering. This is the intra-group collective preference that will be exercised over the social decision-choice space. Coalition formation tends to create inter-group preferences or intra-coalition collective preferences. With respect to the nature of social goal-objective formation, the intra-coalition preferences will be shaped by intra-group *cost-benefit balances* of the social goal-objective elements, as they are transformed into policies and implementations for the welfare implications of the general population. The individual decision to belong to a group of either the public or the private sector advocates, for any decision-choice element, will be shaped by the degree to which one is influenced by either the grand or the party ideology. The direction of decision-choice actions of the individuals will also be affected by the understanding of the knowledge about the national interest and its structure, given the cultural parameters of the social organism. The members of the general population do not belong to the social decision-making core after its formation. They can, however, influence the decision-choice actions of the members of the social decision-making core through coalition formations and other pressure processes as admitted within legal boundaries to create preponderating effects on the direction of the social policy choices by the use of vote-threats in many cases, when the role is reversed and the voting public becomes transformed into the agent of the decision-making core.

In the current discussion, the three coalitions to be considered are $\mathbb{A}, \mathbb{B}$ and $\mathbb{C}$. Coalition $\mathbb{A} = (\mathbb{P}, \mathbb{D})$ is formed between the advocates, $\mathbb{P}$, of the private sector goals and objectives and the decision-making core, $\mathbb{D}$, while the game is played against advocates, Π, of the public sector goals and objectives. The coalition $\mathbb{B} = (\Pi, \mathbb{D})$ is formed between the advocates, Π, of the public sector goals and objectives and the constructed social decision-making core, $\mathbb{D}$, while the game is played against the advocates, $\mathbb{P}$, of the private sector goals and objectives. Similarly, there arises a time when a coalition $\mathbb{C} = (\mathbb{P}, \Pi)$ is formed between the advocates, $\mathbb{P}$, of private sector objectives and advocates, Π, of the public sector objectives while the game is played against the decision-making core, $\mathbb{D}$. There are two intra-group advocates of private sector $\mathbb{P}$ and public sector Π from the general population. There are three possible coalitions of $\mathbb{A} = (\mathbb{P}, \mathbb{D})$, $\mathbb{B} = (\Pi, \mathbb{D})$ and $\mathbb{C} = (\mathbb{P}, \Pi)$.

The environment of the game, with the participating social units, which are provided in a cognitive geometry in Figure 3.8, is structured in opposites and conflicts. The private sector advocates face the combined effort of the coalition of the decision-making core and the public sector advocates as opposites. Similarly, the public sector advocates face the combined effort of the coalition of the decision-making core and the private sector advocates as opposites. The members of the decision-making core face the combined effort of the coalition of the private-sector advocates and the public-sector advocates as

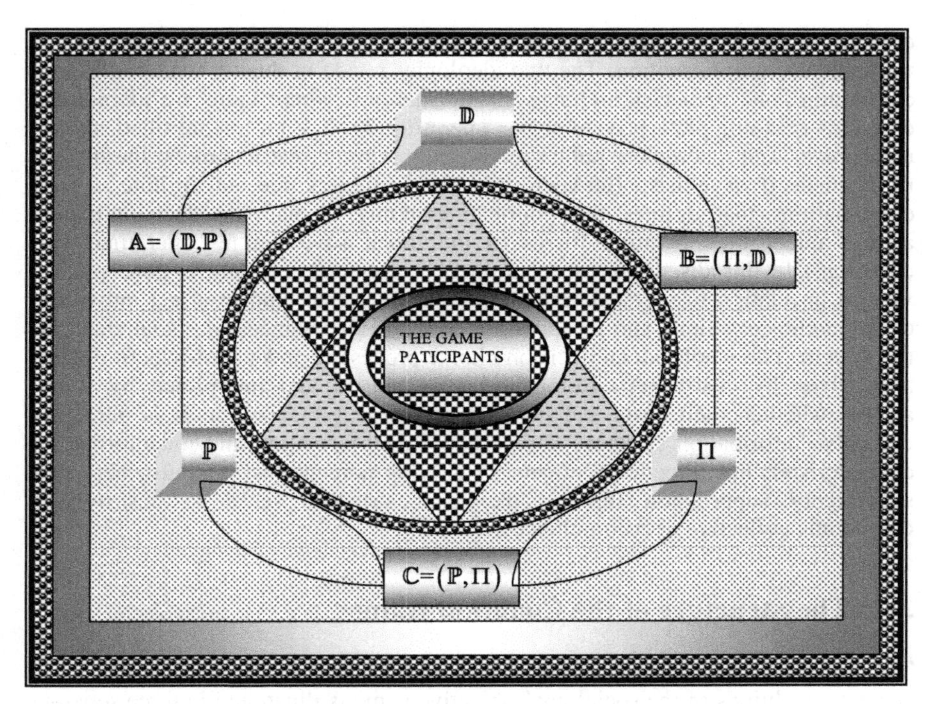

Fig. 3.8 The Cognitive Geometry of the Game Environment of the Participants

opposites. The coalitions are formed around social goal-objective elements whose implementations will change the social cost-benefit configuration and its distribution among the members as well as shape the direction of the national history. These opposites in the political economy must be seen in terms of dualities with continua as well as in unity with the social decision-choice process. It is, precisely, these conflicts in the political opposites and the resolution of them that generate continual qualitative and quantitative transformations which reflect the notions of socio-economic development, social change and nation building.

In these discussions, we have suppressed the conditions of party affiliation and ideology that may influence individual preferences and assessments in relation to the possible cost-benefit balances that may be associated with different social states, national interests and social vision. All social decision-choice actions are undertaken in relation to the individual perceptions of the relative cost-benefit configurations as they relate to the established national interest and social vision as understood by the individuals on the basis of available social information. The grouping of the society into public and private sector advocates is on the basis that the social set up operates on the principles of private-public duality, and on the same principle of individual-community duality under the continuum hypothesis of qualitative social transformations of a never-ending process. Within any continuum there is no permanent social category. All social categories are under transformations depending on the resultants of the operating social forces. The decision-choice principles behind the two groups of the general population are defined by public and private sector ideologies operating under the conditions of available social information. The two ideologies reside in duality with conflicting continuum zone of decision-choice transformations, where there are ideological extremes of a complete private sector operating through an uncontrolled market mechanism, and a complete public sector with extreme restrictions on the market mechanism if it exists. Between the two extremes are an infinite set of combinations that defines the political economy of national decision-choice actions under the operating forces of individual sovereignties operating with coalitions. In this discussion, the notion of a political economy is conceptually viewed in relation to social decision-choice actions that tend to affect individual decision-choice actions and the possible cost-benefit balances of social states of the system. The whole social decision-choice process is under cost-benefit duality and driven by cost-benefit rationality as perceived and interpreted by social agents, whether such costs and benefits are measurable or not. It is here that deceptive information structure composed of disinformation and misinformation characteristics has directional effects by altering perceptions. It is also here that sovereignty suppression through the legal structure seeks to destroy the democratic collective decision-choice system.

3.4 The Structure of the Coalition Game

The politico-economic-game environment as presented in Figure 3.8 with three separate coalitions of the social goal-objective game is illustrated in three separate games in Figure 3.9 panels a, b and c. As it has been pointed out, the basic utility of coalitions is to affect the center of gravity of the democratic collective decision-choice system. Panel (a) presents a private-sector-government coalition to defeat the decision-choice actions of the public sector advocates, in order to include certain goal-objective elements in favor of private sector activities while restricting the goal-objective elements favorable to the public sector socio-economic activities. Panel (b) presents a coalition between the government and the public sector advocates to the game which includes goal-objective elements of public interest but restricts the goal-objective elements of private-sector interest. Similarly, Panel (c) presents the private-public sector coalition to the game against some or all the goal-objective elements of the social decision-making core for inclusion into the social goal-objective set. In all these situations, the main interest of the private-sector advocates is to create favorable conditions for rent-seeking possibilities on the basis of which rent harvesting can occur. The public-sector advocates work on the general principle of public welfare that may be antagonistic to an environment of private-sector rent-seeking and harvesting activities.

The outcome of this complex game of varying structure of coalitions is the emergence of a set of social goals and objectives and the indicated national interest that will define the path of the national socio-economic history as well asthe path of the political economy the quality of which is defined by the degree of achievement of the social vision. The nature of the democratic collective decision-choice process is such that coalitions are formed with the influential and

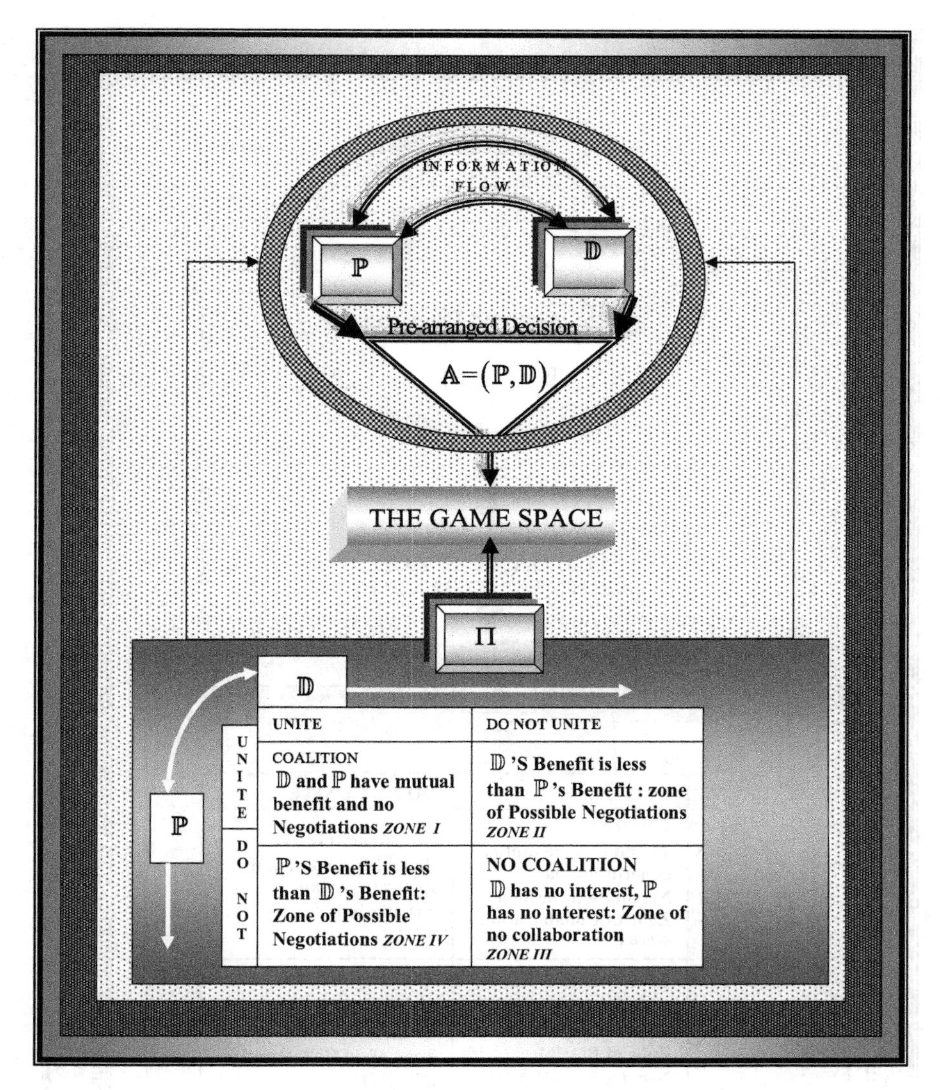

Fig. 3.9a Panel (a) A Game with $\mathbb{P}$-$\mathbb{D}$ Coalition Against Π

powerful members of the decision-making core who then work to deliver the pre-arranged social goals and objectives. The private and the public sector advocates may be split into interest groups who usually constitute small minorities in the voting space. It is usually these minorities that form coalitions with the deciding members through the majority of the decision-making core under the prescribed rules of the democratic decision-making process to create the prevailing set of social goals, objectives and national interest and then work to protect them for harvesting.

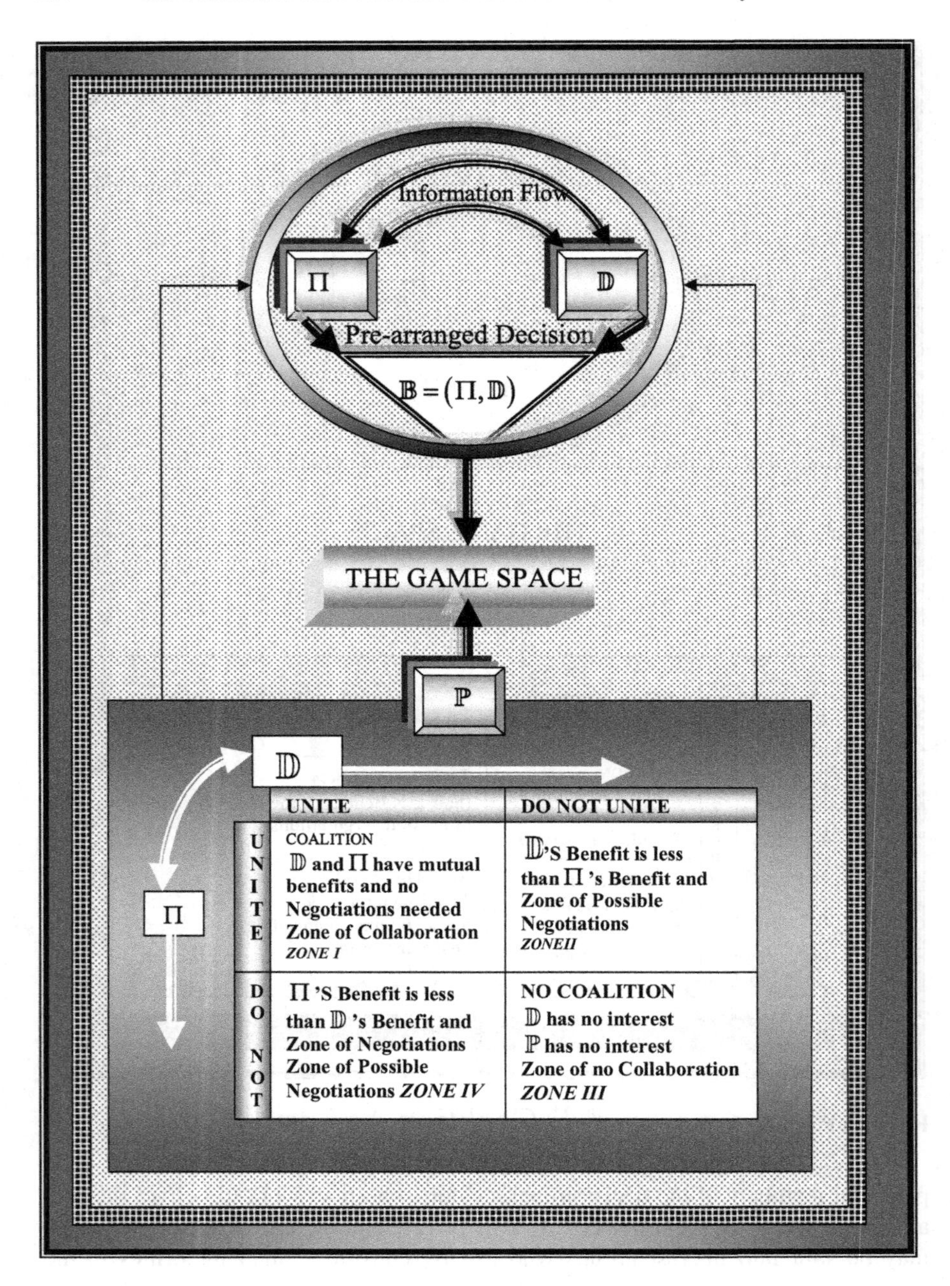

Fig. 3.9b Panel (b): A Game with Π-$\mathbb{D}$ Coalition Against $\mathbb{P}$

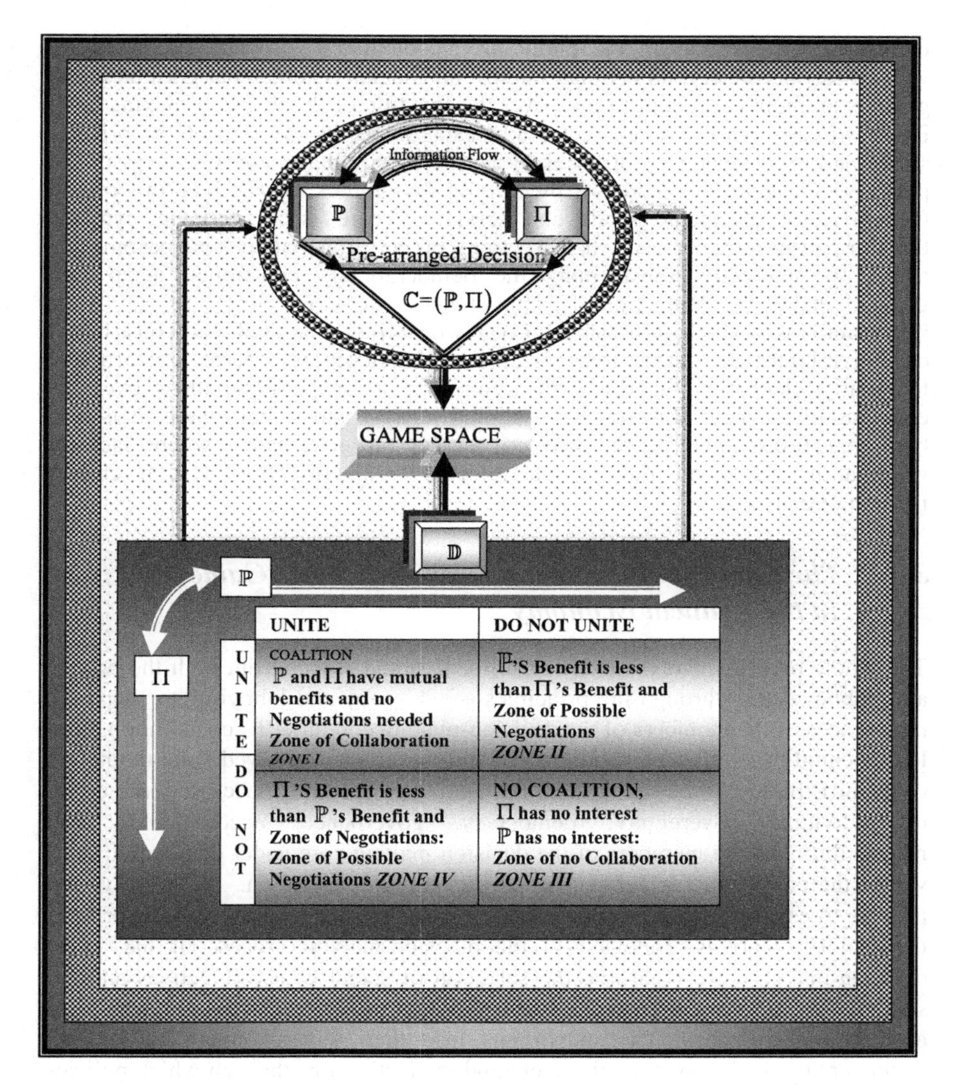

Fig. 3.9c Panel (c) A Game with ℙ-Π Coalition Against 𝔻

The primary reward of the game is the group's social goal-objective elements that are selected to enter into the social goal-objective set on the basis of potential cost-benefit balances for rent-seeking. The secondary reward is the rent harvesting from the favorite conditions that are created by the inclusion of particular elements in the social goal-objective set if the private sector advocates win to include their goals and objectives. The motivating force of the game is the secondary reward which drives the advocates' social goal-objective selection process. Here lie the problems of governance and conflicts in the principal-agent duality. The decision-making core is constructed to manage the general welfare of the society by mediating the conflicts in the public-private duality to produce an increasing welfare for the society, its national interest and social vision. What goes on in the game is an indirect attempt to take over the function of the decision-making core. Every success of the private sector advocates increases the size of the private-sector and is in favor of profit-seeking activities. Similarly, every success of the public-sector advocates increases the size of the public sector defined in terms of the provision of social goods and services. There is a continual dynamic transformation taking place where sections of the principal seek to become the agent, by indirectly transforming the democratically constituted agent into the principal in the democratic decision-choice system. This might be one of the important reasons for non-transparent behavior through the continual construct of deceptive information structure. It is through this deceptive information structure that party ideologies function with vengeance to shape preferences over the elements of the goal-objective set, national interest and social vision. It is also here that transparency gives way to secrecy in governance where lack of information disclosure through an information-classification system creates illusion of democracy.

3.4.1 The Concept of the Social Decision-Choice Game in the Political Economy

The whole process involves exerting influences and forces from both the private and public sectors through the principles of advocacy and lobbying, to shape the preferences of the members of the decision-making core after its constitution. The logic behind this influence peddling and force exertion is derived from the nature of the distribution of the social decision-making power on the accepted principles of individual sovereignty, democracy, governance, accountability and responsibility of the members of the decision making core under democratic rules within the parameters as defined in the prevailing constitution. After the voting process to establish the decision-making core is done, the individual vote or even the majority vote is irrelevant to the decision-making process of the social decision-making core regarding the selection of social goals, objectives and the setting of the national interest and social vision of the society in the political, legal and economic spaces. The individuals lose their sovereignties and become dependent on the behavior of the agent and the institutions through which policies are transmitted. It is here that the slogan of *government of the people, by the people and for the people* acquires no meaning and places the general population in the zone of deception and organizational ignorance, even though the joy that the phrase connotes is irresistibly sweet in the psychology of life of the voting public creating social exuberance that is couched in a mirage of democracy. This mirage of democracy reveals itself in an illusory sense that the voting public is made to believe that it is participating in the collective decision making that affects the individual and collective lives in all spheres of existence. This mirage has a fuzzy covering in the collective decision-choice space that hides the true reality of the democratic collective decision-choice system. This fuzzy covering is amplified by *vague information* structure and *deceptive information structure* that must be processed for an effective participation in the the politico-economic games within the collective decision-choice space. Keep in mind that the deceptive information structure is made up of disinformation sub-structure and misinformation sub-structure which are related to negative campaigns. It is this presence of fuzzy covering over the democratic mirage that provides the utility of the application of fuzzy paradigm and its analytical tools in the theories on the political economy

The individual and collective preferences about social goals and objectives are at complete mercy of the preferences of the members of the decision-making core. The members of the decision-making core create and establish the social goals and objectives as well as the rules, laws and regulations that guide the general decision-making processes in the three structures of politics, law and economics. They also hold and collectively control the instruments of enforcement and compliance to maintain their will and preference orderings. The democratic process is such that the decision-making core acquires a position of dictatorship for the period of the members' tenure. The process is such that the decision-choice actions by the decision-making core on the goal-objective elements may run counter to the will of the majority of the populace. This is attributed to the asymmetry of power distribution in the social decision-choice space after the constitution of the decision-making core, the agent who acquires dictatorial characteristics. The principal works to correct the power asymmetry while the agent works to maintain the power asymmetry in the social decision-choice space through a series of institutions. It is this process of an asymmetric power relation and the correction strategies in the principal-agent duality that invite coalition formations to create influence and exert socio-political force on the members of the decision-making core to choose a particular direction of social decision-choice actions in the goal-objective selection process. The process may be viewed as a continual correction of power imbalances in the democratic decision-choice system based on the decision-choice algorithm of majority rule to create the principal-agent duality.

Let us keep in mind that the game of selecting the members of the decision-making core is such that all the voters, irrespective of who they vote for, peacefully surrender their rights of individual and collective sovereignties of the social decision-making, by the rules of the democratic game, to the winning candidates. After the voting, the voters have no meaningful and *cost-effective* process to either reverse their votes or compel the members of the decision-making core to stick to their platforms or to even adopt the collective preferences of the majority as implied by the rules of democratic decision-choice system. The presented set of ideas and information to get votes may be quite different from or even may be diametrically opposed to the internally held true ideas, beliefs and preferences of the candidates. It is here that a dilemma arises in the democratic collective decision-choice space and corresponding social institutions. It is here that democracy negates itself and acquires characteristics of a dictatorship. The democratic decision-choice process, in establishing a decision-choice agent by transfers of individual and collective sovereignties, reveals itself as a process of willingly establishing a temporary and possibly a permanent dictatorship. A new conflict arises in the collective decision-choice space, after the collective decision-choice action has been undertaken by voting to create the decision-making core. Alternatively, after the creation of the institutional principal-agent duality, categories of conflict states are defined over the continuum of the duality. Each category is identified by its quality, where each quality is specified by the relational characteristics that are established institutionally between the principal and the agent. Besides the inter-preference asymmetry of the principal and the agent, there are two intra-preference asymmetries that intensify the conflicts in the social preference space. One of the two is expressed within the principal to create intra-principal decision-choice conflicts, and the other is expressed within the agent to also create intra-agent decision-choice conflicts. The principal-agent preference differences define conflict conditions of inter-group preferences while the preference differences within the principal and agent define conflict conditions of intra-group preferences. Within the decision-making core, there arises a conflict in the intra-preferences of the members. The resulting reconciliation of intra-group preferences of the members of the decision-making core, regarding the elements in the goal-objective set, may stand in contradistinction to the aggregate preferences of the majority whose voting actions created the decision-making core.

The interest of the decision-making core, as an agent of the public, replaces the interest of the public as the principal, when the decision-choice process, within the agent, is such that the majority of the members of the agent decide against the preference order of the principal. The dilemma in the practice of democratic principles of collective decision-choice process on the principle of majority rule, therefore, arises in the sense that in a process to avoid dictatorship, the implementation leads to a creation of a dictatorial group whose preferences are imposed on the general preferences of the society, irrespective of what the public majority preference order reveals. Furthermore, freedom of voting in a democratic social set-up is also the process to willingly surrender the same freedom in the social decision-choice actions. It is also here that the Arrow's impossibility paradox provides some useful instructions in understanding the problems of non-proportionality in non-weighted democratic decision-choice system [R2.5] [R3.1.23] [4.94] [R4.46] [R15.17]. This is another problem that is characteristic of elicitation of information on the basis of which individual voting decisions are structured, given the individual preference structures.

Besides the fact that the democratic principle of majority rule always creates dictatorship of a class over the minority, the voting principle and process of the practice of democracy may lead to the creation of a dictator decision-making core whose interests may substantially diverge from those of the majority and minority groups. The differences in the principal-agent preferences may be effectively maintained by ideology, information asymmetry, fear, deception, and suppression of many dimensions in the democratic collective decision-choice system. In other words, the practice of democracy may diverge from the philosophical basis of

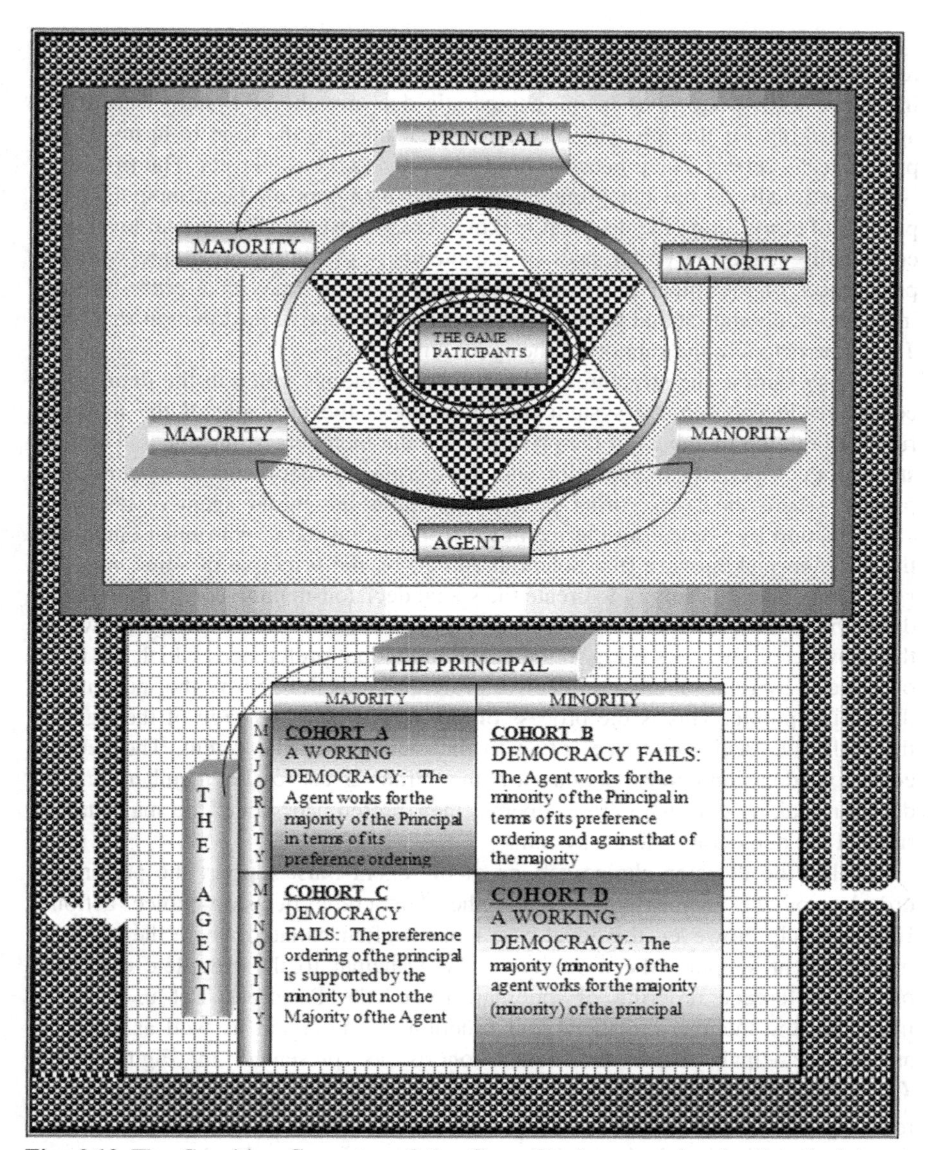

Fig. 3.10 The Cognitive Geometry of the Game Environment for the Principal-Agent Interactions in the Goal-Objective Formation

democracy which may become a mirage in the conceptual desert of nothingness, but turn into a dictatorial decision-choice system that decimates the very foundation of the democratic collective decision-choice system. This may be accomplished directly or indirectly by different strategies to distort perception and shape individual and collective preferences as part of the game. It is here that the problem of financing the activities of the democratic collective decision-choice system presents a difficulty and intensifies the dilemma. The principal-agent relational politico-economic game structure is presented as a cognitive geometry in Figure 3.10. Distortions in the democratic decision-choice system lead to the agent working for the minority of the principal most of the time in terms of the preference ordering. The process of the game is on two basic levels. The first level is to continually shape the preferences of the members of the principal in the process of constituting the decision-making core, the Agent. The second level is to continually shape the preferences of the members of the agent to side with the preferences of the minority of the principal. All of this may take place through information elicitation and manipulation to create deceptive information structure. The game is to win the support of the agent's majority.

Let us examine the structure of the politico-economic game in the democratic collective decision-choice system operating under the principle of the majority rule for both the members of the principal and the members of the agent. The first step in the operationalization of the democratic decision-choice system is a process where the legal voting public constitutes the initial agent of decision-choice action by selecting the members from within them, who must enter into the decision-making core (the potential Agent) by the principle of some *majority rule*. This process is simply to create the social decision-making core. When this is done, the constituted social decision-making core becomes the actual agent, where the constituted members bring in their own individual preferences which are to be overridden by representative collective preferences as defined by the majority of the voting public. This is the process of *sovereignty surrendering* of the right to the collective decision-choice action to the members in the Agent. The general collective decision-choice actions are then transferred to the members in the decision-making core who are to operate their preferences under the majoritarian rules of the democratic choice system.

Ideally, the decision-choice actions of the agent must support the preference ordering of the majority to support the democratic decision-choice actions. Unfortunately, however, this ideal case is most often violated where the majority of the Agent supports the minority of the Principal to the complete disregard of the preference ordering of the majority of the Principal. Let us keep in mind that the members in the Agent function as the Agent as well as function as the Principal. There are four Cohorts as presented in Figure 3.10. $\text{Cohort A} \Leftrightarrow \text{Cohort D}$ when the Agent's majority supports the Principal's majority and the Agent's minority supports the Principal's minority in the democratic collective decision-choice system. In this way, the Agent represents the majority and works for the preference ordering of the Principal's majority. The democratic decision-choice system works if there are few violations in this structural sequence. The order structure is that $\text{Cohort B} \Leftrightarrow \text{Cohort C}$ where the Agent's majority supports the Principal's minority preference positions in the decision-choice actions; and where the Agent's minority supports the principal's majority preference positions. The democratic decision-choice is said to break down in this case. The democratic collective decision-choice system itself is said to fail, if this structural sequence produces greater decision-choice preference points where the Agents' majority continually supports the principal's minority more than the former. The explanation of the existence of the relational Cohort B $\Leftrightarrow$ Cohort C and this behavior has given rise to the theories of rent-seeking, social choice, public choice, corruption and dictatorship.

The gap between information elicitation and the actual preferences not only help to explain the political discontent but also the *initial* cost-benefit balances of any outcome in the social goal-objective set. The right to vote is also a right to surrender one's democratic social decision-making right to the winner among the competing candidates who will become a member of the decision-making core that will hold the power as an agent of social decision-choice actions. The right to vote is an obligation not to commit violence after the results. It is also a right to compete to be elected to become a member of the decision-making core as well as an obligation to adhere to the rules in the democratic-decision-choice system. The right to vote is also an obligation to refrain from direct participation in deliberations and decisions to construct the social goal-objective set in addition to the set of programs and projects that will support it after the creation of the decision-making core. This is, again, a typical example of the *principal-agent problem* in the democratic-choice system of the political economy where the agent's preferences and decisions may run, and usually do run, counter to the principal, particularly in the social decision-making process.

The right to make social decisions and the power to create laws and rules with the instruments of enforcement are bestowed on the members of the social decision-making core after the core is constructed through the first level of the democratic decision-choice game. Democracy, in all its variants, is a process to ensure the resolution of conflicts and differential preferences in the social decision space, under acceptable rules, with rights, responsibilities and obligations without violence, and where *freedom* and *justice* are the cornerstones of social stability in the social decision-choice space. Responsibilities and obligations are the costs of rights while the rights are the benefits of responsibilities and obligations in the democratic collective decision-choice system. The creation of deceptive information structure, information classification on the basis of national security, creates lack of transparency in the democratic decision-choice system, which leads to the destruction of freedom and justice that are claimed to be cornerstones of democracy. In its place, we may have a mirage of a democratic decision-choice system with an effective dictatorial system encompassing law and order, where freedom and justice become casualties of the legal and judicial structures.

The irony of this game of democracy in creating the social decision-making core by voting, is that one's right to vote is also one's right to strip off one's democratic rights and liberty through the activities of the democratically constructed decision-making core.

Once votes are cast to establish the social decision-making core, the voters and all members of the society are completely at the mercy of the compassion, integrity, honesty and good will of the members who are elected into the social decision-making core with their supporting complex interest groups, until there is a new round of elections to affirm or change the composition of the members. At any new voting round, a role reversal occurs where the principal becomes the agent and the members seeking reelection collectively become the principal on behalf of whom decisions are made. And even then, the outcomes of the new round of elections are at the mercy of the incumbents in the decision-making core and their supporting cronies, who stand to earn rent by using and implementing economic and legal restraints on the principal. The dilemma in social decision-choice actions in democratic social formations cannot be resolved by proportionate representation in the constitution of the collective decision-making core because such proportionate representation cannot do away with subjective preferences of the collectivity of the members of the decision-making core which functions in the social decision-choice space as the agent. The intensity of the dilemma rests on the efficiency of the solution to the principal-agent problem in the democratic social formation and its governance. The argument in support of democratic governance and its method of constituting the decision-making core (the agent) must demonstrate that democracy solves the principal-agent problem in social organizations better than any other form of governance and its method of constituting the decision-making core. The differences in dictatorial impositions in all principal-agent problems of the political economy are simply the size of the class that constitutes each agent and the methods of selecting the members.

The explanatory logic of the defining characteristics of the democratic decision-choice process can be found in the fundamental relationships among the political, legal and economic structures of the social unit. When the social decision-making core is constructed in the political structure, the voting agent is transformed to the principal while the winning decision-choice candidates, as the principal, are transformed into constituting the agent. After the establishment of the principal-agent duality, the principal is disarmed in critical social decisions. The social decision-making power is transferred and vested in the unit, (the agent). The disarmament of the decision-making sovereignty of the principal in the decision-choice space means that the preferences of the principal, the populace, are held hostage by the democratic rules and the governing constitution where power transfers occur.

The members of the decision-making core, and particularly the leadership of the social decision-making core, can, and usually do, by forming a core within the core, corrupt the power process, the social trust, accountability, and transparency, and hijack the social decision-making unit to serve their personal ambitions and interests as well as other private interests instead of providing a true service to the interest of the public and the people who provide the living blood of the unit. Having succeeded in corrupting the power process and the channels of utilization, the core within the core uses the power bestowed on the social decision-making unit to increase its power through active changes in the legal structure in order to establish legal privileges to itself and its cronies, and to exercise more coercion on the people, members of the principal, that establish and finance the decision-making core through their votes and taxes. These legal privileges are then transformed into socio-economic privileges at the disadvantage of the masses and the public by continual changes of the social cost-benefit distribution in favor of the minority. The legal and socioeconomic privileges are maintained with the acquired decision-choice power and enforced with the power of the state as an institutional construct. It is the conflict of preferences in the principal-agent duality that establishes a continuum of politico-economic games and brings about a continual social change, as time proceeds, though the dynamics of substitution-transformation processes of quantity and quality in the relational structure of the society. The continuum within the principal-agent duality is a collection of categories of actual and potential social states that may be changed through the game strategies and different formations of the social goal-objective set. There social states are couched in cost-benefit configurations. Each category has a corresponding quantity-quality duality where any actual politico-economic category is derived from a primary politico-economic category, as well as it is a contestant for a future politico-economic category that is to be derived. The quantity-quality duality is seen in terms of principal-agent relations which, in the last analysis, are transformed into social cost-benefit configurations involving income, justice and freedom.

3.4.2 On the Individual-Community Duality, Principal-Agent Duality and the Political Economy

These actual or potential realities and characteristics of the process of social decision making by means of elected representatives find expressions in the relationship between principal and agent, where the principal is the public and the agent is the elected body. This must be related to individual-community duality in the space of interest as seen in terms of social cost-benefit balances and the corresponding preferences of the individuals and the community, the collective. The individual preferences and decision-choice actions are expressed in the private sector. The collective or the community preferences are expressed in the public sector. A conflict arises in the space of interests regarding the interactions of the two sectors. In the private sector the individual seeks the best cost-benefit configuration by avoiding or transferring unnecessary costs. In other words, the individual decision-choice actions do not encounter distributional effects of the cost-benefit configuration. In fact, there is cost shifting from the private sector to the collective in terms of the social unit. This private-sector cost shifting makes individual activities highly desirable and preferable when an advantage can be abstracted and transformed into benefit increases or into cost reduction. In the public sector, the collective also follows the principle of cost-benefit rationality by seeking the best cost-benefit configuration by avoiding unnecessary costs and requiring equality in cost-benefit sharing. The public absolves all social cost; it cannot transfer part of the cost of decision-choice action to the private sector depending on the constitutionality of rules. In the public sector, however, even the collective choice of the best social cost-benefit configuration may result in tensions after the choice. These tensions are generated by the distributional effects of the resulting cost-benefit configuration in relation to the perceived social fairness, justice, freedom and equity by the members of the public.

At the level of competing interests in the individual-community duality, a problem arises as to how private and public interests should be identified. Who should be charged with the responsibility to identify and separate them? Similarly, who should establish rules and regulations to maintain them through the power of enforcement that is bestowed on it by the community? It is interesting to note that the private-public sector interests exist as opposites and in a continuum. The concept of continuum simply implies that every private-sector (public-sector) interest has a support of a public-sector (private-sector) interest to maintain the social unity of the production-consumption system. It is the need for this institutional separation which one may see the relevance of the institution of government as a social decision-choice technology and a tool for social management. In this connection, as viewed from the social decision-choice space, the role of a government as an institution, in the social organization, is to mediate the separation of private-sector and public-sector interests as well as unite them for the stable management of the social set-up. The institution of government must be controlled by a class from within the general population and who will use the power of the government to manage the social set up for stability and progress. Thus, the conflict in individual-community duality in the decision-choice space leads ultimately to the establishment of principal-agent duality that generates a relational good-evil duality also in a continuum.

The formation of the principal-agent duality does not dissolve the individual-community duality. It simply helps to manage the conflicts of interest in the individual-community duality. The conflicts in the principal-agent duality are the result of the management and distribution of interests viewed in terms of fairness and equity of the cost-benefit configuration. As it is, the individual is not only a member of the community, but his or her identity is only meaningful only with and within the community. Similarly the community is not only made up of its members but its identity is only meaningful with its members. The individual has identity within the community and the community has identity from the constituent individuals. The members are the support of the community and the community is the support of the members for stability and progress. Similarly, when the principal-agent duality is established, the principal becomes the support of the agent and the agent becomes the support of the principal in mutual give-and-take relations for social stability and progress. The point in the relational continuum of the good-evil duality, which the society settles at any time, is continually changing. The change depends on the character of the class that comes to inherit the established institution of government. Thus, there are complex interactions between the individual-community duality and the principal-agent duality to generate relational good-evil duality in the social set up. There are intra-individual and intra-community conflicts of interest as well as conflicts between the individual interests and the community interests. Similarly, there are conflicts between the principle's interests and the agent's interests and within themselves.

The dynamics of the behavior of the individual-community duality is complexly connected to the dynamics of the behavior of the principal-agent duality. The give-and-take relationship between the individual-community duality and principal-agent duality has a potential to generate corruption through actual and potential conflicts of interests. The presence of these actual and potential interest leads to the development of a deceptive information structure for lack of transparency. In a democratic decision-choice system, the principal and the agent are the community that is composed of its individual. The agent is calved from the community and seeks validation from the principal while the principal seeks protection from the agent in the political, legal and economic structures. The processes and the nature of the dynamics in the principal-agent duality, as they relate to the behavioral give-and-take process with the individual-community duality, have also given rise to a new class of sales persons in the *political market*. These sales persons are called by different names such as political action groups, lobbyists, advocates, interest groups, political think-tanks, civil societies, non-governmental organizations (NGOs) and many more whose activities are legally established but non-wealth-producing rent seeking.

They are members of the parasitic segment of the production-consumption process through their ability to induce wealth transfers rather than wealth creation through the process of interest distribution by the decision-making core. One of the main objectives of these political salespersons is to either positively or negatively market particular social objectives and goals that must enter the set of social objectives, goals and national interest to the members of the social decision-making core. In other words, these salespersons help to shape the nature of the internal conflicts and the dynamics of the principal-agent duality through their effects on social preferences towards the elements in the goal-objective set as well as the national interest. They also influence the democratic collective decision-choice system by marketing ideologies from the deceptive information structure. They influence the direction, size and structure of the decision-making core as well as the social goal-objective set, either by forging alliance with or by buying of institutions of information dissemination such as mass media. The media, then, becomes a collaborative institution of mass deception in support of the rent-seeking, rent-creation and rent harvesting. Such activities are *legal corruption* that distorts the nature of the true structure of the democratic collective decision-choice system.

In this process, the social goals and objectives as well as the national interest and social vision may not and usually do not reflect the wishes of the citizens (principal) as expressed by the voters in a democratic decision-choice system. No amount of critical reformation of the voting mechanism by improvement of the rules and procedures will resolve the problem of the dilemma in the political game to establish the social goal-objective set that accurately reflects preferences of the voters. The best that can be done is to create an institutional process to minimize the preference distance between the principal and the agent which hopefully will reduce the intensity of the dilemma. The dilemma decreases in intensity as the preferences of the agent approach those of the principal. The coalition game to establish social goals and objectives is made possible by the formation of small pressure groups that exert pressure on the social choice of the members of the social decision-making core at the expense of the preferences of the majority voters. In other words, the social goals and objectives and the corresponding national interest are made to correspond to the interest of the pressure groups, rather than to the true interests that correspond to the outcomes of the democratic decision-choice process.

The social goal-objective game, which can also be viewed as a private-public interest game, as conceived, is carried on through a democratic process that is designed to promote social decision-choice efficiency in the political market by reconciling conflicts in the private-public interests without violence, in that the right to vote is also an obligation to refrain from violence after the results of the voting activity. The national constitution lays down the rules in the social set-up. It acts as an umpire to regulate conflicts in decision-choice activities among the three segments of the private sector, the public sector and the social decision-making core after the members have been elected into the core by votes. The same constitution acts as referee for intra-group and inter-group decisions and social decision making. The constitution in this public-private decision-choice space is a regulator of behavior for fair play without exerting group pressure from either the private or public sector advocates, where the members in the social decision-making core decide with true conscience. The analytical process of the social

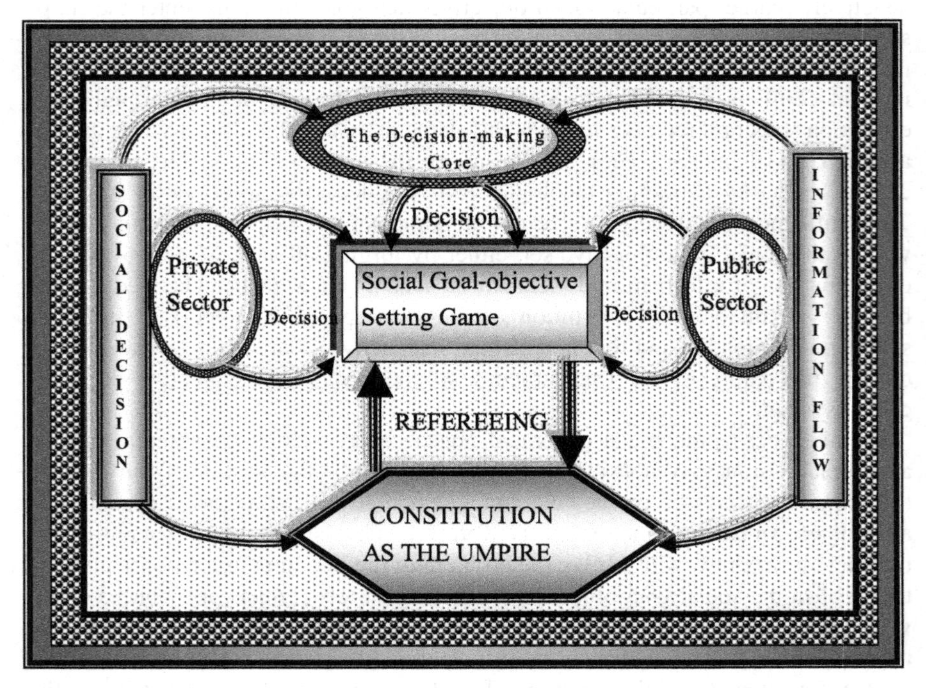

Fig. 3.11 Decision-Information-Interactive Process in the Social-Goal-Objective Formation

formation may be seen in terms of distribution of freedom of the decision-choice power in the political economy between the collective and the individual and between the private and the public, while the members of the social decision-making core mediate for fairness and equity over the economic, political and legal spaces and how such distributed freedoms affect the cost-benefit distribution over the path of the national history. The nature and structure of the game is depicted in Figure 3.11.

The process and the game work through interactive and interrelated social dualities, polarities, unity, continuum, opposites and tension. The essential ones for our analytical interest in examining the structure of the goal-objective game are public-private polarity, individual-community duality, and principal-agent duality as foundations of social organization, management and governance. In each pole of the public-private polarity, we have the residence of individual-community duality with its own internal conflicts and superimposed on it is the principal-agent duality with its internal conflicts. It is a game where the individual-community conflicts interact with the principal-agent conflicts to shape and reshape preferences, and create forces of change that move the society through the quality-quantity space on the basis of cost-benefit distributions with neutrality of time.

The nature of the conflicts in political polarity and dualities is diagrammatically represented in Figure 3.12 with *public-private polarity* that tends to house the *individual-community duality*, and the *principal-agent duality* where the principal represents the general population and the agent represents the government (the decision-making core). The interactive elements of the polarity-duality structures define the action environment for the democratic decision-choice process and the governance decision-choice process in the political economy organized around a democratic decision-choice system for both the principal and agent. If the community's preference characteristic set dominates the individualistic characteristic set, then the social system moves in an increasing size of the public sector. In this way, the decision-choice processes tend to favor increasing public sector elements in the social goal-objective set. Social institutions are then created in the legal, political and economic structures in support of such social preference configuration with public sector dominance. Similarly, there is an increasing size of the private sector if the individualistic preference characteristic set dominates that of the community with the result that the decision-choice processes tend to favor increasing private sector elements in the social goal-objective set.

The collective decision-choice processes, in this case, tend to favor increasing private sector elements in the social goal-objective set. Social institutions are created in the legal, political and economic structures in support of such social preference configurations with private sector dominance. The dominance includes balances between the individual and collective freedom and property rights in the social decision-choice space. As either the information structure changes or preferences shift so also the private-public sector combination varies in response in the continuum. Thus the private-public sector combination is never permanent, with its position constantly responding to the nature of the information set and collective and individual preferences in the private-public-sector combination .

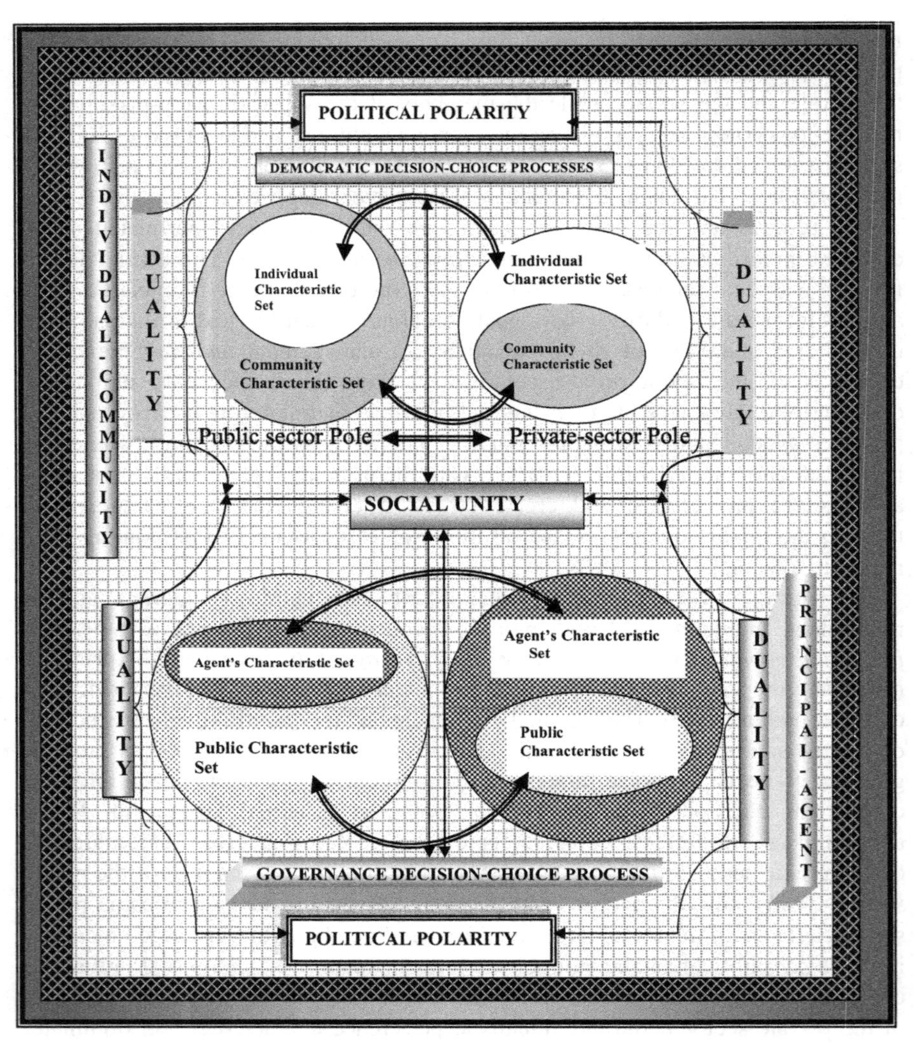

Fig. 3.12 Venn-Diagram of Social Unity of Public-Private Sector Polarity with either Individual-Community Duality or Principal-Agent Duality in Decision-Choice Conflicts

Let us now consider the public-private polarity and its interaction with principal-agent duality. If the social structure is such that the agent accepts the principal's preference characteristic set to dominate the personal characteristic set of the members of the decision-making core then the social system moves in an increasing intensity in the application of democratic principles of governance where the agent represents the population in accordance with the rules of the democratic collective decision-choice system. In this way, the preference characteristic set of the members of the decision-making core is asymptotically close to the preferences of the principal, such that the decision-choice processes tend to favor increasing public sector or private sector elements in the social goal-objective set in accordance with the direction of the preferences of the principal. Social institutions are then created by the decision-making core in the legal, political and economic structures in support of such social preference configurations with either public sector or private sector dominance. Similarly, there is an increasing private-public sector conflict if the agent's preference characteristic set dominates that of the populace (the principal) where the decision-choice processes tend to favor personal preferences of the members of the decision-making core.

In this situation, either private or public-sector ideology comes to shape the relative structure of the elements of the social goal-objective set which may contain increasing goal-objective elements of private sector for rent seeking, creation and harvesting. Alternatively, it may contain goal-objective elements of the public sector that will go to improve the general welfare of the population but not specifically to some individual welfare. Social institutions are created in the legal, political and economic structures by the members of the decision-making core in support of such social preference configurations with either private or public sector dominance. The dynamics of the social decision-choice process is explainable through the conceptual system of polarity, duality, conflict, continuum and unity in the quantity-quality-time space. It is here, that the fuzzy paradigm and its analytical methods can help in the problem formulation and soft computing under a given constitution and a democratic decision-choice system. It will also assist in a useful understanding of the changing qualitative characteristics as the social system traverses between poles and within dualities under the continuum principle. In a simple nutshell, the political game, in all the three sectors is about power distribution in shaping the private-public sector proportions in a nation, and how such proportions affect resource distribution, income-effort distribution, fairness and justice.

Even under a strict adherence to the constitution, as an umpire, to create channels of efficient refereeing and fairness of the game of socio-political decisions, the members of the decision-making core decide not on the basis of social good that reflects *majoritarian* views and preferences. On the contrary, they decide on the basis of their preferences and the maintenance of their incumbency. In other words, the representative government, constituting the unit of social decision making, may fail to serve the governed in terms of the collective preferences defined by the majority. Instead of governance based on a set of socially held goals and objective, we may have rulers that serve their ambition and the ambitions of some specific interest groups rather than the interests of the populace and the general public. The system may be such that the democratic rules of the collective decision-choice process may assert the dominance of the private pole through the behavior of the principal-agent duality to the complete neglect of the role of the individual-community duality.

Generally, we may tend to expect that the internal dynamics of the behavior of principal-agent duality, with periodic role reversals, that are meant to produce self-organizing social order, is constrained by the internal dynamics of the individual-community duality in the role reversal processes for greater degree of efficiency of the democratic decision-choice process. Usually, however, this is not the case. Here, comes the complexity that demands a non-classical process of reasoning. The analytical framework is such that the optimizing decision-choice behavior of the agent is constrained by the internal conflicts of differential preferences of the principal. Similarly, the optimizing decision-choice behaviors of the individuals are constrained by the conflicts of the collective preferences of the community. The two are superimposed on each other under a defective information-knowledge structure composed of vagueness that produces fuzzy decision-choice variables, and limited and asymmetric information that produces random decision-choice variables. The complex nature leads us to deal with both fuzzy-random and random-fuzzy decision-choice variables in multiple dimensional interactions in the quality-quantity-time space of the self-organizing and self-correcting social systems. The classical paradigm and its laws of thought are, thus, confronted with increasing difficulty in the synthesis even given the basic analytics.

The complexity of the political economy of the national interest and the supporting goal-objective set are such that the resultant is produced in the direction of maximum socio-economic power that the individual or a group holds. In this respect, the set of social goals and objectives, in addition to national interest, may be established to meet the demands of pressure groups that have a power advantage but not necessarily what is socially and collectively desirable for the national progress and the people's welfare. The contending political forces are to distribute greater benefits to the members with greater sociopolitical power and greater cost to the members with weaker sociopolitical power. This process of distribution of decision-making power leads to the emergence of institutions of power struggle that produces increasing numbers of political pressure groups in the political economy. The rise of political salespersons, lobbyists, institutions of interest and political pressure groups is founded on the acknowledgement of the fact that the created power of the state and its supporting institutions may be hijacked by some members of the constitutionally created decision-making core to serve their interests and the interests of their cronies; and that the direction of the social decisions may be influenced by applying political pressure that is legally sanctioned on the members of the decision-making core without producing violence.

The recognition of this situation leads to the formation of pressure groups on the basis of ideology and specific interests rather than on the basis of social welfare optimum and the will of the people. In this way, the democratic collective decision-choice system has a built in seed of its internal self-destruction through the internal forces that operate in the principal-agent duality in continuum, and under the principle of intense conflicts in the private-public-sector conditions of the provision of goods and services in the society where the fundamental ethical postulate is individualism which is taken as a prior in social decision-choice actions. Alternatively stated, democracy tends to destroy itself from within. In this respect, some questions posed by Keirstead are useful to reflect on:

> *Has the political system the flexibility necessary to assimilate economic changes and to operate economic controls without loss of freedom or responsibility? Are the developing institutions and ideas of capitalism consistent with a healthy democracy and free political institutions? Have the attitudes, the social beliefs and myths and ideals associated with what we call capitalism the kind of moral quality which survives?* [R15.24, p.436].

The seed of destruction of democracy may also be seen from the organized institutions by groups through coalition formation to acquire power in the political structure to serve their group interests in the legal and economic structures without consideration of the societal sovereignty. It is on this recognition of power-relational games which affect the national-interest setting, and social goal-objective formation that makes it useful to reflect on the idea:

> *It is doubtful if any notions of national interest can emerge. National policy is simply a compromise of group interests. Sometimes the compromise is in the form of a conspiracy against the rest of the public. .. All national policy is necessarily a compromise, but there is a difference between compromises of the general interest because of conflicting or qualifying elements, and compromises or conspiracies between powerful groups for their own special advantage at the expense of the general good. There is a difference in principle between the power exercised by government and that exercised by organized pressure groups. The power exercised by government is responsible power. If it is abused, if the general interest is not promoted, there is a constitutional recourse against those who have abused their position* [R15.24, p. 442].

This difference in the use of power is destroyed if an interest group controls the government and acquires the vehicle to exercise the power of the government to serve its interests. Here comes the paradox in democracy where the great enemy of individual freedom is the state, it is claimed, and yet the protection of individual freedom seem to call for an increase in the power of the state which is set up by the democratic collective decision-choice system in relation to the goal-objective setting to support the national interest social vision and collective good.

Chapter 4
Coalition Formations and the Social Goal-Objective Setting

At the end of Chapter 3 of this monograph, we discussed social power and the exercise of the decision-making power in relation to social goal-objective formations in support of the national interests and social vision. The control of this decision-making power and the exercise of it may substantially depend on organized groups to claim power in the political structure and alter the conditions of individual and collective freedoms as may be seen in the economic and legal structures. The organized groups are called coalitions which are formed on the basis of common social preferences to will some form of social decision-making power. The direction to which the coalitions move the social decision-choice process is called the *coalition effect*. The coalition effect is part of the *systemic causation* that shapes the dynamics of the democratic collective decision-choice system from within. Three major social blocks are advanced. They are the public-sector-interest advocates, the private-sector-interest advocates and the social decision-making core. The public-sector and private-sector advocates constitute the initial decision-making group that creates the social decision-making core who then assumes the public responsibilities of governance and management of the social setup. These groups may be subdivided into other smaller groups with common sub-interest.

We shall now turn our attention to the process of coalition formation to affect the power distribution in the democratic collective decision-choice systems. Let us consider the decision problem of constructing the optimal social goal-objective set as has been defined and how it relates to the coalition formation in a democratic decision-choice system. Three coalitions are formed based on aggregate interests in relation to institutions of private, public and government as are specified in Figures 3.8 and 3.9 (a, b, and c) of Chapter 3. Each of these coalitions has an opposition to whom they have to play the game against in order for their subjective goals and objectives to be included into the social goal-objective set. Without coalition formation each advocate group will have to play the game by itself against the other two, where in this case, the game becomes a three-dimensional one without alliances. The goal-object-setting game does not become three-dimensional even though this is possible. There are situations when coalitions are not only necessary but are the most effective way to work for the inclusion of particular goals and objectives into the social goal-objective set, as

K.K. Dompere, *Social Goal-Objective Formation, Democracy and National Interest*, Studies in Systems, Decision and Control 4,
DOI: 10.1007/978-3-319-05173-4_4, © Springer International Publishing Switzerland 2014

well as to set the direction of the national interest, history and the system's welfare and its distribution. The good aspects of the coalition effect in the goal-objective setting may be overshadowed by the evil side of asserting the preferences of the coalitions, at the expense of the majority, in the place of the true goal-objective set of the nation, and in relation to the national interest and social vision on the principle of majority rule in a democratic collective decision-choice system.

4.1 The Basic Structure and the Geometry of Coalitions

Each coalition formation requires some degree of *intra-group negotiations* on preference compromises by the members as to what goals and objectives the members of the coalition agree on, in order to collectively play the game for their inclusion into the social goal-objective set. Such intra-group negotiations involve trade-offs and priority ranking on the basis of degrees of collective preferences as seen from cost-benefit balances. This may be intense. The rankings of degree of collective preferences may be constructed from weighted utilities where the weights may be obtained from the nature of the distribution of participation rates of the group members. They also may be obtained by priority scoring that may be organized into a system of weights. The participation rates may be numerically measured in terms of money or time, or both, or an appropriate unit. The reward of negotiations within the coalition is the design of a common coalition goal-objective set that provides the coalition with a platform to design strategies for inter-group negotiations. Each coalition is called an interest group with common characteristics that unite its members. Given the coalition's constructed goal-objective set, the coalition is ready to play the political market game against the opponents in order to establish the social goal-objective set for social action.

The negotiation processes for the group and coalition formations proceed on two fronts of intra and inter coalition compromises. The intra-coalition negotiation is intra-coalition preference formation over the identified goals and objectives that the members collectively consider as essential, or have high degrees of social relevance to the coalition members in terms of their perception of social good or desirable to the members. The intra-coalition negotiation for each of the interest group, Π, $\mathbb{P}$, and $\mathbb{D}$ may proceed in the following steps: 1) elicitation of all goals and objectives that the coalition members consider as socially important; 2) construction of zone $\mathbb{A}$ of completely agreed objectives; 3) construction of zone $\mathbb{B}$ of negotiable goals and objectives and 4) construction of zone $\mathbb{C}$ of irreconcilable goals and objectives. Intra-coalition negotiation does not arise in zones $\mathbb{A}$ and $\mathbb{C}$. The set, $\mathbb{A}$, is augmented by the negotiated goals and objectives from the set, $\mathbb{B}$, to obtain the coalition goal-objective set, $\overleftrightarrow{\mathbb{A}}$, while those goal-objective elements that fail the test of negotiation go to augment the set, $\mathbb{C}$,into the final non-negotiable coalition goal-objective set, $\overleftrightarrow{\mathbb{C}}$, which shows elements of complete disagreement without a coalition common preference. The zone, $\mathbb{B}$ is where cooperation and trade-offs occur to increase the admissible set of the coalition objectives. After negotiation, we have $\mathbb{B} = \overleftrightarrow{\mathbb{A}} \cap \overleftrightarrow{\mathbb{C}} = \varnothing$, that is there is no element remained for negotiation. The processes of intra and inter coalition formation are viewed to be independent of each other.

Each interest group enters the political market with a well-defined goal-objective set that reflects the members' preferences with a *fuzzy conditionality* [R7.14][R7.16] [R15.13][R15.14]. The inter-coalition negotiations emerge when the three coalitions of interest groups, Π, $\mathbb{P}$ and $\mathbb{D}$ enter into the political market to play the social goal-objective formation game. Each group enters with a defined set of social goal-objective set that is agreed upon by its members. The defined goal-objective sets may be identified as, $\mathbb{G}_{\Pi}$, for the public sector advocates, $\mathbb{G}_{\mathbb{P}}$, for the private sector advocates and, $\mathbb{G}_{\mathbb{D}}$, for the decision-making core respectively. The specific interest groups' goal-objective sets are assumed to be nonempty subsets of the general social goal-objective set. The intersection of them will constitute the social goal-objective set that the three interest groups agree on. Let such a set be, $\mathbb{Q}$ and hence

$$\begin{cases} \mathbb{Q} = \mathbb{G}_{\Pi} \cap \mathbb{G}_{\mathbb{D}} \cap \mathbb{G}_{\mathbb{P}} \\ \text{where } \# \mathbb{Q} \geq 0 \end{cases} \tag{4.1}$$

The structure of the interactions of the goal-objective sets of the interest groups may be represented in a Venn diagram as in Figure 4.1. From these interactive set structures, we may also define other common goal-objective structures. The intensity of preference relations as defined over sets $\mathbb{H}, \mathbb{F}$ and $\mathbb{E}$ will dictate the nature and direction of coalition formations among interest groups. As such, a set of relational algebra may be constructed as:

$\mathbb{H} = \mathbb{G}_{\mathbb{D}} \cap \mathbb{G}_{\Pi}$, $\mathbb{F} = \mathbb{G}_{\Pi} \cap \mathbb{G}_{\mathbb{P}}$ and $\mathbb{E} = \mathbb{G}_{\mathbb{P}} \cap \mathbb{G}_{\mathbb{D}}$ where $\#\mathbb{H} \geq 0$, $\#\mathbb{F} \geq 0$ and $\#\mathbb{E} \geq 0$, where (#) represent the cardinality of the goal–objective elements. Theoretically, it is possible that $\#\mathbb{Q} = 0$ from the beginning of the game, but practically, negotiations will lead to $\#\mathbb{Q} \neq 0$. The set, $\mathbb{H}$, contains the common goal-objective elements of the members in both $\mathbb{D}$ and Π. The set, $\mathbb{F}$, contains the common goal-objective elements of the members in both Π and $\mathbb{P}$ while the set, $\mathbb{E}$, contains the common goal-objective elements of the members in both, $\mathbb{D}$ and $\mathbb{P}$. The coalition between any two of Π, $\mathbb{P}$ and $\mathbb{D}$ is referred to as a supra-coalition. On the basis of the group preference relations, the direction and the nature of a supra-coalition formation among the decision-making core, $\mathbb{D}$, the private sector, $\mathbb{P}$, and the public sector, Π, interest groups or advocates may be examined for negotiations. The relative intensity of preferences in terms of coalition formation will be shaped by the relative cost-benefit balances as assessed by the groups over the elements in $\mathbb{F}$, $\mathbb{H}$ and $\mathbb{E}$. Alternatively stated, the nature of the supra-coalition formation will depend on the social cost-benefit configuration and its distribution as assessed by the interest groups.

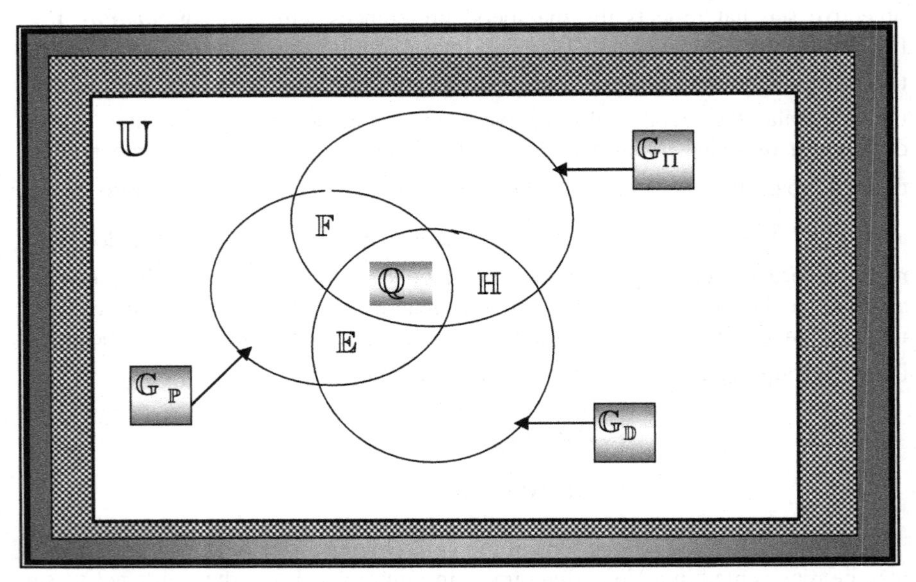

Fig. 4.1 Coalition Goal-objective Sets Showing Agreement-disagreement sets in the Universal Goal-objective set $\mathbb{U}$

The cost-benefit assessments of the coalition-formation process are undertaken relative to the goal-objective sets that are common to the advocates of private and public sectors and the decision-making core. In this respect, there are four possible cases to be considered in addition to the one illustrated in Fig. 4.1. This is the case where $\mathbb{Q} = \varnothing$ which simply implies that the three groups of private, public and the decision-making core have no common set of social goals and objectives. There are cases where there are common goals and objectives that satisfy the inter-group preference when $\mathbb{Q} \neq \varnothing$ as in the following cases:

case1) $\mathbb{E} = \varnothing$, $\mathbb{H} \neq \varnothing$ and $\mathbb{F} \neq \varnothing$;

case2) $\mathbb{H} = \varnothing$, $\mathbb{E} \neq \varnothing$ and $\mathbb{F} \neq \varnothing$;

case3) $\mathbb{F} = \varnothing$, $\mathbb{H} \neq \varnothing$ and $\mathbb{E} \neq \varnothing$, and

case 4) $\mathbb{E} = \varnothing$, $\mathbb{H} = \varnothing$ and $\mathbb{F} = \varnothing$. Cases 1, 2, and 3 are such that two groups have a goal-objective set that satisfies their collective preferences. Case 4 is such that the policy has no common elements for the three groups and hence their social preferences diverge and there is nothing to negotiate on. These cases are illustrated with Venn Diagrams in Fig 4.2.

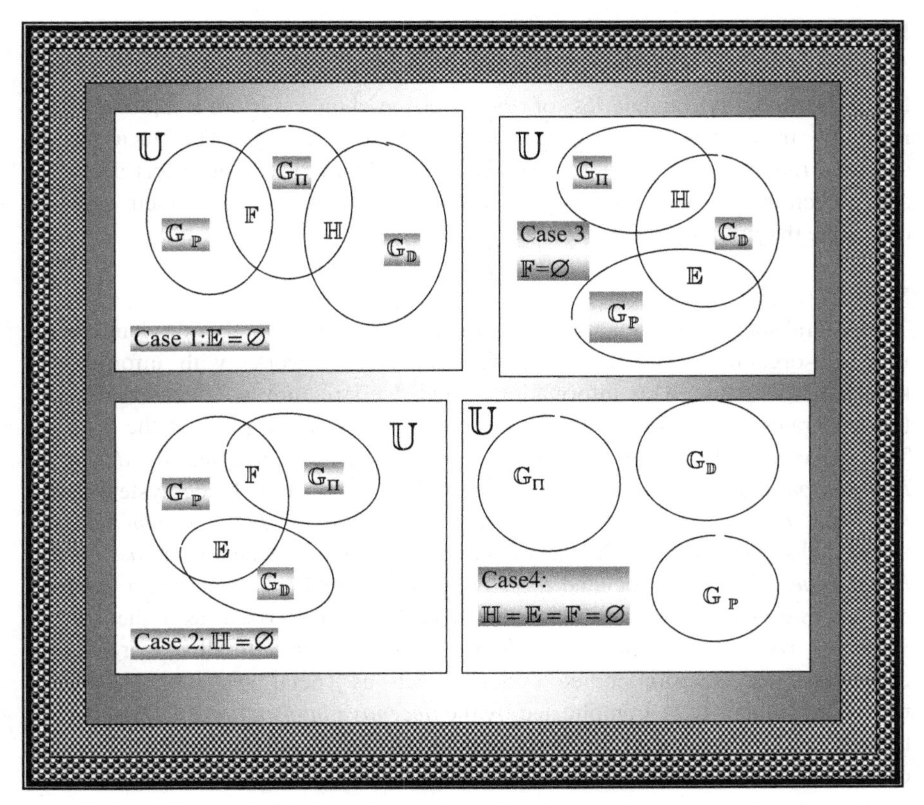

Fig. 4.2 Goal-Objective Conditions for Possible Coalition Formation with $\mathbb{Q} = \varnothing$

The process of inter-group coalition formation takes place on the basis of group preferences as defined and agreed upon by their members. The direction and the nature of the supra-coalition formations among the decision-making core, $\mathbb{D}$, the private sector advocates, $\mathbb{P}$, and public sector advocates, Π, are driven by the group cost-benefit perceptions and estimates on the basis of which preference-relational functions are defined to create the force of resultant coalitions. The task now is to examine the nature and structure of the preference-relational functions, coalition-relational functions and the unity of relations among the groups. The game is seen in terms of political games in economic and power spaces in the democratic decision-choice system. The game takes place in the institutional set-up that defines the operative structure of the democratic social formation. The decision-choice action and its outcome involve decision-choice risk associated with the democratic social decision-choice system. This is the *systemic risk* of the democratic social formation and the corresponding decision-choice process.

Definition 4.1: Systemic Risk (Verbal)
A systemic risk ρ is defined as the risk associated with the system's departure from the stability boundaries of the decision-choice system's optimal path resulting from deficient information-knowledge structure. The measure of systemic risk is an aggregation of stochastic and fuzzy risks that reflect decision-choice actions in both the possibility and probability spaces which together constitute the *fuzzy-stochastic risk.*

Note 4.1
The general social decision-choice system, whether democratic or non-democratic, is a self-organizing and self-correcting system that works with information-knowledge structure. This information-knowledge structure is defective in a way that generates stochastic and fuzzy uncertainties as integral parts of the process. The defective information structure may come to be corrupted by *deceptive information structure* which is particular to social decision-choice systems. The *stochastic uncertainty*, by itself generates a *probabilistic belief system* with an associated *stochastic risk.* The *fuzzy uncertainty*, by itself, generates a *possibilistic belief system* with an associated *fuzzy risk.* They combine to generate a *complex risk decision-choice* environment that may be characterized as either *fuzzy-stochastic risk* or *stochastic-fuzzy risk.* This risk is systemic to all self-organizing and self-exited decision-choice systems such as social systems. The fuzzy uncertainty comes to be complicated by the *deceptive information structure* which thus increases the value of the fuzzy risk and the total systemic risk of the social decision-choice system.

The deceptive information structure is made up of two components of *disinformation* and *misinformation* sub-structures to shape preferences in intended direction by the originator contrary to the information recipient. For detailed exposition see [R15.13]. The defective information structure and the deceptive information structure combine to define the complex environment of the social decision-choice framework in which the decision-choice agents operate. Such an environment may be presented with a pyramidal logic as in Figure 4.3. In the diagram, defective information structure and vague and incomplete information sub-structures constitute one pyramidal logical block while deceptive information structure and misinformation and disinformation sub-structures constitute another pyramidal logical block that is superimposed to define the environment for the decision-choice framework that cognitive agents operate from the center of the social system. Notice the implied concepts of logical trinity, duality and continuum where the degree of deception has a supporting degree of transparency and the degree of defectiveness has a supporting degree of perfection which may be seen as linguistic variables and represented as fuzzy sets.

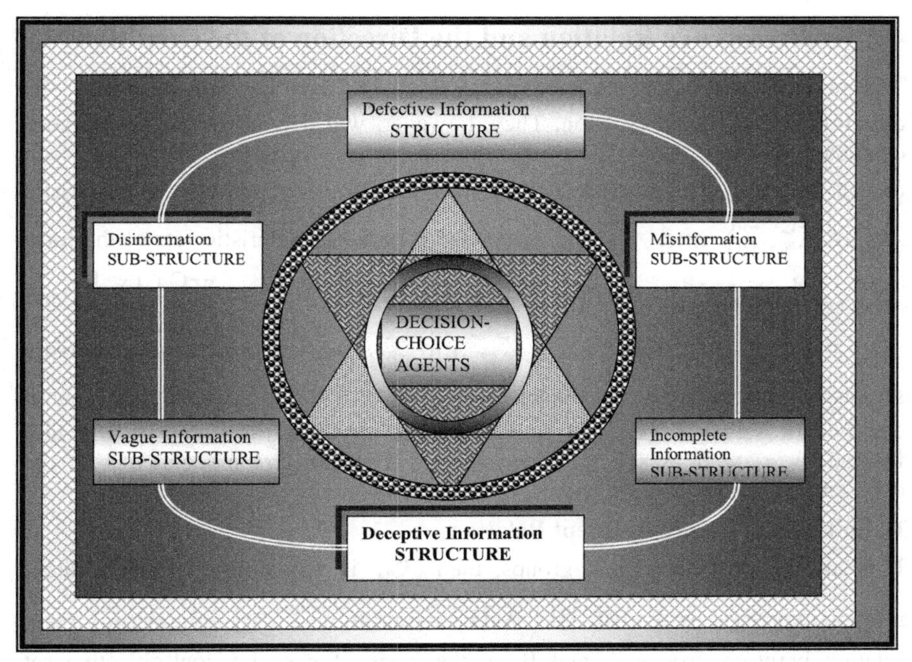

Fig. 4.3 Pyramidal Logical Geometry of the Interactions of Defective information and Deceptive Information Structures and their Substructures that Define the Environment of Decision-Choice Actions.

Definition 4.2: Preference-Relation Function

A relation, $\Re$, is said to be a goal-objective preference-relational function defined over sets $\mathbb{E}$, $\mathbb{F}$ and $\mathbb{H}$, for negotiation, if , for any pair of the goal-objective sets such as $\mathbb{E}$ and $\mathbb{H}$ there exist:

a) $\Re(\mathbb{H}) \succ \Re(\mathbb{E})$ or $\Re(\mathbb{E}) \succ \Re(\mathbb{H})$ or $\Re(\mathbb{H}) \sim \Re(\mathbb{E})$

b) If $\Re(\mathbb{H}) \succ \Re(\mathbb{E}) \Rightarrow \Re(\mathbb{E}) \nsucc \Re(\mathbb{H})$

c) If $\Re(\mathbb{H}) \succcurlyeq \Re(\mathbb{E})$ and $\Re(\mathbb{E}) \succcurlyeq \Re(\mathbb{F}) \Rightarrow \Re(\mathbb{H}) \succcurlyeq \Re(\mathbb{F})$ or $\Re(\mathbb{F}) \not\succcurlyeq \Re(\mathbb{H})$

The symbol $\succ$ implies more socially valuable and hence preferred, $\sim$ implies equally valuable and hence indifference in preference, $\succcurlyeq$ implies socially valuable or equally valuable, and $\not\succcurlyeq$ not socially or equally valuable, as the negotiable sets are compared within the mutually established goal-objective set in support of the national interest or the set of national interest.

4.2 Preference Relation and the Direction of the Coalition

The preference relations defined over $\mathbb{F}$, $\mathbb{H}$ and $\mathbb{E}$ will determine the direction of coalition formation. They, however, will depend on the cost-benefit characteristics and their relative intensities as perceived by the groups. For each one of the sets $\mathbb{F}$, $\mathbb{H}$ and $\mathbb{E}$, we may specify associated cost-characteristic sets as $\mathbb{X}_{\mathbb{E}}^{C}$, $\mathbb{X}_{\mathbb{F}}^{C}$ and $\mathbb{X}_{\mathbb{H}}^{C}$ and associated benefit-characteristic sets of the form $\mathbb{X}_{\mathbb{E}}^{B}$, $\mathbb{X}_{\mathbb{F}}^{B}$ and $\mathbb{X}_{\mathbb{H}}^{B}$ such that $\mathbb{X}_{\mathbb{E}} = \mathbb{X}_{\mathbb{E}}^{C} \cup \mathbb{X}_{\mathbb{E}}^{B}$; $\mathbb{X}_{\mathbb{F}} = \mathbb{X}_{\mathbb{F}}^{C} \cup \mathbb{X}_{\mathbb{F}}^{B}$, and $\mathbb{X}_{\mathbb{H}} = \mathbb{X}_{\mathbb{H}}^{C} \cup \mathbb{X}_{\mathbb{H}}^{B}$ where $\mathbb{X}_{\mathbb{F}}$, $\mathbb{X}_{\mathbb{H}}$ and $\mathbb{X}_{\mathbb{E}}$ are the complete characteristic sets associated with the goal-objective sets $\mathbb{F}$, $\mathbb{H}$ and $\mathbb{E}$ respectively. These cost-benefit sets will be used to indicate the possible directions of coalition formations as well as specify a *measure of systemic risk* associated with the democratic social formation and decision-choice system.

Definition 4.3: Coalition-Intent Relation

If **A** and **B** are two interest groups, then ${}_{A}\mathfrak{C}_{\overrightarrow{B}}$ is a coalition formation-intent relation expressed by **A** to **B**. $\mathfrak{U}\langle A,B\rangle$ is said to be a unity (symmetric) intent relation between groups **A** and **B** on the basis of a social goal-objective set iff ${}_{A}\mathfrak{C}_{\overrightarrow{B}}$ and ${}_{B}\mathfrak{C}_{\overrightarrow{A}}$ (That is, there is a mutually symmetric intent relation to create a coalition and hence $\mathfrak{C}_{A\rightleftharpoons B}$)

Definition 4.4: Coalition Realization Function

$\mathfrak{C}(\bullet)$ is a coalition formation realization function defined over interest groups A and B if ${}_{A}\mathfrak{C}_{\overrightarrow{B}}$, ${}_{B}\mathfrak{C}_{\overrightarrow{A}}$ and $\mathfrak{U}\langle A,B\rangle$ are realized with an agreement relational value, $\mathcal{A}(A,B)=1$, where $\mathcal{A}(A,B)\in[0,1]$ is the degree of agreement realization for coalition formation.

Proposition 4.1: Preference Ordering

Given three interest groups $\mathbb{P}$, $\mathbb{D}$ and Π with goal-objective sets $\mathbb{G}_{\mathbb{P}}$, $\mathbb{G}_{\mathbb{D}}$ and $\mathbb{G}_{\Pi}$ respectively and goal-objective intersection sets, $\mathbb{E}$, $\mathbb{H}$ and $\mathbb{F}$ such that the set interaction $\mathbb{E}\cap\mathbb{H}=\mathbb{E}\cap\mathbb{F}=\mathbb{H}\cap\mathbb{F}=\mathbb{H}\cap\mathbb{E}\cap\mathbb{F}=\varnothing$ where if $\succ$ implies more valuable and $\sim$ implies equally valuable, then:

$$\begin{cases} \mathfrak{R}_{\mathbb{P}}(\mathbb{E}) \succ \mathfrak{R}_{\mathbb{P}}(\mathbb{F}) \Rightarrow {}_{\mathbb{P}}\mathfrak{C}_{\overrightarrow{\mathbb{D}}} \\ \mathfrak{R}_{\mathbb{P}}(\mathbb{F}) \succ \mathfrak{R}_{\mathbb{P}}(\mathbb{E}) \Rightarrow {}_{\mathbb{P}}\mathfrak{C}_{\overrightarrow{\Pi}} \\ \mathfrak{R}_{\mathbb{P}}(\mathbb{F}) \sim \mathfrak{R}_{\mathbb{P}}(\mathbb{E}) \Rightarrow {}_{\mathbb{P}}\mathfrak{C}_{\overrightarrow{\mathbb{D}\sim\Pi}} \end{cases} \tag{4.2}$$

$$\begin{cases} \mathfrak{R}_{\mathbb{D}}(\mathbb{E}) \succ \mathfrak{R}_{\mathbb{D}}(\mathbb{H}) \Rightarrow_{\mathbb{D}} \underset{\rightarrow}{\mathfrak{C}}_{\mathbb{P}} \\ \mathfrak{R}_{\mathbb{D}}(\mathbb{H}) \succ \mathfrak{R}_{\mathbb{D}}(\mathbb{E}) \Rightarrow {}_{\mathbb{D}}\underset{\rightarrow}{\mathfrak{C}}_{\Pi} \\ \mathfrak{R}_{\mathbb{D}}(\mathbb{E}) \sim \mathfrak{R}_{\mathbb{D}}(\mathbb{H}) \Rightarrow {}_{\mathbb{D}\rightharpoondown}\mathfrak{C}_{\Pi\sim\mathbb{P}} \end{cases} \tag{4.3}$$

$$\begin{cases} \mathfrak{R}_{\Pi}(\mathbb{F}) \succ \mathfrak{R}_{\Pi}(\mathbb{H}) \Rightarrow {}_{\Pi}\underset{\rightarrow}{\mathfrak{C}}_{\mathbb{P}} \\ \mathfrak{R}_{\Pi}(\mathbb{H}) \succ \mathfrak{R}_{\Pi}(\mathbb{F}) \Rightarrow {}_{\Pi}\underset{\rightarrow}{\mathfrak{C}}_{\mathbb{D}} \\ \mathfrak{R}_{\Pi}(\mathbb{F}) \sim \mathfrak{R}_{\Pi}(\mathbb{H}) \Rightarrow {}_{\Pi}\underset{\sim}{\mathfrak{C}}_{\mathbb{P}\sim\mathbb{D}} \end{cases} \tag{4.4}$$

These preference orderings may then be expressed in the structure of coalition formations.

Proposition 4.2: Coalition Formation

$$\text{if} \begin{cases} \text{a)} \begin{cases} {}_{\mathbb{P}}\underset{\rightarrow}{\mathfrak{C}}_{\mathbb{D}} \text{ and } {}_{\mathbb{D}}\underset{\rightarrow}{\mathfrak{C}}_{\mathbb{P}} \Rightarrow \mathfrak{R}_{\mathbb{P}}(\mathbb{E}) \succ \mathfrak{R}_{\mathbb{P}}(\mathbb{F}) \text{ and } \mathfrak{R}_{\mathbb{D}}(\mathbb{E}) \succ \mathfrak{R}_{\mathbb{P}}(\mathbb{H}) \\ \text{then } \exists\, \mathfrak{U}\langle \mathbb{D},\mathbb{P}\rangle \text{ that leads to } \mathfrak{C}(\mathbb{D},\mathbb{P}) \end{cases} \\ \text{b)} \begin{cases} {}_{\mathbb{P}}\underset{\rightarrow}{\mathfrak{C}}_{\Pi} \text{ and } {}_{\Pi}\underset{\rightarrow}{\mathfrak{C}}_{\mathbb{P}} \Rightarrow \mathfrak{R}_{\mathbb{P}}(\mathbb{F}) \succ \mathfrak{R}_{\mathbb{P}}(\mathbb{E}) \text{ and } \mathfrak{R}_{\Pi}(\mathbb{F}) \succ \mathfrak{R}_{\Pi}(\mathbb{H}) \\ \text{then } \exists\, \mathfrak{U}\langle \Pi,\mathbb{P}\rangle \text{ that leads to } \mathfrak{C}(\Pi,\mathbb{P}) \end{cases} \\ \text{c)} \begin{cases} {}_{\Pi}\underset{\rightarrow}{\mathfrak{C}}_{\mathbb{D}} \text{ and } {}_{\mathbb{D}}\underset{\rightarrow}{\mathfrak{C}}_{\Pi} \Rightarrow \mathfrak{R}_{\Pi}(\mathbb{H}) \succ \mathfrak{R}_{\Pi}(\mathbb{F}) \text{ and } \mathfrak{R}_{\mathbb{D}}(\mathbb{H}) \succ \mathfrak{R}_{\mathbb{P}}(\mathbb{E}) \\ \text{then } \exists\, \mathfrak{U}\langle \Pi,\mathbb{D}\rangle \text{that leads to } \mathfrak{C}(\Pi,\mathbb{D}) \end{cases} \\ \text{d)} \begin{cases} {}_{\mathbb{P}}\underset{\rightarrow}{\mathfrak{C}}_{\mathbb{D}\sim\Pi}\ {}_{\mathbb{D}}\underset{\rightarrow}{\mathfrak{C}}_{\Pi\sim\mathbb{P}} \text{ and } {}_{\Pi}\underset{\rightarrow}{\mathfrak{C}}_{\mathbb{D}\sim\mathbb{P}} \Rightarrow \mathfrak{R}_{\mathbb{P}}(\mathbb{E}) \sim \mathfrak{R}_{\mathbb{P}}(\mathbb{F}) \sim \mathfrak{R}_{\mathbb{P}}(\mathbb{H}), \\ \mathfrak{R}_{\mathbb{D}}(\mathbb{E}) \sim \mathfrak{R}_{\mathbb{D}}(\mathbb{F}) \sim \mathfrak{R}_{\mathbb{D}}(\mathbb{H}) \text{ and} \mathfrak{R}_{\Pi}(\mathbb{E}) \sim \mathfrak{R}_{\Pi}(\mathbb{F}) \sim \mathfrak{R}_{\Pi}(\mathbb{H}) \\ \text{then the goal-objective set will be } \mathbb{E} \cup \mathbb{F} \cup \mathbb{H} \cup \mathbb{Q} \end{cases} \end{cases}$$

4.3 Negotiations and Coalition Formations under Cost-Benefit Rationality and Approximate Reasoning

So far we have not said anything about the problem of negotiation and coalition formation among the three interest groups around the formation of the social goal-objective set. We have only provided the relational structure of possible coalitions without the driving force of the coalitions and the possible directions of the coalitions. Let us now turn our attention to this problem. In this section, we shall deal only with the three interest groups of the public, Π, the private, $\mathbb{P}$, sectors and the decision-making core $\mathbb{D}$. We shall then concentrate on inter-coalition negotiation under the assumption that intra-coalition formations and the corresponding preferences of the three have been established either exactly or inexactly. Inter-coalition negotiations to arrive at the social goal-objective set do not arise in the set, $\mathbb{Q}$, where its elements are accepted by the private, public and decision-making groups. Agreements are difficult in inter-coalition negotiation outside the set, $\mathbb{F} \cup \mathbb{E} \cup \mathbb{H} \cup \mathbb{Q}$, where there are complete non-interest interactions. The necessity for coalition formations is dissolved under conditions of inter-group indifference ranking leading to the goal-objective set, $\mathbb{F} \cup \mathbb{E} \cup \mathbb{H} \cup \mathbb{Q}$. Inter-coalition negotiations are limited to zones of possible conflicts but with possible agreements toward cooperation and common fronts after the agreed-upon inter-coalition social goal-objective set has been established. To carry on the inter-negotiation analysis, we shall introduce some important relational concepts and operators on the basis of cost-benefit rationality.

4.3.1 Cost-Benefit Characteristics in Negotiations and Coalition Formations in the Social-Goal Objective Formation

In dealing with disagreements, agreements and negotiations towards establishing the social goal-objective set in support of national interest as made up of the set of domestic and international interests, an important question arises as to what is the process that lead to intra-group and inter-group agreements in establishing equilibrium of the social goal-objective set. To answer this question, we may proceed by viewing the identity as defined by a characteristic set. The characteristic set places each element in the potential goal-objective set as a duality. The duality is defined by subsets of cost characteristics and benefit characteristics which establish a cost-benefit configuration in differential proportions as seen by different evaluators. The cost-benefit configuration is seen in a continuum of opposites that provides the goal-objective element its identity, and allows incremental trade-offs to occur in negotiations. The cost-benefit characteristics, therefore, present inter-group conflict assessments as well as constitute tradeoff possibilities. The inter-group conflict assessments and the tradeoff possibilities define a framework of negotiation techniques in the social goal-objective formation relative to the national interest. The process of negotiation decisions that is being advanced is that every decision-choice variable and every state variable is defined by its cost-benefit characteristics where choices and trade-offs are specified in terms of cost-benefit balances by the decision-choice agent. Such cost-benefit balances need not be quantitatively measurable or be in monetary values. The cost-benefit configuration is defined in a quality-quantity plane with neutrality of time.

Let Ω be the possibility set of goals and objectives with generic element ω. Each element, $\omega \in \Omega$ is defined by a characteristic set, $\mathbb{X}$, which may be partitioned at any time into subsets of cost characteristics $\mathbb{X}^{C}$ and a benefit characteristic subset $\mathbb{X}^{B}$ such that $\mathbb{X}=\left(\mathbb{X}^{C}\cup\mathbb{X}^{B}\right)$. For the individual groups, Π, $\mathbb{D}$ and $\mathbb{P}$, the evaluated cost characteristic subsets of $\mathbb{X}$ may be written as $\mathbb{X}_{\Pi}^{C}$, $\mathbb{X}_{\mathbb{P}}^{C}$ and $\mathbb{X}_{\mathbb{D}}^{C}$, and the benefit characteristic subset may be written as $\mathbb{X}_{\Pi}^{B}$, $\mathbb{X}_{\mathbb{P}}^{B}$ and $\mathbb{X}_{\mathbb{D}}^{B}$.

4.3.1.1 The Structure of the Cost-Benefit Partition

For any given objective $\omega\in\Omega$ the following cost-benefit set conditions hold

$$\left.\begin{array}{l} 1.\ \mathbb{X}=\left(\mathbb{X}_{\Pi}^{B}\cup\mathbb{X}_{\Pi}^{C}\right)=\left(\mathbb{X}_{\mathbb{P}}^{B}\cup\mathbb{X}_{\mathbb{P}}^{C}\right)=\left(\mathbb{X}_{\mathbb{D}}^{B}\cup\mathbb{X}_{\mathbb{D}}^{C}\right) \\ 2.\ \left(\mathbb{X}_{\Pi}^{B}\cap\mathbb{X}_{\Pi}^{C}\right)=\left(\mathbb{X}_{\mathbb{P}}^{B}\cap\mathbb{X}_{\mathbb{P}}^{C}\right)=\left(\mathbb{X}_{\mathbb{D}}^{B}\cap\mathbb{X}_{\mathbb{D}}^{C}\right)=\varnothing \end{array}\right\}\text{Complete Partition} \tag{4.5}$$

$$\left.\begin{array}{l} 3\ \text{a)}\ \left(\mathbb{X}_{\Pi}^{B}\cap\mathbb{X}_{\mathbb{D}}^{C}\right)\neq\varnothing,\ \text{b)}\ \left(\mathbb{X}_{\Pi}^{C}\cap\mathbb{X}_{\mathbb{P}}^{B}\right)\neq\varnothing \\ \quad \text{c)}\ \left(\mathbb{X}_{\mathbb{D}}^{C}\cap\mathbb{X}_{\mathbb{P}}^{B}\right)\neq\varnothing \end{array}\right\}\text{Cost-Benefit Assessment conflicts} \tag{4.6}$$

$$\left.\begin{array}{l} 4.\ \text{a)}\ \left(\mathbb{X}_{\Pi}^{B}=\mathbb{X}_{\mathbb{D}}^{B}\right),\ \text{b)}\ \left(\mathbb{X}_{\Pi}^{B}=\mathbb{X}_{\mathbb{P}}^{B}\right),\ \text{c)}\ \left(\mathbb{X}_{\mathbb{D}}^{B}=\mathbb{X}_{\mathbb{P}}^{B}\right) \\ \quad \text{d)}\ \left(\mathbb{X}_{\Pi}^{C}=\mathbb{X}_{\mathbb{D}}^{C}\right),\ \text{e)}\ \left(\mathbb{X}_{\Pi}^{C}=\mathbb{X}_{\mathbb{P}}^{C}\right),\text{f)}\ \left(\mathbb{X}_{\mathbb{D}}^{C}=\mathbb{X}_{\mathbb{P}}^{C}\right) \end{array}\right\}\text{Cost-Benefit Similarities} \tag{4.7}$$

Eqn. (4.6) simply states that some characteristics may be seen as benefits to some groups while the same characteristics may be seen as costs by other groups. It affirms the notion of unity in duality in the social set-up. We may now specify *cost-benefit relative complement* (difference) as assessed by the interest groups $(\Pi\mathbb{D})$, $(\Pi\mathbb{P})$ and $(\mathbb{P}\mathbb{D})$ where the uses of subscripts mean cost-benefit assessment difference between the indicated groups, and $\left({}^{\bullet}\!/\!{}_{\bullet}\right)$ represents relative complement or set difference. For example if $\mathbb{X}$ is the characteristic set then:

$$\mathbb{X}_{\Pi\mathbb{D}}^{C}=\left(\mathbb{X}_{\Pi}^{C}\Big/\mathbb{X}_{\mathbb{D}}^{C}\right)=\left\{x\mid x\in\mathbb{X}_{\Pi}^{C}\text{ and }x\notin\mathbb{X}_{\mathbb{D}}^{C}\ \forall\ x\in\mathbb{X}\right\} \tag{4.8}$$

in that there are some characteristics in $\mathbb{X}$ that are evaluated as cost by group, Π ,while the same characteristics are assessed as benefit by group $\mathbb{D}$. Thus, from the the cost-benefit space we may write:

$$\left.\begin{array}{ll} \mathbb{X}^{C}_{\Pi\mathbb{D}} = \left(\mathbb{X}^{C}_{\Pi} \big/ \mathbb{X}^{C}_{\mathbb{D}}\right); & \mathbb{X}^{C}_{\mathbb{D}\Pi} = \left(\mathbb{X}^{C}_{\mathbb{D}} \big/ \mathbb{X}^{C}_{\Pi}\right) \\ \mathbb{X}^{C}_{\Pi\mathbb{P}} = \left(\mathbb{X}^{C}_{\Pi} \big/ \mathbb{X}^{C}_{\mathbb{P}}\right); & \mathbb{X}^{C}_{\mathbb{P}\Pi} = \left(\mathbb{X}^{C}_{\mathbb{P}} \big/ \mathbb{X}^{C}_{\Pi}\right) \\ \mathbb{X}^{C}_{\mathbb{P}\mathbb{D}} = \left(\mathbb{X}^{C}_{\mathbb{P}} \big/ \mathbb{X}^{C}_{\mathbb{D}}\right); & \mathbb{X}^{C}_{\mathbb{D}\mathbb{P}} = \left(\mathbb{X}^{C}_{\mathbb{D}} \big/ \mathbb{X}^{C}_{\mathbb{P}}\right) \end{array}\right\} \text{Cost Assessment Differences} \tag{4.9}$$

$$\left.\begin{array}{ll} \mathbb{X}^{B}_{\Pi\mathbb{D}} = \left(\mathbb{X}^{B}_{\Pi} \big/ \mathbb{X}^{B}_{\mathbb{D}}\right); & \mathbb{X}^{B}_{\mathbb{D}\Pi} = \left(\mathbb{X}^{B}_{\mathbb{D}} \big/ \mathbb{X}^{B}_{\Pi}\right) \\ \mathbb{X}^{B}_{\Pi\mathbb{P}} = \left(\mathbb{X}^{B}_{\Pi} \big/ \mathbb{X}^{B}_{\mathbb{P}}\right); & \mathbb{X}^{B}_{\mathbb{P}\Pi} = \left(\mathbb{X}^{B}_{\mathbb{P}} \big/ \mathbb{X}^{B}_{\Pi}\right) \\ \mathbb{X}^{B}_{\mathbb{P}\mathbb{D}} = \left(\mathbb{X}^{B}_{\mathbb{P}} \big/ \mathbb{X}^{B}_{\mathbb{D}}\right); & \mathbb{X}^{B}_{\mathbb{D}\mathbb{P}} = \left(\mathbb{X}^{B}_{\mathbb{D}} \big/ \mathbb{X}^{B}_{\mathbb{P}}\right) \end{array}\right\} \text{Benefit Assessment Differences} \tag{4.10}$$

Note 4.2

If $\mathbb{A}$ and $\mathbb{B}$ are two characteristic sets in a general characteristic space, $\mathbb{U}$, then their difference or the relative complement between $\mathbb{A}$ and $\mathbb{B}$ is the set of the form $\left(\mathbb{A}/\mathbb{B}\right) = \{x \mid x \in \mathbb{A} \text{ and } x \notin \mathbb{B} \ \forall\ x \in \mathbb{U}\}$ and similarly the relative complement between $\mathbb{B}$ and $\mathbb{A}$ is $\left(\mathbb{B}/\mathbb{A}\right) = \{x \mid x \in \mathbb{B} \text{ and } x \notin \mathbb{A} \ \forall\ x \in \mathbb{U}\}$. These are illustrated in Figure 4.4 with Venn diagrams. The relative complement allows us to specify either benefit or cost characteristic associated with a particular group assessment.

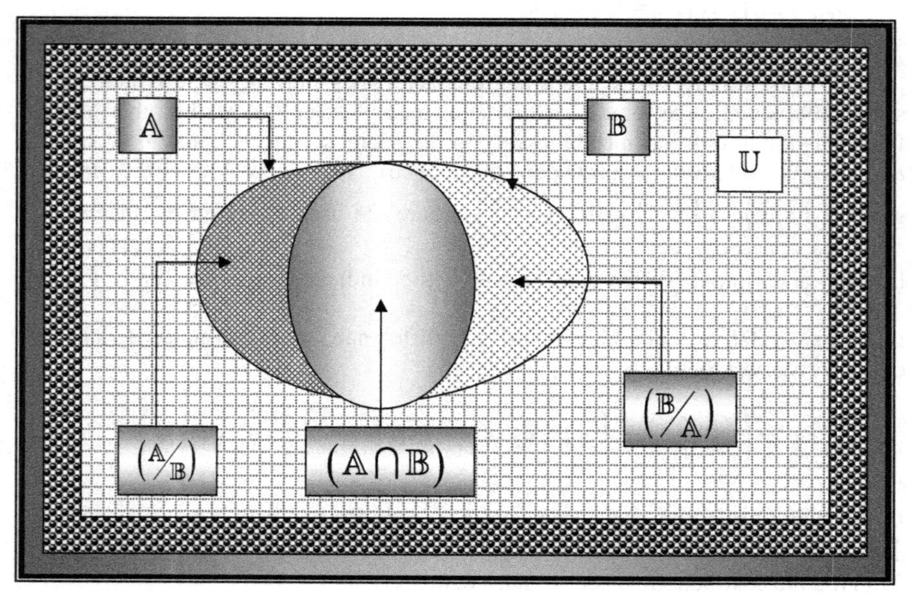

Fig.4.4 Venn diagram of Relative Complement of Characteristic Sets $\mathbb{A}$ and $\mathbb{B}$

To see the nature of the assessment differences and the conflicts over a social goal-objective element, we shall introduce into the analysis the concept of *symmetric cost-benefit assessment differences* between groups. The structure being developed here has applications to all negotiations to resolve differences and conflicts. Conflicts arise in the general decision-choice space because of differences in the cost-benefit evaluations and how such cost-benefit differences are assessed against individual and collective interests. Negotiations are about cost-benefit tradeoffs among differential interest groups. Let ∇ be the *symmetric difference* operator for two benefit-cost characteristic subsets of $\mathbb{A}$ and $\mathbb{B}$ then $\mathbb{A}\nabla\mathbb{B} = \left\{x \mid x \in (\mathbb{A} \cup \mathbb{B}) \text{ and } x \notin (\mathbb{A} \cap \mathbb{B})\right\} = \left(\mathbb{A}/\mathbb{B}\right) \cup \left(\mathbb{B}/\mathbb{A}\right)$. Thus the symmetric difference is the union of the individual relative complements of any two sets. The nature of the symmetric difference is illustrated in Figure 4.5.

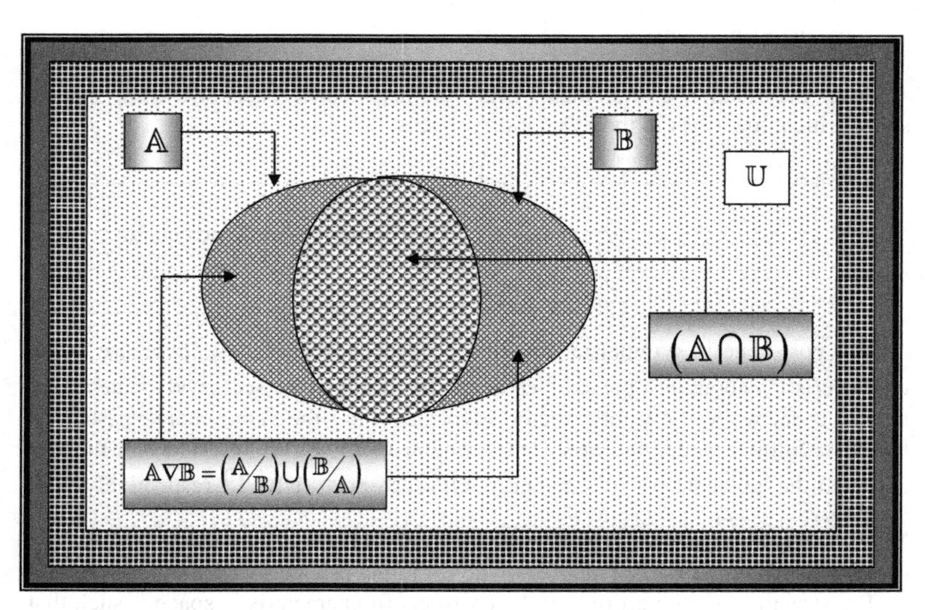

Fig. 4.5 Venn diagram of Symmetric Difference of Characteristic Sets $\mathbb{A}$ and $\mathbb{B}$

We can use the symmetric difference to distinguish between inter-group common cost characteristics and common benefit characteristics in terms of their evaluation structures for negotiation and coalition formation.

$$\left.\begin{aligned} \mathbb{X}_{\Pi}^{C} \nabla \mathbb{X}_{\mathbb{D}}^{C} &= \left(\mathbb{X}_{\Pi}^{C} \big/ \mathbb{X}_{\mathbb{D}}^{C} \right) \cup \left(\mathbb{X}_{\mathbb{D}}^{C} \big/ \mathbb{X}_{\Pi}^{C} \right) \\ &= \left(\mathbb{X}_{\Pi\mathbb{D}}^{C} \cup \mathbb{X}_{\mathbb{D}\Pi}^{C} \right) \\ \mathbb{X}_{\Pi}^{C} \nabla \mathbb{X}_{\mathbb{P}}^{C} &= \left(\mathbb{X}_{\Pi\mathbb{P}}^{C} \cup \mathbb{X}_{\mathbb{P}\Pi}^{C} \right) \\ \mathbb{X}_{\mathbb{P}}^{C} \nabla \mathbb{X}_{\mathbb{D}}^{C} &= \left(\mathbb{X}_{\mathbb{P}\mathbb{D}}^{C} \cup \mathbb{X}_{\mathbb{D}\mathbb{P}}^{C} \right) \end{aligned}\right\} \text{Common Coalition Cost Characteristics} \tag{4.11}$$

$$\left.\begin{aligned} \mathbb{X}_{\Pi}^{B} \nabla \mathbb{X}_{\mathbb{D}}^{B} &= \left(\mathbb{X}_{\Pi}^{B} \big/ \mathbb{X}_{\mathbb{D}}^{B} \right) \cup \left(\mathbb{X}_{\mathbb{D}}^{B} \big/ \mathbb{X}_{\Pi}^{B} \right) \\ &= \left(\mathbb{X}_{\Pi\mathbb{D}}^{B} \cup \mathbb{X}_{\mathbb{D}\Pi}^{B} \right) \\ \mathbb{X}_{\Pi}^{B} \nabla \mathbb{X}_{\mathbb{P}}^{B} &= \left(\mathbb{X}_{\Pi\mathbb{P}}^{B} \cup \mathbb{X}_{\mathbb{P}\Pi}^{B} \right) \\ \mathbb{X}_{\mathbb{P}}^{B} \nabla \mathbb{X}_{\mathbb{D}}^{B} &= \left(\mathbb{X}_{\mathbb{P}\mathbb{D}}^{B} \cup \mathbb{X}_{\mathbb{D}\mathbb{P}}^{B} \right) \end{aligned}\right\} \text{Common Coalition Benefit Characteristics} \tag{4.12}$$

From the relative complements and symmetric differences we may specify some critical identities in the evaluation process.

$$\left.\begin{aligned} \mathbb{X}_{\mathbb{P}\Pi}^{C} = \mathbb{X}_{\Pi\mathbb{P}}^{B} \neq \varnothing &\Rightarrow \left(\mathbb{X}_{\mathbb{P}}^{B} \big/ \mathbb{X}_{\Pi}^{B} \right) = \left(\mathbb{X}_{\Pi}^{C} \big/ \mathbb{X}_{\mathbb{P}}^{C} \right) \\ \mathbb{X}_{\Pi\mathbb{D}}^{C} = \mathbb{X}_{\mathbb{D}\Pi}^{B} \neq \varnothing &\Rightarrow \left(\mathbb{X}_{\Pi}^{C} \big/ \mathbb{X}_{\mathbb{D}}^{C} \right) = \left(\mathbb{X}_{\mathbb{D}}^{B} \big/ \mathbb{X}_{\Pi}^{B} \right) \\ \mathbb{X}_{\mathbb{P}\mathbb{D}}^{C} = \mathbb{X}_{\mathbb{D}\mathbb{P}}^{B} \neq \varnothing &\Rightarrow \left(\mathbb{X}_{\mathbb{P}}^{C} \big/ \mathbb{X}_{\mathbb{D}}^{C} \right) = \left(\mathbb{X}_{\mathbb{D}}^{B} \big/ \mathbb{X}_{\mathbb{P}}^{B} \right) \end{aligned}\right\} \text{Cost-Benefit Symmetric equality} \tag{4.13}$$

The cost-benefit symmetric equality projects a notion that there are some characteristics about an objective $\omega \in \Omega$ that the private sector advocates view as real costs while the public sector advocates view the same characteristics as benefits. Similar situations arise between the members of the decision-making core and the advocates of private and public sectors $\mathbb{P}$ and Π respectively. The value $\mathbb{X}_{\mathbb{P}\Pi}^{C} = \mathbb{X}_{\Pi\mathbb{P}}^{C}$ implies that the total cost characteristic set of the private sector is viewed as a benefit characteristic set to the private sector.

The structure of the partition of the cost-benefit characteristic space is such that the characteristic set $\mathbb{X}$ is partitioned, relative to each group, into cost-benefit characteristics with *possible group conflicts* as generated by differences in the cost and benefit assessments associated with each group. These assessment differences produce symmetric assessment differences and relative complements that lead to the establishment of critical cost-benefit differences which then generate disagreements in conditions of relevance and irrelevance of social goal objective assessments. We shall use the structure of the relative complement and symmetric differences to define *degrees of agreement* and *disagreement* about elements in the goal-objective set. The assessment process leads to three cost-identification matrices and three benefit-identification matrices for the three groups. These identification matrices will be of the form:

$$\mathbb{X}^{\mathrm{C}} = \left(\begin{array}{c|ccc} \Omega & \Pi & \mathbb{P} & \mathbb{D} \\ \hline \omega_1 & \mathrm{X}_{\Pi}^{\mathrm{C}_1} & \mathrm{X}_{\mathbb{P}}^{\mathrm{C}_1} & \mathrm{X}_{\mathbb{D}}^{\mathrm{C}_1} \\ \omega_2 & \mathrm{X}_{\Pi}^{\mathrm{C}_2} & \mathrm{X}_{\mathbb{P}}^{\mathrm{C}_2} & \mathrm{X}_{\mathbb{D}}^{\mathrm{C}_2} \\ \vdots & & \ddots & \\ \omega_i & \mathrm{X}_{\Pi}^{\mathrm{C}_i} & \mathrm{X}_{\mathbb{P}}^{\mathrm{C}_i} & \mathrm{X}_{\mathbb{D}}^{\mathrm{C}_i} \\ \vdots & & \ddots & \\ \omega_n & \mathrm{X}_{\Pi}^{\mathrm{C}_n} & \mathrm{X}_{\mathbb{P}}^{\mathrm{C}_n} & \mathrm{X}_{\mathbb{D}}^{\mathrm{C}_n} \end{array} \right), \quad \mathbb{X}^{\mathrm{B}} = \left(\begin{array}{c|ccc} \Omega & \Pi & \mathbb{P} & \mathbb{D} \\ \hline \omega_1 & \mathrm{X}_{\Pi}^{\mathrm{B}_1} & \mathrm{X}_{\mathbb{P}}^{\mathrm{B}_1} & \mathrm{X}_{\mathbb{D}}^{\mathrm{B}_1} \\ \omega_2 & \mathrm{X}_{\Pi}^{\mathrm{B}_2} & \mathrm{X}_{\mathbb{P}}^{\mathrm{B}_2} & \mathrm{X}_{\mathbb{D}}^{\mathrm{B}_2} \\ \vdots & & \ddots & \\ \omega_i & \mathrm{X}_{\Pi}^{\mathrm{B}_i} & \mathrm{X}_{\mathbb{P}}^{\mathrm{B}_i} & \mathrm{X}_{\mathbb{D}}^{\mathrm{B}_i} \\ \vdots & & \ddots & \\ \omega_n & \mathrm{X}_{\Pi}^{\mathrm{B}_n} & \mathrm{X}_{\mathbb{P}}^{\mathrm{B}_n} & \mathrm{X}_{\mathbb{D}}^{\mathrm{B}_n} \end{array} \right), \tag{4.14}$$

The terms $\mathrm{X}_j^{\mathrm{C}_i}$'s and $\mathrm{X}_j^{\mathrm{B}_i}$'s are column vectors associated with the cost and benefit characteristic of the form:

$$\left. \begin{array}{l} \mathbb{X}_j^{\mathrm{C}_i} = \left\{ \left[x_1^{\mathrm{c}}, x_2^{\mathrm{c}} \cdots x_i^{\mathrm{c}} \cdots x_n^{\mathrm{c}} \right]' \mid x \in \mathbb{X},\ j \in \{\Pi, \mathbb{P}, \mathbb{D}\} \right\} \\ \mathbb{X}_j^{\mathrm{B}_i} = \left\{ \left[x_1^{\mathrm{b}}, x_2^{\mathrm{b}} \cdots x_i^{\mathrm{b}} \cdots x_m^{\mathrm{b}} \right]' \mid x \in \mathbb{X},\ j \in \{\Pi, \mathbb{P}, \mathbb{D}\} \right\} \end{array} \right\} \tag{4.15}$$

We may note that $n + m = \#\mathbb{X}$ for $\mathbb{P}, \mathbb{D}$, and Π. The cost-benefit characteristics are then aggregated for each $\omega \in \Omega$ and then ranked by an appropriate cost-benefit criterion for the optimal construction of the individual goal-objective sets $\mathbb{G}_{\Pi}, \mathbb{G}_{\mathbb{P}}$, and $\mathbb{G}_{\mathbb{D}}$ by the respective groups $\Pi, \mathbb{P}$ and $\mathbb{D}$. If a goal-objective element is single-period duration then equation 4.14 will be for that one period. On the other hand, if it is of multiple-period duration then there will be as many matrixes as there are time periods where such time periods will have differential weight that will reflect intergenerational cost-benefit shifting.

4.3.1.2 Agreement-Disagreement Relations and Cost-Benefit Characteristics

The cost-benefit characteristics are initially in real values and whether prices exist for their aggregate construct will depend on the nature of the goal-objective element. The view, here, is that one cannot conceive of the use of *consumer surplus* as a nominal value if the real cost and benefit elements are not identified and related to the conditions of the existing market in terms of prices and quantities. In general, we shall end up with a composite aggregate of a mixture of quantitative and non-quantitative elements and monetary and non-monetary elements with differential social weight assessments of the impact of each goal-objective element on the welfare of the society and the defined structure of the national interest.

Definition 4.5: Disagreement relation

A function, $\mathcal{D}(\bullet)$ is said to be a degree of disagreement relational function between two groups, if given any cost-benefit preference relation held by interest groups $(\Pi, \mathbb{P}, \mathbb{D})$ we can write:

$$\left.\begin{array}{l}\mathcal{D}(\Pi,\mathbb{P})=\mathcal{D}(\mathbb{P},\Pi)\in[0,1]\Rightarrow \mathbb{X}^{C}_{\mathbb{P}\Pi}=\mathbb{X}^{B}_{\Pi\mathbb{P}}\neq\varnothing \text{ ie, } \forall\omega\in\left({}^{\mathbb{G}_{\mathbb{P}}}/_{\mathbb{G}_{\Pi}}\right) \text{ and } \omega\in\left({}^{\mathbb{G}_{\Pi}}/_{\mathbb{G}_{\mathbb{P}}}\right)\\ \mathcal{D}(\Pi,\mathbb{D})=\mathcal{D}(\mathbb{D},\Pi)\in[0,1]\Rightarrow \mathbb{X}^{C}_{\Pi\mathbb{D}}=\mathbb{X}^{B}_{\mathbb{D}\Pi}\neq\varnothing \text{ie, } \forall\omega\in\left({}^{\mathbb{G}_{\Pi}}/_{\mathbb{G}_{\mathbb{D}}}\right) \text{ and } \omega\in\left({}^{\mathbb{G}_{\mathbb{D}}}/_{\mathbb{G}_{\Pi}}\right)\\ \mathcal{D}(\mathbb{P},\mathbb{D})=\mathcal{D}(\mathbb{D},\mathbb{P})\in[0,1]\Rightarrow \mathbb{X}^{C}_{\mathbb{P}\mathbb{D}}=\mathbb{X}^{B}_{\mathbb{D}\mathbb{P}}\neq\varnothing \text{ie, } \forall\omega\in\left({}^{\mathbb{G}_{\mathbb{P}}}/_{\mathbb{G}_{\mathbb{D}}}\right) \text{ and } \omega\in\left({}^{\mathbb{G}_{\mathbb{D}}}/_{\mathbb{G}_{\mathbb{P}}}\right)\end{array}\right\}, \text{for } \omega\in\Omega$$

Note: 4.3

The *degree of disagreement* as defined in terms of cost-benefit measures may be based on the collective preferences of interest groups $(\Pi, \mathbb{P}, \mathbb{D})$ and on conditions that characterize the costs and benefits of $\omega \in \Omega$. As defined, $\mathcal{D}(\bullet) = 0$ means no disagreement or complete agreement and $\mathcal{D}(\bullet) = 1$ means complete disagreement or no agreement, while $\mathcal{D}(\bullet) \in (0,1)$ means partial agreement defined on an open interval of zero and one. In the partial disagreement, we speak of degrees of agreement and disagreement where such degrees are specified in terms of fuzzy membership characteristic functions. It is also the *possible agreement zone* (PAZ). The membership characteristic functions are defined in terms of the number of benefit characteristics, $\mathbb{X}^B \subseteq \mathbb{X}$ relative to number of cost characteristics where $(\#\mathbb{X}^B \leq \#\mathbb{X})$, and $(\#\mathbb{X}^C \leq \#\mathbb{X})$ with $(\#\mathbb{X}^C + \#\mathbb{X}^B = \#\mathbb{X})$ for any group and for each $\omega \in \Omega$. At any moment of time, the *degree of agreement* in the negotiation process is $\mathcal{A}(\cdot) = (1 - \mathcal{D}(\cdot))$. From the definition of disagreement relation we may also define the agreement relation.

Definition 4.6: Agreement relation

A relation, $\mathcal{A}(\cdot) = (1 - \mathcal{D}(\cdot))$, is said to be an agreement relational function between two groups, if given any cost-benefit preference relation held by the interest groups $(\Pi, \mathbb{P}, \mathbb{D})$, then

$$\left.\begin{array}{l}\mathcal{A}(\Pi,\mathbb{P})=\mathcal{A}(\mathbb{P},\Pi)\in[0,1]\Rightarrow \mathbb{X}^{C}_{\mathbb{P}\Pi}=\mathbb{X}^{C}_{\Pi\mathbb{P}}\neq\varnothing \text{ ie, } \forall\omega\in\left({}^{\mathbb{G}_{\mathbb{P}}}/_{\mathbb{G}_{\Pi}}\right) \text{ and } \omega\in\left({}^{\mathbb{G}_{\Pi}}/_{\mathbb{G}_{\mathbb{P}}}\right)\\ \mathcal{A}(\Pi,\mathbb{D})=\mathcal{A}(\mathbb{D},\Pi)\in[0,1]\Rightarrow \mathbb{X}^{C}_{\Pi\mathbb{D}}=\mathbb{X}^{C}_{\mathbb{D}\Pi}\neq\varnothing \text{ie, } \forall\omega\in\left({}^{\mathbb{G}_{\Pi}}/_{\mathbb{G}_{\mathbb{D}}}\right) \text{ and } \omega\in\left({}^{\mathbb{G}_{\mathbb{D}}}/_{\mathbb{G}_{\Pi}}\right)\\ \mathcal{A}(\mathbb{P},\mathbb{D})=\mathcal{A}(\mathbb{D},\mathbb{P})\in[0,1]\Rightarrow \mathbb{X}^{C}_{\mathbb{P}\mathbb{D}}=\mathbb{X}^{C}_{\mathbb{D}\mathbb{P}}\neq\varnothing \text{ie, } \forall\omega\in\left({}^{\mathbb{G}_{\mathbb{P}}}/_{\mathbb{G}_{\mathbb{D}}}\right) \text{ and } \omega\in\left({}^{\mathbb{G}_{\mathbb{D}}}/_{\mathbb{G}_{\mathbb{P}}}\right)\end{array}\right\}, \text{for } \omega\in\Omega$$

From the agreement relation, we have in the negotiation process the condition $\mathbb{X}^{B}_{\mathbb{P}\Pi} = \mathbb{X}^{B}_{\Pi\mathbb{P}} \neq \varnothing$, $\mathbb{X}^{B}_{\Pi\mathbb{D}} = \mathbb{X}^{B}_{\mathbb{D}\Pi} \neq \varnothing$ and $\mathbb{X}^{B}_{\mathbb{P}\mathbb{D}} = \mathbb{X}^{B}_{\mathbb{D}\mathbb{P}} \neq \varnothing \}$ Benefit agreement

Proposition 4.3

If $\mathcal{A}(\mathrm{A,B})$ and $\mathcal{D}(\mathrm{A,B})$ are degrees of agreement and disagreement relations respectively for individuals A and B with respect to a goal-objective element $\omega \in \Omega$ then

$$\lim_{\mathcal{D}(\mathrm{A,B}) \to 0} \mathcal{A}(\mathrm{A,B}) \to 1 \Rightarrow \text{agreement},$$

$$\lim_{\mathcal{D}(\mathrm{A,B}) \to 1} \mathcal{A}(\mathrm{A,B}) \to 0 \Rightarrow \text{disagreement}$$

$$\lim_{\mathcal{A}(\mathrm{A,B}) \to 0} \mathcal{D}(\mathrm{A,B}) \to 1 \Rightarrow \text{ disagreement}$$

and $\lim_{\mathcal{A}(\mathrm{A,B}) \to 1} \mathcal{D}(\mathrm{A,B}) \to 0 \Rightarrow \text{Agreement}$ The limiting process takes place through a negotiation function with a cost-benefit conversion moment.

The value, $\mathcal{D}(\bullet)$, is the degree of disagreement that can be made small through the negotiation process while at the same time increasing or decreasing the agreement relational value $\mathcal{A}(\bullet)$. The disagreement relational function, $\mathcal{D}(\bullet)$ and the agreement relational function may be implicitly defined as $\mathcal{D}(\mathfrak{R}_{A}, \mathfrak{R}_{B})$ and $\mathcal{A}(\mathfrak{R}_{A}, \mathfrak{R}_{B})$ where the, $\mathfrak{R}$'s are defined over the appropriate social goal-objective sub-sets of $\mathbb{F}$, $\mathbb{H}$ and $\mathbb{E}$ with corresponding cost-benefit characteristic sets. The nature of the coalition formation and the direction of the formation, among the private sector advocate, $\mathbb{P}$, the public sector advocates, Π, and the decision-making core, $\mathbb{D}$, will depend essentially on the groups' perception of the cost-benefit configuration associate with the negotiating elements for entries into the social goal-objective set. Let us keep in mind that the private sector advocates support private-sector widening and deepening. The public-sector advocates support public-sector widening and deepening. The role of the decision-making core is to mediate for fairness and equity under true democratic decision-choice system which will provide us with a solution point in the continuum of the private-public sector duality. To this solution, we shall take epistemic reflections on the nature of the private-public sector duality in the chapter 5.

Chapter 5
Fuzzy Rationality and Negotiated Equilibria of the Social Goal-Objective Set in the Political Economy

5.1 Information Structure, Society and Social Institutions

As has been discussed in [R7.13] [R7.14] [R15.1] [R15.19] [R15.29] [R15.33], every decision-choice action by cognitive agents has a knowledge support which in turn has information support. The decision-choice actions in either competitive or non-competitive situation are driven by the individual and collective motivations which affect the nature of the information-knowledge structures. The information available to decision-choice agents is processed into knowledge which then becomes an input into the decision-choice modules. Information on all human decision-choice actions is defective in the sense of carrying the characteristics of vagueness and incompleteness. It is, here, and elsewhere referred to as *defective information structure* which characterizes all knowledge and decision-choice systems. The vagueness relates to quality of information and the incompleteness relates to the quantity of information which will include the classification of information on the basis of national security. In the political economy, the defective information structure may be contaminated with a complexity of *deceptive information structure* which is made up of disinformation and misinformation. Disinformation relates to stripping the mind off what is known in order to create a cognitive vacuum, while misinformation relates to filling the cognitive vacuum with contaminated information to create confusion in reasoning in order to influence cognitive calculations of the benefit-cost imputations. The deceptive information structure may be related to propaganda. It is here that the concepts of the battle of ideas and the winning of minds acquire meaning in the decision-choice processes in social set ups since most decision-choice actions are obtained by computing with words.

K.K. Dompere, *Social Goal-Objective Formation, Democracy and National Interest*, Studies in Systems, Decision and Control 4,

DOI: 10.1007/978-3-319-05173-4_5,

The cognitive algorithms composed of the laws of thought and the calculus of reasoning depend on the initial assumption of the nature of the information structure which is to be processed into a knowledge structure as an input into the decision-choice actions. When one assumes away the elements of vagueness except incompleteness, one deals with the component of the quantity of information as measured in volume. to the neglect of the quality information as measured in degrees of credibility One then can use the methods of classical logic and mathematics with classical decision-choice rationality to resolve the individual conflicts in the collective decision-choice space under the democratic principle of citizens' sovereignty in the collective decision making, if the sources of the information input are credible. This is what the traditions in decision-choice theories and economic science have done by following the classical logic and mathematical theory of information processing in assuming the conditions that makes it possible to use the implied techniques. The validity of the conclusions depends on the credibility of the source of information. The methods of the classical paradigm lose their analytical power if the quantity of the information set is compromised with deceptive information structure from the source, especially in the social systems. Alternatively, we may combine both the defective information structure and the deceptive information structure into a unit as one

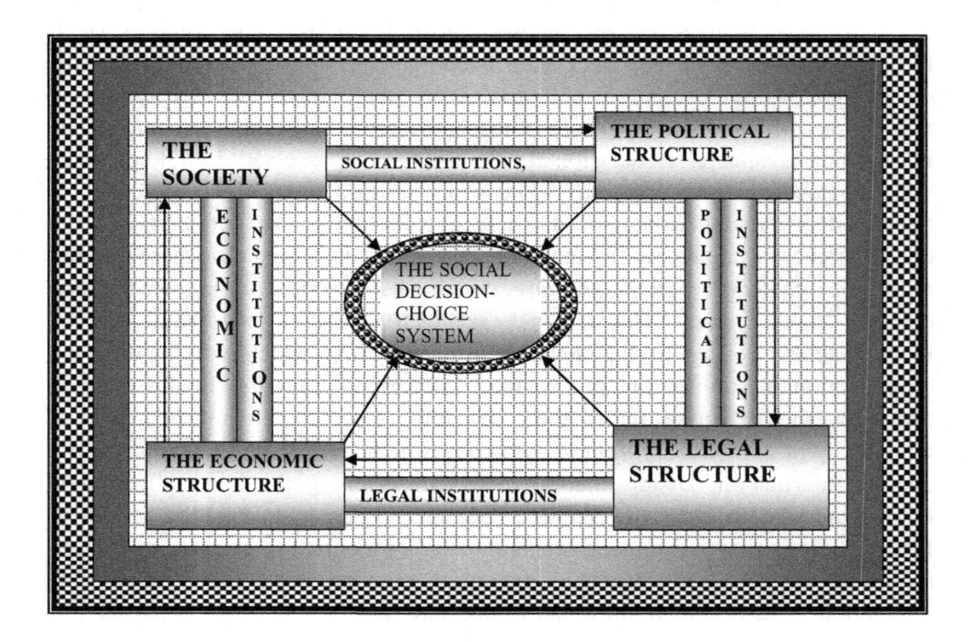

Fig. 5.1 The Analytical Geometry of Social Organizational and Institutional Links with the Social Decision-Choice System

information structure that contains qualitative and quantitative characteristics. We may then exit from the classical rationality with its logic and mathematics, and employ methods of fuzzy paradigm and its rationality. In this framework, the set of social institutions $\mathbb{Z}$ composed of the sets of political, legal and economic institutions $\mathbb{Z}^P, \mathbb{Z}^L, \mathbb{Z}^E$ respectively are taken as given where $\mathbb{Z} = \left(\mathbb{Z}^P \cup \mathbb{Z}^L \cup \mathbb{Z}^E\right)$. The society, its three structures and the linking institutions are shown in Figure 5.1. The interesting thing about the $\mathbb{Z}$ is that $\left(\mathbb{Z}^P \cap \mathbb{Z}^L \cap \mathbb{Z}^E\right) \neq \varnothing$, $\left(\mathbb{Z}^P \cap \mathbb{Z}^L\right) \neq \varnothing$, $\left(\mathbb{Z}^P \cap \mathbb{Z}^E\right) \neq \varnothing$ and $\left(\mathbb{Z}^L \cap \mathbb{Z}^E\right) \neq \varnothing$.

For the analytics of cost-benefit relationships of negotiations and coalition formations, it is useful to examine the use of methods of fuzzy rationality and its analytical methods in obtaining the path of the moving equilibrium of the social goal-objective set under defective and deceptive information structures. The analytical frame may be seen in terms of explanatory or prescriptive rationality [R7.14] [R7.15].

5.2 Fuzzy Computability of the Negotiated Equilibrium of the Social Goal-Objective Set of the Political Economy

Each negotiator, either as a group or an individual, enters into the negotiation game with $\left(\mathcal{A}_0(\cdot), \mathcal{D}_0(\cdot)\right)$ as an initial agreement and disagreement position of the negotiators in relation to their relative assessments of the cost-benefit configurations and preferences where $\mathcal{A}_0$ and $\mathcal{D}_0$ are the initial agreement and disagreement respectively. Thus $\mathcal{P}_{\Pi_0} = \left(\mathcal{A}_0(\cdot), \mathcal{D}_0(\cdot)\right)_\Pi$, $\mathcal{P}_{\mathbb{P}_0} = \left(\mathcal{A}_0(\cdot), \mathcal{D}_0(\cdot)\right)_\mathbb{P}$ and $\mathcal{P}_{\mathbb{D}_0} = \left(\mathcal{A}_0(\cdot), \mathcal{D}_0(\cdot)\right)_\mathbb{D}$ are the initial .negotiating positions of $\Pi, \mathbb{P}$ and $\mathbb{D}$ respectively.

For each negotiating group and for each goal-objective element $\omega \in \Omega$ the decision to accept $\omega \in \Omega$ to be included in the social goal-objective set is to reconcile the conflicts in the assessments of cost characteristic subset $\mathbb{X}^C \subseteq \mathbb{X}$ and benefit characteristic subset $\mathbb{X}^B \subseteq \mathbb{X}$ and the trade-off possibilities. Thus the evaluation of $\omega \in \Omega$ is the evaluation of relative value of $x \in \mathbb{X}^B$ and $x \in \mathbb{X}^C$. This evaluation is done through the *cost-benefit preference ordering* over $\mathbb{X}$ that is mapped onto the membership characteristic functions, where each characteristic element is placed in either fuzzy cost subset or fuzzy benefit subset. The trade-off possibility is defined in terms of giving up some benefits associated with one element $\omega_i \in \Omega$ for acquiring a different set of benefits associate with another element $\omega_j \in \Omega$. As specified, the giving up of the benefits associated with $\omega_i \in \Omega$ becomes the cost or the price for the benefit $\omega_j \in \Omega$. The trade-offs take place through a negotiation process which will establish the element that will constitute the agreed upon social goal-objective set. The opportunity cost concept is operating in the negotiation process in such a way that the end of the game is driven towards social cost-constrained of social benefit maximum on the basis of *majoritarian principle* of democratic collective decision-choice process. Such an optimization is through fuzzy logic and mathematics to produce fuzzy decision-choice rationality in the construct of the social goal-objective set.

The concepts, techniques and methods of the fuzzy paradigm that are utilized here, consist of duality, opposites, unity, continuum, fuzzification defuzzification and fuzzy optimization. The duality involves negative and positive characteristic sets as opposites. The unity connects them in a relational setting where the opposites are seen in continual give-and-take relations in a continuum. The continuum presents a situation where the positive characteristic set has a negative support and the negative characteristic set has positive support at all analytical points. The negative and positive characteristic sets are projected or defined by defective and deceptive information structures that create stochastic and fuzzy uncertainties which in turn create penumbral regions of collective and individual decision-choice actions due to linguistic vagueness and ambiguities in thought, in the social decision-choice space. The movements through the penumbral regions of choice require subjective interpretation which must be incorporated into the analytical structure. The complete formulation of the collective decision-choice problem and the corresponding solution come under the fuzzy process. In the fuzzy process the choice and state variables are seen as fuzzy or linguistic variables which are created through the *fuzzification process* of the negative and positive characteristic sets.

The fuzzification process allows us to create the fuzzy variables from linguistic and non-linguistic quantities with the concept of the degree to which an element belongs to the negative or the positive set. The linguistic and non-linguistic quantities are given fuzzy coverings where the positive set becomes *fuzzy positive set* and the negative set becomes a *fuzzy negative set*. The fuzzy variable has two components of an element and a degree of belonging to a set, called the membership characteristic function. The fuzzification process thus creates vague or inexact symbolism at the level of information representation for both logical and mathematical operations [R6.130]. The defuzzification process strips the fuzzy variable of its vagueness, ambiguity and subjectivity to create exact-value equivalences with *fuzzy conditionality* which simply tells us the optimal degree that is subjectively attached to the certainty value. In the cognitive calculus, the fuzzy negative set and the fuzzy positive set are taken to create a connected fuzzy decision space with its membership characteristic function that must be defuzzified through an analytical process. One of the methods of the defuzzification-process of the fuzzy decision space is the *fuzzy optimization* where the positive characteristic set is taken as a decision-choice objective to be maximized while the negative characteristic set serves as a constraint on the choice an element in the positive characteristic set [R7.12][R7.30] [R7.36].

Alternatively, however, the negative characteristic set may be taken as an objective to be minimized while the positive characteristic set is taken as a constraint of the decision-choice action on an element of the negative characteristic set . This analytical process of thought satisfies the logical system of duality, opposites, unity, and continuum. Thus every benefit has its cost support and every cost has its benefit support in cost-benefit duality, and corresponding to it we have categories of dualities such as relevance-irrelevance duality, good-evil duality, female-male duality, sweet-bitter duality and many more. The position taken in this monograph is that all human action is decision-choice based, and that the decision-choice actions relate to benefit-cost configurations that provide cognitive ordering where every benefit has a cost support and every cost has a benefit support and that every decision-choice action is a process of balancing benefits against costs as opposites in a duality and continuum. All cost-benefit information structures composed of qualitative and quantitative characteristics are viewed as either defective or deceptive or both as have been previously explained and hence we may speak of fuzzy cost-benefit sets.

Definition 5.1: Fuzzy Cost-Benefit Sets

The fuzzy benefit and cost sets in the cost-benefit space are of the form:

$$\widehat{\mathbb{X}}^{\mathrm{B}} = \left\{ \left(x, \mu_{\mathbb{X}^{\mathrm{B}}}(x) \right) \mid x \in \mathbb{X}, \mu_{\mathbb{X}^{\mathrm{B}}}(x) \in (0,1), \left({}^{d\mu_{\mathbb{X}^{\mathrm{B}}}} \big/ {}_{d\#\mathbb{X}^{\mathrm{B}}} \right) \geq 0 \right\}$$

$$\widehat{\mathbb{X}}^{\mathrm{C}} = \left\{ \left(x, \mu_{\mathbb{X}^{\mathrm{C}}}(x) \right) \mid x \in \mathbb{X}, \mu_{\mathbb{X}^{\mathrm{C}}}(x) \in (0,1), \left({}^{d\mu_{\mathbb{X}^{\mathrm{C}}}} \big/ {}_{d\#\mathbb{X}^{\mathrm{B}}} \right) \leq 0 \right\}$$

where the cup identifies fuzzy set.

The term $\mu_{\mathbb{X}}(x)$ specifies the degree to which the element x belongs to the cost or benefit characteristic set of the group in accord with the corresponding defective information structure. . The above definitional equations constitute a duality in a continuum where for each element, the benefit characteristics increase as the cost characteristics decrease thus the benefit membership characteristic function is upward sloping while the cost membership characteristic function is downward sloping. The appropriate membership characteristic functions are exponential-type, or Z-type or S-type that are continuous and differentiable in the interval of $(0,1)$ where the continuous and differentiability conditions satisfy the continuum principle in duality.

The problem of cost-benefit reconciliation in goal-objective selection and negotiation processes may be seen in terms of a decision to either maximize a benefit characteristic set that is constrained by a cost characteristic set or minimize a cost characteristic set that is constrained by a benefit characteristic set. The problem is to find $x \in \widehat{\mathbb{X}}^{B} \cap \widehat{\mathbb{X}}^{C}$ for each $\omega \in \Omega$ on the basis of which a particular social goal-objective element is evaluated as *relevant* or *irrelevant* social objective relative to some notion of national interest and general social welfare. Relevance and irrelevance are seen in terms of cost-benefit distributions over the elements in the goal-objective set and social states. Every goal-objective element exists in the relevance-irrelevance duality for selection relative to the social vision and national interest. The relevance and irrelevance are operationalized by transforming them into cost-benefit configuration for each goal-objective set for individual vote or support. This support, however, depends on the quality and quantity of information, given the information processing capacity of the voting individual. In a well organized system, we may assume that the individual processing capacity is the same for all even if preferences are different. It is the preferences of decision-choice agents that order decision-choice elements on the basis of equally available cost-benefit information structure which merely defines the decision-choice environment. Let us define the linguistic variables of relevance and irrelevance.

Definition 5.2: Relevance

Let $\mathbb{R} \subset \Omega$ be the relevant social goal-objective set; then we define a relevant social goal-objective set as a fuzzy set of benefit characteristics in the form

$$\mathbb{R} = \left\{ \left(\omega, x, \mu_{\mathbb{X}^{B}}(x) \right) \mid \omega \in \Omega, x \in \mathbb{X}, \mu_{\mathbb{X}^{B}}(x) \geq \mu_{\mathbb{X}^{C}}(x) \right\}$$

The $\mathbb{R} \subset \Omega$ is the set of all goal-objective elements whose benefit aggregate valuations are greater than cost aggregate valuations and the irrelevant objective as the complement of $\mathbb{R}$, that is $\mathbb{R}'$. The value $x^* \in \mathbb{X}$ for which $\mu_{\mathbb{X}^{B}}(x) = \mu_{\mathbb{X}^{C}}(x)$ is the support of each relevant goal-objective element $\omega \in \Omega$, which is the reconciliation between cost-benefit characteristic associated with x^*.The decision-choice problem of relevance or irrelevance is $\Delta = \left(\mathbb{R} \cap \mathbb{R}' \right)$ whose membership function is $\mu_{\Delta}(x) = \left(\mu_{\mathbb{X}^{B}}(x) \wedge \mu_{\mathbb{X}^{C}}(x) \right)$. In other words, the social benefit has a full cost support or the social cost is justified by its social benefit as in Figure 5.2. When one chooses a goal-objective element, one accepts it benefit as well as its cost support where the cost is the price of the benefit, irrespective of the domain of their measurements [R3.1.14] [R3.1.23] [R3.1.24] [R3.2.8] [R3.2.12].

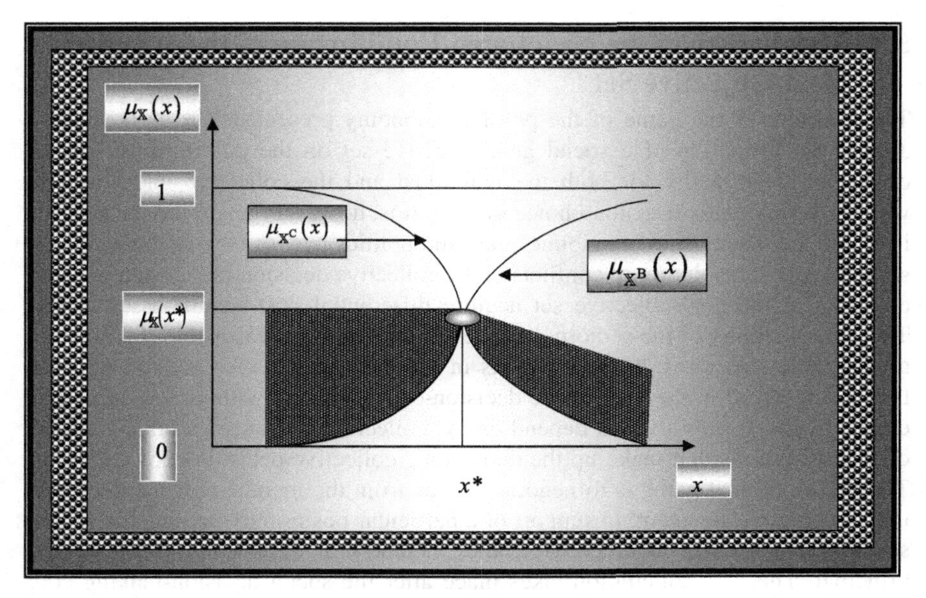

Fig. 5.2 Fuzzy Cost-Benefit Optimal Definition of Relevance

Definition 5.3: Negotiation Function

A relational function, $\mathcal{N}\left(\mathcal{P}_{A}\mathcal{P}_{B}\right)$ is said to be negotiation activity function between groups A and B, or individual A and B if there exist $\mathcal{A}(\text{A,B}),\ \mathcal{D}(\text{A,B}) \in (0,1)$ where either A and B may be any pair of interest groups $\Pi,\ \mathbb{D}$ and $\mathbb{P}$ and where $\mathcal{P}_{A}$ and $\mathcal{P}_{B}$ are the positions of A and B. and areas of negotiation are Zones II and IV in Figure 3.3.1 panels a, b, and c).

Proposition 5.1

The interest groups A and B are in the *zone of agreement* or *reasonable agreement* around a goal-objective subset $\mathbb{E}$ or $\mathbb{H}$ or $\mathbb{F}$ if there exists an $\mathcal{E}$-neighborhood set defined around 0, $N_{\varepsilon}(0) \ni \mathcal{D}(\bullet) \in N_{\varepsilon}(0)$ and $\left|\mathcal{D}(\bullet)-0\right| \geq 0$ and there exists an $\mathcal{E}$-neighborhood set defined around one, $N_{\varepsilon}(1) \ni \mathcal{A}(\bullet) \in N_{\varepsilon}(1)$ and $\left|\mathcal{A}(\bullet)-1\right| \geq 0$. A crisp negotiation equilibrium is established when $\mathcal{D}(\cdot)=0$ and $\mathcal{A}(\cdot)=1$ where, A and B are defined over $\Pi,\ \mathbb{D}$ and $\mathbb{P}$. Any agreement with $\left|N_{\varepsilon}(1)\right| \geq 0$ is a fuzzy equilibrium with a degree of acceptance defined as $\mu_{N_{\varepsilon}(\bullet)}(0)=1-\left|\varepsilon\right|$.

5.3 The Nature of the Negotiated Equilibrium of the Social Goal-Objective Set

The structure of the game of the political economy presented here is developed around the formation of a social goal-objective set on the guiding principle of cost-benefit rationality for both the individual and the collective under an un-weighted democratic decision-choice system. It is, however, general in the manner in which the structures of economics, law and politics may or may not bring social stability by the resolution of conflicts in the collective decision-choice space as the elements in the goal-objective set acquire differential cost-benefit assessments. The social stability of the socio-political economy requires reasonable stabilities in all the three structures, as the conflicts in the individual preferences shift, or as they are resolved in the democratic decision-choice system without violence. The degree of social stability will depend on the collective tolerance and the strengths of the institutions that make up the democratic collective decision-choice system. The socio-political game is to negotiate votes from the members of the decision-making core, as the agent, in support of a particular position by private and public sector advocates that together with other members of the society constitute the principal. The vote solicitation takes place after the social decision-making core has been constructed by some principles of social acceptance. The members of the social decision-making core are assumed to be sensitive to the preferences of the voting public as they understand them. The negotiation process begins with it the notion that the economic structure is divided into public sector, $\mathbb{P}$ and private sector $\mathbb{B}$ with corresponding advocates in support, on the basis of ideology that relates to cost-benefit assessments.

In this respect, we may conceive of any social system with a possible set $\mathbb{S}$ that is composed of varying proportions of private-public sector configuration where $\mathbb{P} \neq \varnothing$ and $\mathbb{P} \subset \mathbb{S}$ is the public sector characteristic set, and $\mathbb{B} \neq \varnothing$ with $\mathbb{B} \subset \mathbb{S}$ is the private sector characteristic set, all are measured in the value of provision of aspects of social goods and services. These sets may be viewed as institutional characteristics for organizing production, income distribution and consumption. In fact, we shall also assume that the social system's characteristic set, $\mathbb{S}$, is measured by the provision of goods and services in the society. If we assume that the private and public sectors are mutually exclusive and collectively exhaustive then we have $\mathbb{S} = \mathbb{P} \bigcup \mathbb{B}$ and $\mathbb{P} \bigcap \mathbb{B} = \varnothing$, and hence there are no government-private sector economic entities. From the structure, we may define proportionality values of public sector proportion $\pi = \left({}^{\#\mathbb{P}}/_{\#\mathbb{S}} \right)$ and private sector proportion, $\beta = \left({}^{\#\mathbb{B}}/_{\#\mathbb{S}} \right)$. The system is said to assume total public sector if $\pi = \left({}^{\#\mathbb{P}}/_{\#\mathbb{S}} \right) = 1$ and $\beta = 0$. Alternatively, it is said to be completely private sector if $\beta = \left({}^{\#\mathbb{B}}/_{\#\mathbb{S}} \right) = 1$ and $\pi = 0$. Any point of the socio-political economy is described by a duplex (π, β). The case where we have $(\pi, 0)$ may be associated with perfect communism where the ownership of the means of production is in the hands of the public. The case where $(0,\beta)$ may also be associated with perfectly competitive capitalism where the means of production is in the private hands.

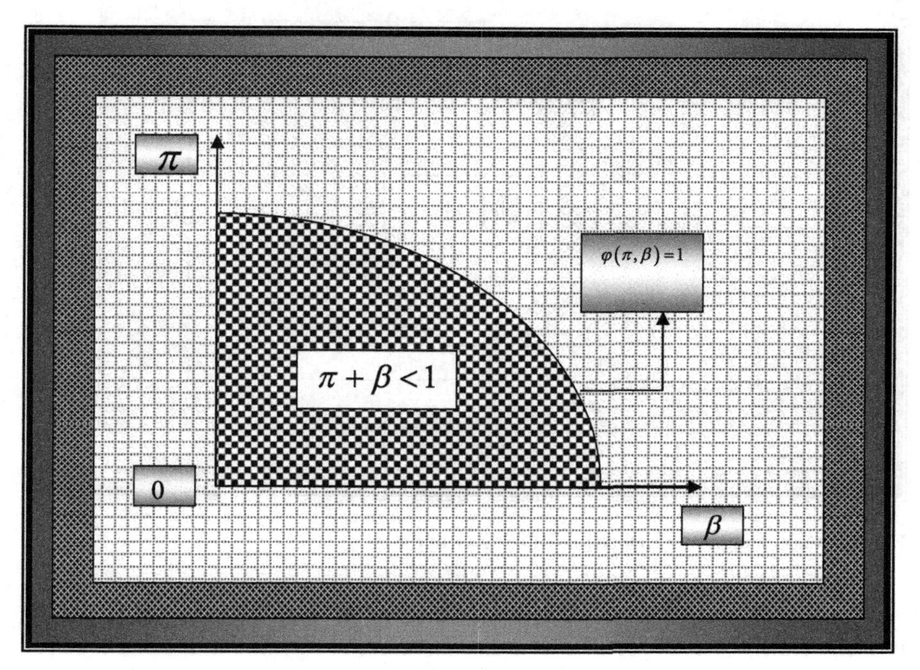

Fig. 5.3 Private-Public Sector Efficiency Frontier

There is, thus, a functional relationship between π and β that establishes private-public sector efficiency frontier $\varphi(\pi,\beta)=1$ for trade-off possibilities for decision-choice by the decision-making core regarding the preferred private-public sector combination. The frontier is specified in Figure 5.3. We may view **S** as the size of the political economy, **P** the size of the public sector and hence π is its proportion and, **B** the size of the private sector and hence β is its share in the political economy. The term $\varphi(\bullet)$ is structural transformation function where $\varphi(\cdot)=1$ defines efficiency frontier where $\beta=1-\pi$. The condition $\varphi(\pi,\beta)<1$ is feasible but inefficient public-private sector combination in duality. This may be the case where part of the social production is external in the sense that it is neither produced by domestic private or public sectors. At all times, the system operates on the frontier such that $\beta+\pi=1$.

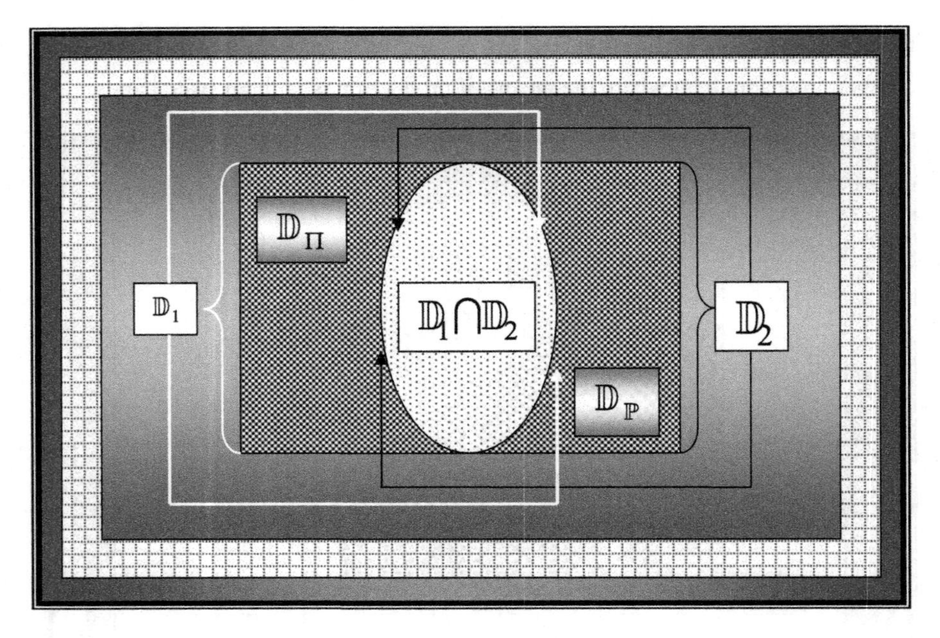

Fig. 5.4 Venn Diagram of Decision-Making Groups

The private and public sectors are measured in the same units to give meaning to their *relative size. Three measurable candidates are available to us. They are: 1) employment*-proportions relative to total employees or population, 2) expenditure-proportions relative to gross national product, and 3) the value of the provision of goods and services produced. Because of deficit spending, tax-revenue as a share in income distribution may be in appropriate. We have specified the existence of public sector advocates Π and the private sector advocates, $\mathbb{P}$. The game in the decision-choice process is such that the set of decision-making core (the elected officials) is divided into $\mathbb{D}_1$ as the set of the members of $\mathbb{D}$ that supports or will support the public sector advocate and $\mathbb{D}_2$, as the set of those who support or will support the private sector advocates. The division is such that $\mathbb{D}_1 \neq \varnothing$, $\mathbb{D}_2 \neq \varnothing$ where $\mathbb{D} = \mathbb{D}_1 \cup \mathbb{D}_2$ and $\mathbb{D}_1 \cap \mathbb{D}_2 \neq \varnothing$ with $\mathbb{D}_\Pi$ and $\mathbb{D}_\mathbb{P}$ defined as relative complements and $\mathbb{D}_\Pi \cap \mathbb{D}_\mathbb{P} = \varnothing$ providing a crisp ideological position with fuzzy conditionality values. The structure of the decision-making sets is provided in Figure 5.4 to illustrate the categories of opposing ideologues and the independent decision makers.

Here, an important problem arises in terms of the definition of a government. Should the definition include federal, state, regional districts, county, city and others in the case of federal governmental system? The situation is a little more complicated than we think when we talk about expressions of individual preferences over central, state, district, city and other sub-governmental entities. These sub-governments must be seen in terms of division of collective sovereignty as it is related to power structure in the social decision-choice space Some individuals may prefer smaller central government but larger sizes of non-central governments and other way round. This is the problem of distribution of preferences over sub-governmental entities where a reduction of the size of the central government does not necessary mean a reduction in the size of the total government as measured in terms of provision of goods and services. Should we, in theory and practice, consider preference orderings over increased sizes of non-central government provisions and the reduction in the central government provisions of goods and services as a decrease in the public sector? There are two paths of preference ordering that are analytically open to us. One path is weighted preferences among the sizes of different governmental structures by the individual and the other is un-weighted preferences over the same sizes. We must keep in mind that there may be some individuals who prefer small (large) size of the central government and large (small) size of non-central government relative to provision of goods and services. There might be some members in both the principal and agent that prefer small sizes for both central and non-central governmental provisions. The analytical problem is how to integrate these preferences to obtain the size of the public sector that is managed by the government in the political economy.

The sets in the Figure 5.4 may be specified as $\mathbb{D}_{\Pi}=\{x \mid x\in \mathbb{D}_1 \text{ and } x\notin \mathbb{D}_2\}$ where $\mathbb{D}_{\Pi}$ must be partitioned into available governmental agencies as measured in the same unit where the weights, $w_{ij}, i\in \mathbb{I}$ and $j\in \mathbb{J}$ for the individual $i\in \mathbb{I}$ and government agency, $j\in \mathbb{J}$ with $\mathbb{D}_{\Pi_{ij}}=\left\{x \mid x\in \mathbb{D}_{\Pi_j} \text{ and } j\in \mathbb{J}\right\}$ Similarly, $\mathbb{D}_{\mathbb{P}}=\left\{x \mid x\in \mathbb{D}_2 \text{ and } x\notin \mathbb{D}_1\right\}$. The set $\mathbb{D}_{\mathbb{L}}=\left(\mathbb{D}_1 \cap \mathbb{D}_2\right)$ is interpreted as independent decision makers that examine the situation in a case by a case and support or reject the elements at each decision-choice situation.

The categories are such that the following decision groups are identified $\mathbb{D}_{\Pi}=\left\{x \mid x\in \mathbb{D}_1 \text{ and } x\notin \mathbb{D}_2\right\}$ and similarly $\mathbb{D}_{\mathbb{P}}=\left\{x \mid x\in \mathbb{D}_2 \text{ and } x\notin \mathbb{D}_1\right\}$. The set $\mathbb{D}_{\mathbb{L}}=\left(\mathbb{D}_1 \cap \mathbb{D}_2\right)$ is interpreted as a set of independent decision makers who examine case by case of the decision-choice elements and decide to support or reject in accordance with their evaluations under the best available information. Corresponding to the decision-independence set is the decision-ideological members that may be specified as a symmetric difference between $\mathbb{D}_1$ and $\mathbb{D}_2$ or the union of the symmetric complement of $\mathbb{D}_\Pi$ and $\mathbb{D}_\mathbb{P}$ as in equation (5.1) and with a Venn diagram illustration in Figure 5.4:

$$\mathbb{D}_\mathbb{I} = \mathbb{D}_1 \nabla \mathbb{D}_2 = \{x \mid x \in \mathbb{D}_\Pi \bigcup \mathbb{D}_\mathbb{P} \text{ and } x \notin \mathbb{D}_1 \bigcap \mathbb{D}_2 = \mathbb{D}_\mathbb{L}\} \tag{5.1}$$

Equation (5.1) defines a set of political and decision ideologues relative to private and public sector activities.

The greater the size of the independent members, $\mathbb{D}_\mathbb{L}$ relative to the size of the decision-making core $\mathbb{D}$ the greater is the instability of the democratic decision-choice system and the management of the political economy, but the greater is the probability of an efficient decision-choice process as well as the greater is the unpredictability of the decision-choice outcomes of the activities of the decision-making core. In this way, the democratic collective decision-choice system dwells on an unstable balance depending on the behavior of the members of the independence set. This unstable role of the members of the independence set in the democratic decision-choice system is over shadowed, if either $\#\mathbb{D}_\Pi > \#(\mathbb{D}_\mathbb{P} \cup \mathbb{D}_\mathbb{L})$ or $\#\mathbb{D}_\mathbb{P} > \#(\mathbb{D}_\Pi \cup \mathbb{D}_\mathbb{L})$ that presents the presence of super-majority of one advocacy. Let us return to the structure of the decision-choice game of voting through preferences to establish private-public sector proportionalities for the provision of social goods and services in the implementation of the elements in the goal-objective set in the support of the national interest and social vision.

The decision-making game to form alliance may be conceived in terms of pyramidal logic. The initial structure for coalition formation may be presented as in Figure 5.5. The point $\mathbb{E}$ is the center of gravity in the decision-making where there is an agreement among the public and private advocates and the decision-making core on the key elements of social policy that relate to public-private sector duality on the goal-objective set. The point $\mathbb{E}_1$ specifies the agreement between the private-sector advocates and the decision-making core on a private-public sector combination for which a coalition may be formed to play the game against the public-sector advocates. The point $\mathbb{E}_2$ specifies a position of agreement by private-public sector advocates that will allow effective coalition to be formed in support of a proportional combination of proportional private-public sector structure in the social policy process. Similarly, the point $\mathbb{E}_3$ defines a position that satisfies the condition for coalition between the members of decision-making core and the advocates of the public sector.

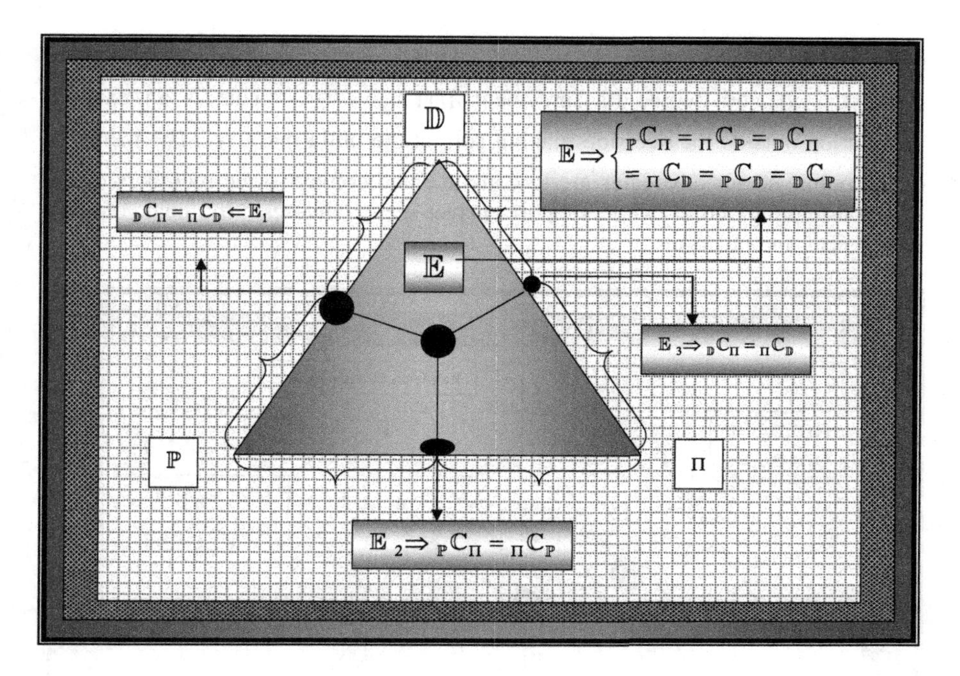

Fig. 5.5 Initial State of the Coalition Game Formation

Available to us, is an internal movement induced by *categorial conversion* to alter the size of $\mathbb{D}_{\Pi}$ or $\mathbb{D}_{\mathbb{P}}$ in support of an increased public or an increased private participation in the provision of goods and services in the political economy. The categorial conversion takes place, in this respect, through negotiations to change the size of either $\mathbb{D}_{\Pi}$ or $\mathbb{D}_{\mathbb{P}}$. The negotiation process regarding the size of the public sector may be represented in a diagrammatic form as in Figure 5.6 where the game stage is set between the public-sector advocates and the decision-making core. Since $\mathbb{D}_{\Pi}$ supports the increase in the size of the public sector, there is no disagreement of the public-sector ideologues in the decision-making core with the public-sector advocates and hence there is no need for negotiation. The implication here is that the disagreement function assumes the value of zero $\mathcal{D}(\Pi,\mathbb{D}_{\Pi})=0$ while $\mathcal{A}(\Pi,\mathbb{D}_{\Pi})=1$ with the negotiation function assuming the value of zero, thus $\mathcal{N}(\Pi,\mathbb{D}_{\Pi})=0$. This is shown as a directional decision line $\xrightarrow{\mathbb{D}_{\Pi}\Pi}$ containing a point B_1. It simply states that negotiation is not needed along this decision line. Active negotiation is required along the decision line $\xrightarrow{\mathbb{D}_{\mathbb{P}}\Pi}$ containing the point B_3 in order to win some decision votes from $\mathbb{D}_{\mathbb{P}}$, the supporters of the private sector (or private sector ideologue). Here, active negotiations are required between members in sets $\mathbb{D}_{\mathbb{P}}$ and Π to move some members in $\mathbb{D}_{\mathbb{P}}$ into $\mathbb{D}_{\Pi}$ to increase the decision support for Π. The negotiation is represented by a function $\mathcal{N}(\mathbb{P}_{\Pi},\mathbb{D}_{\mathbb{D}_{\mathbb{P}}})>0$ with positive disagreement $\mathcal{D}(\Pi,\mathbb{D}_{\mathbb{P}})>0$ and degree of agreement $\mathcal{A}(\Pi,\mathbb{D}_{\mathbb{P}})\in(0,1)$. This is an external negotiation outside the decision-making core at the initial point B_3. The members of the public-sector advocates are to convince and convert some members of the decision-making core who support privet-sector interest to change their preferences and positions.

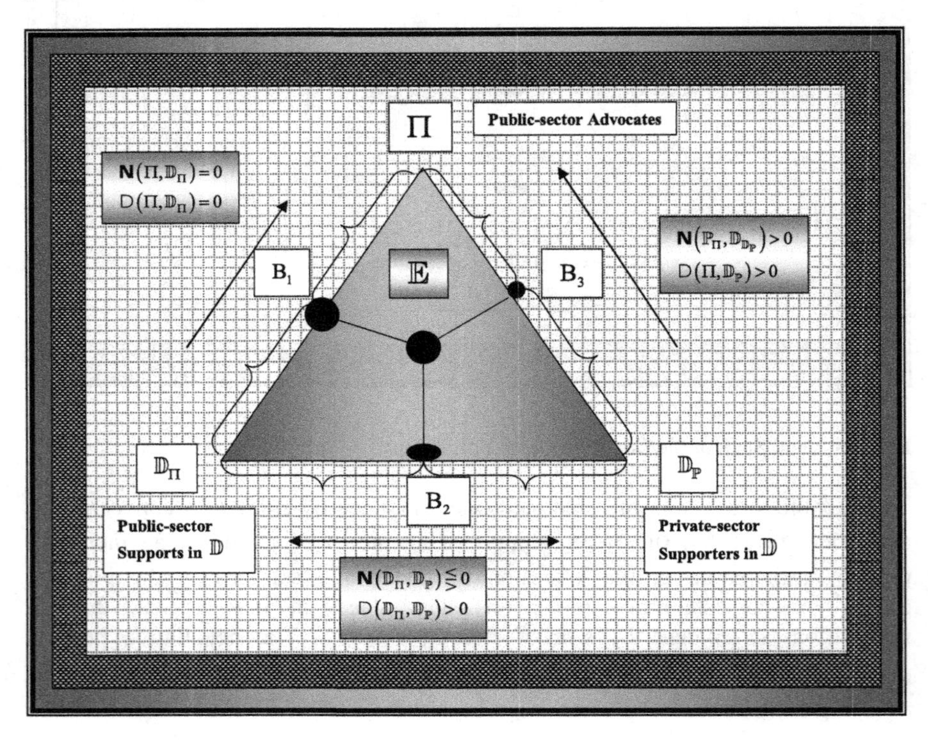

Fig. 5.6 Geometry of Public Sector Negotiation Activities

There are internal negotiations between the members of $\mathbb{D}_{\Pi}$ and those of $\mathbb{D}_{\mathbb{P}}$ within the decision-making core. The initial position for negotiations between $\mathbb{D}_{\Pi}$ and $\mathbb{D}_{\mathbb{P}}$ is such that $\mathcal{N}(\mathbb{D}_{\Pi},\mathbb{D}_{\mathbb{P}})>0$, and $\mathcal{D}(\mathbb{D}_{\Pi},\mathbb{D}_{\mathbb{P}})>0$. The internal negotiation is intense where members of $\mathbb{D}_{\Pi}$ work to convince some members in $\mathbb{D}_{\mathbb{P}}$ to join them to support the position of the public-sector interest while the members of in $\mathbb{D}_{\mathbb{P}}$ do the same thing with some members in $\mathbb{D}_{\Pi}$ in support of the private-sector interest. The decision conversional process is through the establish rules of the socio-political game where political sales persons may be used within the legal structure. We must keep in mind that all these negotiations take place in the political structure in order to change the legal structure to which the economic structure is always at its mercy.

The above analysis of the politico-economic game involves the public sector and public sector advocates playing the game against the private sector advocates for public-sector interest. A similar strategic politico-economic system's dynamics is taking place simultaneously in the private sector with private sector advocates playing key and opposing role against the public-sector advocates in favor of private-sector interest. The stage of the conflict and the theater of the game and its progress are presented in Figure 5.7. Again, since $\mathbb{D}_{\mathbb{P}}$ supports an increasing participation of the private sector, no negotiation is required between $\mathbb{P}$ and $\mathbb{D}_{\mathbb{P}}$ and hence $\mathcal{N}(\mathbb{D}_{\Pi},\mathbb{D}_{\mathbb{P}})=0$ since $\mathcal{A}(\mathbb{D}_{\Pi},\mathbb{D}_{\mathbb{P}})=1$ and $\mathcal{D}(\mathbb{D}_{\Pi},\mathbb{D}_{\mathbb{P}})=0$ along the

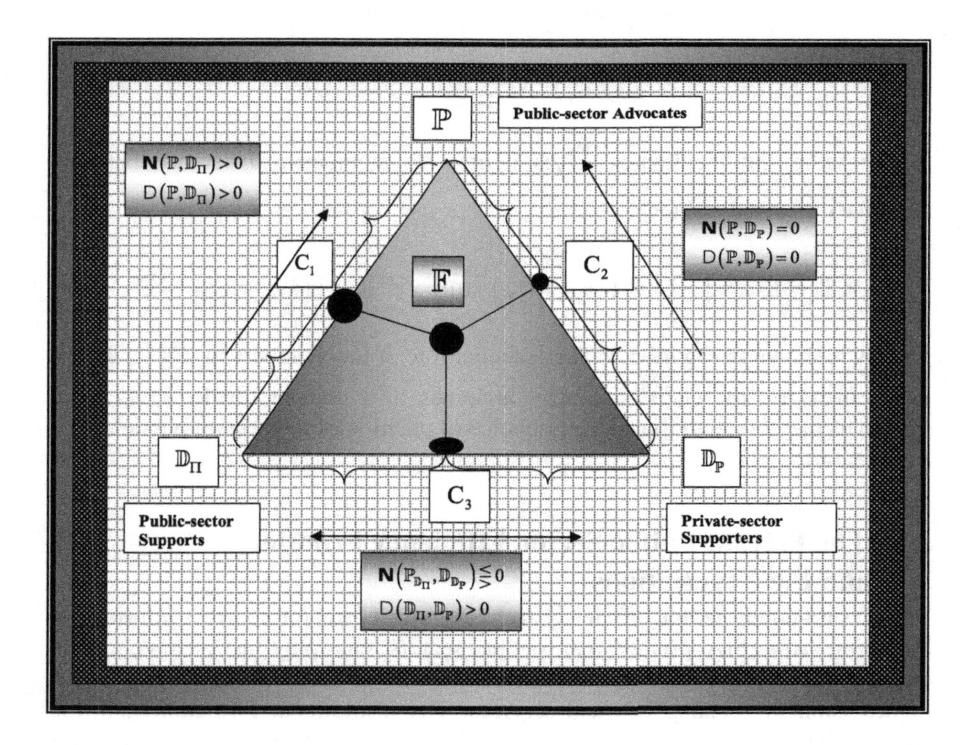

Fig. 5.7 Geometry of the Private-Sector Negotiation Activities

negotiation-decision line $\xrightarrow{\mathbb{D}_{\mathbb{P}}\mathbb{P}}$ containing the point C_2. The decision-negotiation line $\xrightarrow{\mathbb{D}_{\Pi}\mathbb{P}}$ containing the point C_1 projects a need for active negotiation to win votes from $\mathbb{D}_{\Pi}$ in support of $\mathbb{P}$ and hence there is an active negotiation between the members in $\mathbb{D}_{\Pi}$ and $\mathbb{P}$ through external pressures of negotiation. The points C_i's on negotiation decision lines $\xrightarrow{(\cdot)}$ define the initial position where if $\mathcal{A}(\mathbb{D}_\Pi, \mathbb{D}_\mathbb{P}) \neq 1$ then $\mathcal{A}(\mathbb{D}_i, \mathbb{D}_j) \in (0,1)$ as well as $\mathcal{D}(\mathbb{D}_i, \mathbb{D}_j) \in (0,1)$

The politico-economic game, as presented, defines the decision-making core as divided into private-sector ideologues and public-sector ideologues and independent minded members that can support private or public sector activities with their votes under critical social cost-benefit valuations. The set of public-sector ideologues $\mathbb{D}_\Pi$ may be specified as a relative difference between $\mathbb{D}_1$ and $\mathbb{D}_2$ and similarly for $\mathbb{D}_\mathbb{P}$. Let us keep in mind the following set relations

$$\left.\begin{aligned} \mathbb{D}_\Pi &= \left(\mathbb{D}_1 / \mathbb{D}_2 \right) = \{x \mid x \in \mathbb{D}_1 \text{ and } x \notin \mathbb{D}_2\} \\ \mathbb{D}_\mathbb{P} &= \left(\mathbb{D}_2 / \mathbb{D}_1 \right) = \{x \mid x \in \mathbb{D}_2 \text{ and } x \notin \mathbb{D}_1\} \end{aligned}\right\} \tag{5.2}$$

The set of all ideologues $\mathbb{D}_3$ is defined as symmetric difference between $\mathbb{D}_1$ and $\mathbb{D}_2$ and written as:

$$\mathbb{D}_3 = \mathbb{D}_1 \nabla \mathbb{D}_2 = \{x \mid x \in (\mathbb{D}_1 \bigcup \mathbb{D}_2) \text{ and } x \notin \mathbb{D}_1 \bigcap \mathbb{D}_2\} = \mathbb{D}_\Pi \bigcup \mathbb{D}_\mathbb{P} \tag{5.3}$$

This is shown in Figure 5.4. The essential role of the members in $\mathbb{D}_1 \bigcap \mathbb{D}_2$ is their effects on the decision-choice outcomes in a democratic decision-choice system as they relate to the decision-choice of position of the private-public sector combinations.

5.4 The Private-Public Sector Efficiency Frontier and the Negotiations

The private-public sector efficiency frontier projects some important information about the role that private and public sectors can play in socio-economic institutional formation and policy configuration in the political economy in terms of provision of social goods and services. The institutional formations and the social policy configurations, in the last analysis, are about the provision of goods and services within the production-consumption duality as a natural process and superimposed on it is a private-public duality as a social process. Given the efficiency frontier that defines the set of private-public sector combination of supply of social goods and services, we can write two feasible sets of private-public sector combinations for the relative participation of private and public sectors in the political economy within the institutional setup for the application of policy administration of social programs to answer the question as to how much should the decision-making core privatize the public sector interest in the production-consumption space . This is the private-public-sector combination decision-choice problem in the political economy. Let the political economy viewed in terms of the set of private-public sector combination ratios be specified as:

$$\mathbb{E} = \{\varphi(\pi,\beta) \in [0,1] \mid \pi \in \Pi \text{ and } \beta \in \mathbb{P} \ni \pi + \beta \leq 1\} \quad (5.4)$$

$$\mathbb{E}' = \{\varphi(\pi,\beta) \notin (0,1) \mid \pi \in \Pi \text{ and } \beta \in \mathbb{P} \ni \pi + \beta \geq 1\} \quad (5.5)$$

$$\mathbb{F} = \mathbb{E} \cap \mathbb{E}' = \{\varphi(\pi,\beta) \in [0,1] \mid \pi \in \Pi \text{ and } \beta \in \mathbb{P} \ni (\pi + \beta) = 1\} \quad (5.6)$$

Equation (5.4) of the set $\mathbb{E}$ specifies the feasible region of the general politico-economic space with the set of both efficiency and inefficiency elements. It is the attainable social institutional combination set of private and public sectors. ? Equation (5.5) of the set $\mathbb{E}'$ specifies the unattainable region of the general politico-economic space with the set of both efficiency and unattainable elements. It is the unattainable institutional combination set. Equation (5.6) of the set $\mathbb{F}$ specifies the feasible region of the general politico-economic space with the set of both the attainable and efficiency elements. It is the efficient and attainable private-public-sector combination set. The curvature of the of the function of the efficiency frontier exhibits trade-off possibilities between private-public sector combinations with $\varphi(\pi,\beta) = 1$ where $\frac{\partial\varphi}{\partial\pi}d\pi + \frac{\partial\varphi}{\partial\beta}d\beta = 0$ implying that $\left(\frac{\partial\varphi}{\partial\beta}\Big/\frac{\partial\varphi}{\partial\pi}\right) = -\frac{d\pi}{d\beta}$ The elements cardinality of the set $\mathbb{F}$ is infinite ($(\#\mathbb{F} = \infty)$

In terms of private-public institutional set-up $\left(\partial\varphi/\partial\pi\right)$ may be viewed as the marginal productivity of the public sector in the provision of goods and services while $\left(\partial\varphi/\partial\beta\right)$ may be viewed as the marginal productivity of the private sector in the provision of goods and services and $\left(-d\pi/d\beta\right)$ is the condition of the social private-public trade-off process. The nature of the public-private sector combinational structure, in terms of duality, polarity, continuum and trade-offs is presented as a cognitive geometry in Figure 5.8.

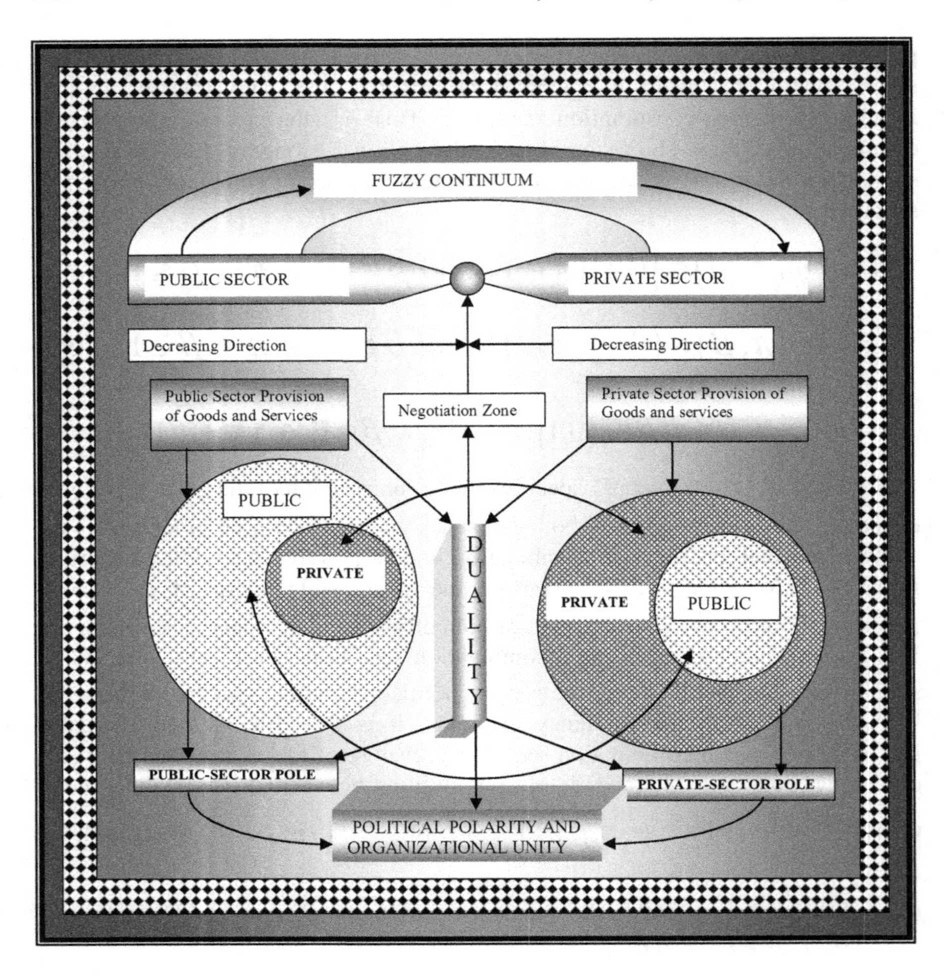

Fig. 5.8 An Epistemic Geometry of Public-Private Sector Duality in a Fuzzy Continuum Defined in Linguistic Quantities of Big and Small for Decision-Choice Actions

The trade-off process exhibits increasing costs in private-public sector combinations in every decision-choice action. It is the presence of this increasing cost of trade-off possibilities that economic studies on the relative private-public sector efficiency have always been inconclusive. The conditions of inconclusiveness rest on the nature of private-public sector duality. Like the cost-benefit duality, the public-private sector continuum suggests that any size of the private sector has a public-sector support. Similarly, any size of the public-sector has private-sector support in any political economy. Alternatively stated, both the private and public sectors reside in categorial unity where they have a give-and-take relation supplying what is lacking in the existence of each other. They constitute an inseparable unity of the social set up. The graphical representation of eqn.(5.4 -5.6) and private-public sector transformation structure of Figure 5.8 is shown in Figure 5.9. The organizational question is where will the optimal combination be on the set of proportions? Is this optimal combination permanent? Is it unique to all societies and generations? Does it depends on cultures, social preferences, generational character and changes with them? The answers to all these questions are essential in understanding the character and evolving the evolving morphology of the political economy.

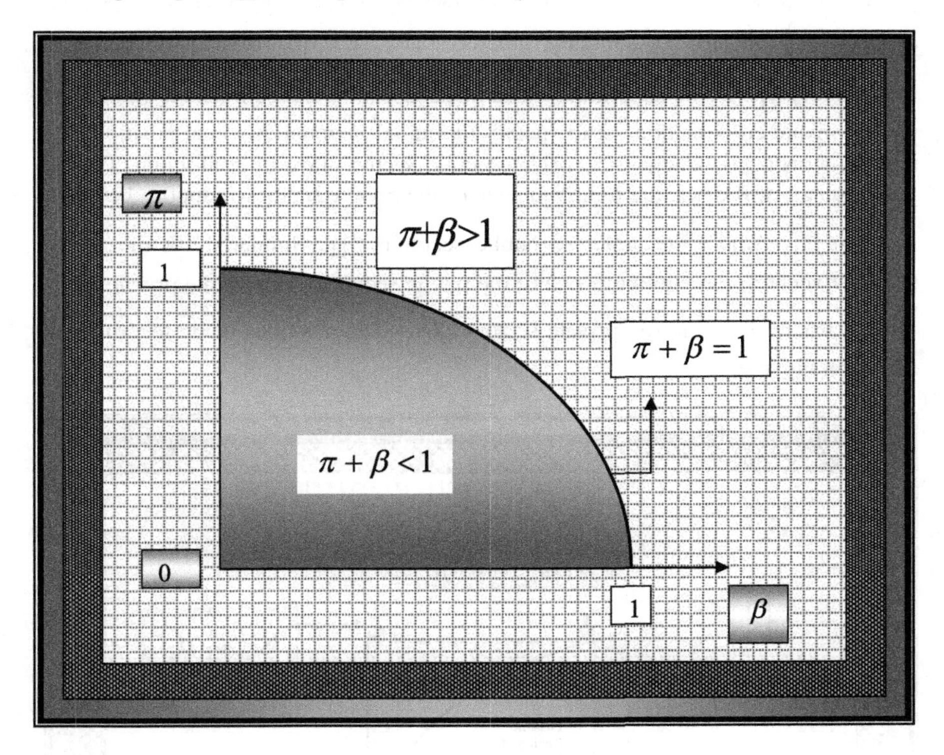

Fig. 5.9 Private-Public Sector Efficiency Frontier

Given the private-public sector frontier, the social decision-choice problem is to select a point $\rho=(\pi,\beta)$ that is optimal in the sense of satisfying the conflicting preferences of public-private sector advocates and interests relative to its social set up and generational conditions. The point of disagreement without violence may deviate from the optimal social welfare. Such an optimal social welfare will vary from social system to social system and for a given social system the optimal social welfare will vary over generations due to cultural dynamics and differences in the collective preferences as well as shifts in inter-generational preferences. The point of acceptance is carried on through negotiations as tatical game between private and public sector advocates under power relation. The negotiation game is such that the public sector advocates either enter the negotiation game with initial reservation proportion $\rho_{\Pi} = \left(\pi_{\Pi}^{r}, \beta_{\Pi}^{r}\right)$ where $\pi_{\Pi}^{r} \gg \beta_{\Pi}^{r}$ which is the most preferred private-public sector combination. The private sector advocates enter with an initial position $\rho_{\mathbb{P}} = \left(\pi_{\mathbb{P}}^{r}, \beta_{\mathbb{P}}^{r}\right)$ where $\beta_{\Pi}^{r} \gg \pi_{\Pi}^{r}$ which is the most preferred private-public sector combination. The superscript *(r)* represents reservation position and $(\gg)$ means substantially greater than. The two reservation positions $\rho_{\mathbb{P}}$ and ρ_{Π} may be superimposed as linear decision-choice lines on the feasible public-private combination set to specify the initial negotiation-game position. The specification acknowledges the private-public sector unity in continuum. For the public sector advocates, the minimum size of the private sector to support the public sector is β_{Π}^{r} where the preferred public sector is π_{Π}^{r}. Similarly, for the private sector advocates, the minimum size of the public sector to support the private sector is $\pi_{\mathbb{P}}^{r}$ where the preferred private sector is $\beta_{\mathbb{P}}^{r}$. The positions of these values are shown in Figure 5.10.

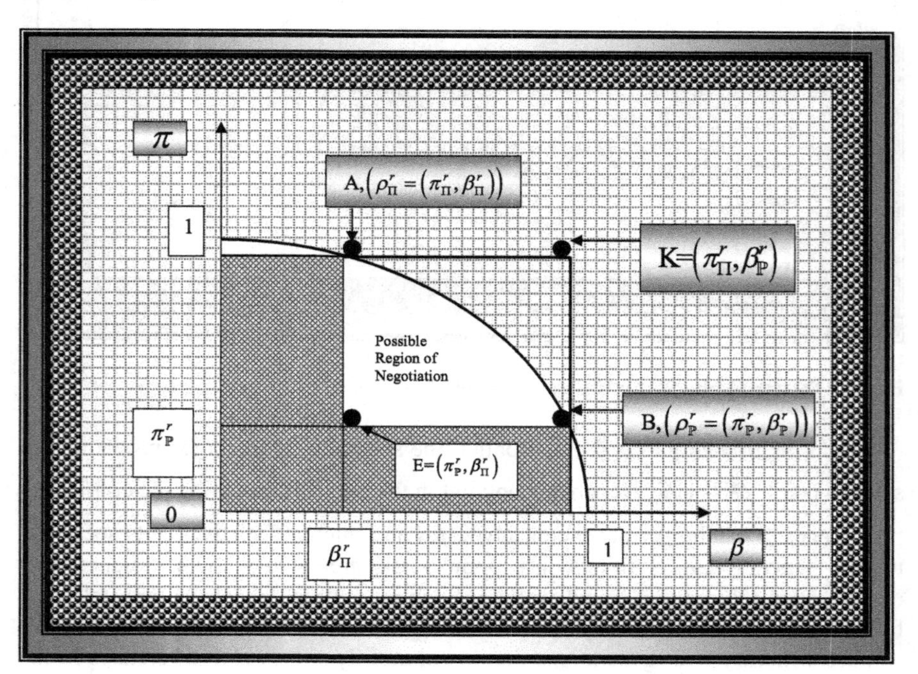

Fig. 5.10 The Negotiation Space for Private-Public Sector Combination

The reservation proportions are established by political ideologies regarding the relative size of private and public sectors in the political economy and hopefully supported by the efficiency assessment by the members in the independence set. Complete public sector implies that $\pi = 1$ and complete private sector implies that $\beta = 1$. The initial preferred public-sector size of the public-sector advocates and preferred private-sector size by the private advocates at $\mathrm{K}=\left(\pi_{\Pi}^{r}, \beta_{\mathbb{P}}^{r}\right)$ does not belong to the feasible set and unsustainable in the organization of the political economy given the competing ideologies and efficiency assessments. The point $\mathrm{E}=\left(\pi_{\mathbb{P}}^{r}, \beta_{\Pi}^{r}\right)$ expresses the initial size $\pi_{\mathbb{P}}^{r}$ of the public sector that is preferred by the private sector advocate and the initial size β_{Π}^{r} of private sector that is preferred by the public sector advocates. The point E while feasible is substantially inefficient and not preferable by both the public and private sector advocates as a solution to the problem of the private-public sector combination regarding the conflicts in the social institutional structure if the political economy. Any public-private sector combination will be called institutional configuration in that any institution in the political economy is either private or public institution or both. The public-private efficiency frontier, feasible and non-feasible set will also be called efficient, feasible and non-feasible institutional configurations. Any socio-political organization of economic production may be characterized as a set of institutional configurations of the private-public sector combination from which a decision-choice action is to be exercised. The decision-choice set therefore may be specified as:

$$\mathbb{S} = \left\{(\pi,\beta) \mid \pi \in \Pi \text{ and } \beta \in \mathbb{P} \text{ and } \pi + \beta \le 1\right\} \text{ Efficient and feasible set} \quad (5.7)$$

$$\mathbb{S}' = \left\{(\pi,\beta) \mid \pi \in \Pi \text{ and } \beta \in \mathbb{P} \text{ and } \pi + \beta \ge 1\right\} \text{ Efficient and unattainable set} \quad (5.8)$$

The efficiency frontier is simply $\left(\mathbb{S} \cap \mathbb{S}'\right)$.

5.4.1 The Private-Public Sector Duality, Individual-Collective-Decision Duality and the Mediating Role of the Government

The general decision-choice process of the political economy may be viewed in terms of trinities and dualities. At the level of the trinity, we have the government, private and the public on one hand and on the other hand, we have the government, individual and the community as social collectivity. At the level of production-consumption space, we have private-public sector duality with tension in a continuum. At the level of decision-choice space, we have the individual-community duality with continuum under tension. At the level of the social organism, the government exists as a mediating instrument to create fairness, peace and stability for the political economy in relation to freedoms and sovereignty. The quality of the mediation, fairness, peace and social stability in the political economy will depend on who comes to use the institutions of the government to assist in creating these qualitative characteristics of the social set up. The group whose is charged with the responsibility of using the instrument of government is what we have referred to as the social decision-making core which is distinct from the government. The general relational structure among the private, public, individual and community is presented as a cognitive geometry in Figure 5.11.

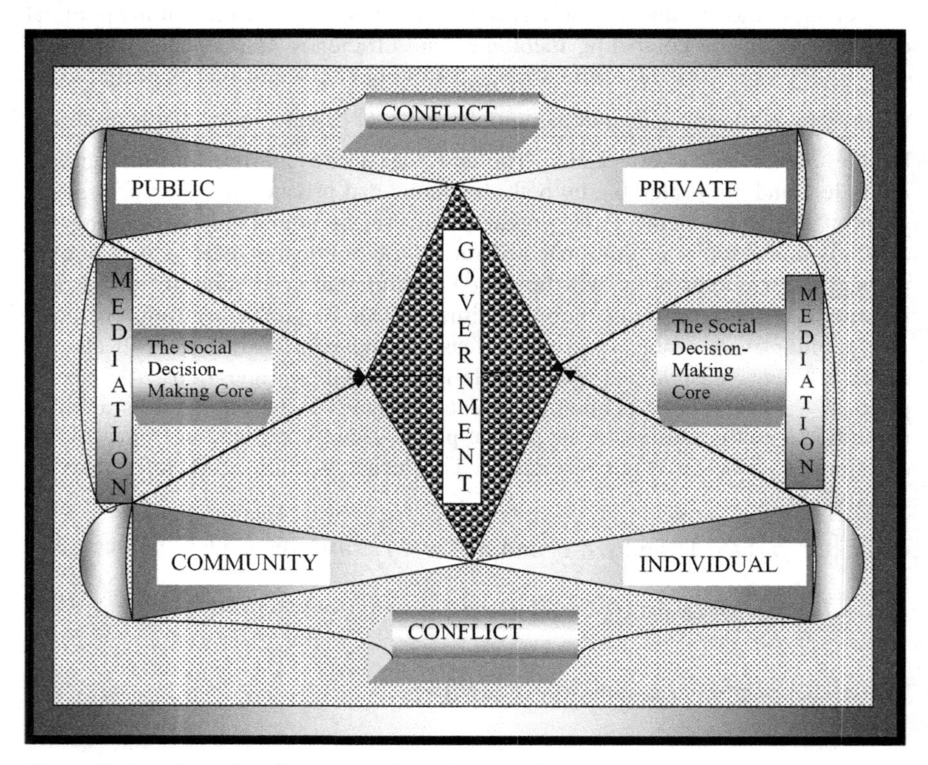

Fig. 5.11 The Cognitive Geometry of the Relational Structure of Trinity, Dualities in the Decision and Organizational Spaces of the Political Economy

Figure 5.11 presents the relational structure of the conflicts in the social decision-choice space involving allocation and distribution of costs and benefit in the social production under individual differential preferences in social production-consumption activities. The benefit-cost distribution may be done by the government through either the private sector or the public sector or both from the organizational point of view. This distribution must be related to the sovereignty questions of the individual or the collective. The sovereignty question involves the distribution of decision-choice powers in the politico-economic space. The nature of the decision-choice actions will depend on who has the power to use the government as an instrument of mediation. The determination of who has the power may be done by collective sovereignty or imposition. The group that acquires the power is what we have called the social decision-making core which is distinct from the government as instrument of mediation. The members of the decision-making core come from the general population with their own preferences, prejudices, and bias in the collective decision-choice process. It is these characteristics of the decision-making core that affect the balancing decisions of the government and provide energy for further conflict and social transformations. Mediation in the conflict zones of the political economy requires negotiation game to which we turn our attention.

5.5 The Decision-Making Core and the Negotiation Process

Given the feasible institutional configuration of the political economy, the choice of a particular institutional configuration that will contain private-public sector combination will depend on the decision-choice action of the decision-making core, the set of the elected officials that acts as the agent to the society which in turn acts as the principal that simultaneously contains the agent. Such decision-choice actions are through expression of votes where the principle of majority votes is the decision-choice rationality. To apply this principle, we define possible voting proportions as $\delta_{\Pi} = \left({}^{\#\mathbb{D}_{\Pi}}\!/_{\#\mathbb{D}} \right)$ and $\delta_{\mathbb{P}} = \left({}^{\#\mathbb{D}_{\mathbb{P}}}\!/_{\#\mathbb{D}} \right)$ with the proportion of independence $\delta_{\mathbb{I}} = \left({}^{\#(\mathbb{D}_1 \cap \mathbb{D}_2)}\!/_{\#\mathbb{D}} \right)$ such that $\left(\delta_{\Pi} + \delta_{\mathbb{P}} + \delta_{\mathbb{I}} \right) = 1$ and $\mathbb{D}_{\mathbb{I}} = \left(\mathbb{D}_1 \cap \mathbb{D}_2 \right)$. With decision-making core partitioned into private and public sector ideologues $\mathbb{D}_{\mathbb{P}}$ and $\mathbb{D}_{\Pi}$ and independent members $\mathbb{D}_{\mathbb{I}}$, we may restate the reservation proportions as conditional functional positions that may be written as:

$$\left. \begin{aligned} \rho_{\Pi} &= \rho_{\Pi}\left(\pi^{r}_{\Pi}, \beta^{r}_{\Pi} \mid \delta^{\mathbb{I}}_{\Pi}, \delta^{\mathbb{I}}_{\mathbb{P}} \right) \\ \rho_{\mathbb{P}} &= \rho_{\mathbb{P}}\left(\pi^{r}_{\mathbb{P}}, \beta^{r}_{\mathbb{P}} \mid \delta^{\mathbb{I}}_{\mathbb{P}}, \delta^{\mathbb{I}}_{\Pi} \right) \end{aligned} \right\} \tag{5.9}$$

The voting size in support of either the private or public sector will depend on the shifting preferences of the members in $\mathbb{D}_{\mathbb{I}}$. To examine this we partition $\mathbb{D}_{\mathbb{I}}$ into supporters of public sector, $\mathbb{D}^{\mathbb{I}}_{\Pi}$ and the supporter of private sector, $\mathbb{D}^{\mathbb{I}}_{\mathbb{P}}$ such that $\mathbb{D}_{\mathbb{I}} = \left(\mathbb{D}^{\mathbb{I}}_{\Pi} \cup \mathbb{D}^{\mathbb{I}}_{\mathbb{P}} \right)$, $\delta^{\mathbb{I}}_{\Pi} = \left({}^{\#\mathbb{D}^{\mathbb{I}}_{\Pi}}\!/_{\#\mathbb{D}} \right)$, $\delta^{\mathbb{I}}_{\mathbb{P}} = \left({}^{\#\mathbb{D}^{\mathbb{I}}_{\mathbb{P}}}\!/_{\#\mathbb{D}} \right)$ and

$\delta_{\mathbb{I}} = \left(\delta_{\Pi}^{\mathbb{I}} + \delta_{\mathbb{P}}^{\mathbb{I}}\right)$. Now as $\delta_{\Pi}^{\mathbb{I}} \uparrow \Rightarrow \delta_{\mathbb{P}}^{\mathbb{I}} \downarrow \Rightarrow \delta_{\Pi} \uparrow$ and similarly, $\delta_{\mathbb{P}}^{\mathbb{I}} \uparrow \Rightarrow \delta_{\Pi}^{\mathbb{I}} \downarrow \Rightarrow \delta_{\mathbb{P}} \uparrow$. The term $\delta_{\Pi}^{\mathbb{I}}$ is public sector augmenter while $\delta_{\mathbb{P}}^{\mathbb{I}}$ is private sector augmenter and $\delta_{\mathbb{I}}$ is the equilibrating factor in terms of parametric shifter in eqn. (5.9) in the negotiating and political decision-choice process in the absence of any super-majority of the participating political parties. If $\delta_{\Pi} > \left(\delta_{\mathbb{P}} + \delta_{\mathbb{I}}\right)$ implies the public sector will always win. Similarly, if $\delta_{\mathbb{P}} > \delta_{\Pi} + \delta_{\mathbb{I}}$ implies the private sector will always win. If however, $\delta_{\Pi} < \left(\delta_{\mathbb{P}} + \delta_{\mathbb{I}}\right)$ or $\delta_{\mathbb{P}} < \left(\delta_{\Pi} + \delta_{\mathbb{I}}\right)$ then the dynamics of $\delta_{\mathbb{I}}$ determines the outcome of the social decision-choice actions.

Now let us define a general social welfare function of the society that depends on the negotiating proportional values of the public and public sector advocates as:

$$\mathrm{W} = \mathrm{W}\left(\pi, \beta\right), \ \frac{d\pi}{d\beta} < 0, \tag{5.10}$$

The problem may be stated as:

$$\begin{gathered}
\underset{\pi,\beta}{\operatorname{opt}} \mathrm{W}\left(\pi_{\Pi}, \beta_{\mathbb{P}}\right), \ \frac{d\pi}{d\beta} < 0 \\
\text{s. t. } \begin{cases} \pi = \pi_{\Pi}^{r} = \pi_{\mathbb{P}}^{r} < 1 \text{ Public sector agreement constraint} \\ \beta = \beta_{\Pi}^{r} = \beta_{\mathbb{P}}^{r} < 1 \text{ Private sector agreement constraint} \\ \pi + \beta \leq 1, \text{ Institutional constraint} \end{cases}
\end{gathered} \tag{5.11}$$

The problem is simply to determine the private-public sector combination proportions that optimize the general welfare of the society in terms of provision of the social goods and services.

The structure of possible solution is presented in Figure 5.12 where three possible social indifference curves are shown for the case where there is complete agreement with the public sector advocates; the case where there is a complete agreement with private sector advocates and the case for the agreement of equal private-public sector proportions on the efficiency frontier. The optimal solution provides us with the optimal public-sector proportion π^* and the corresponding optimal private-sector proportion β^* such that $\left(\pi^* + \beta^*\right) = 1$. The optimal solution tells us about the optimal private-public sector proportions. It does not tell us what goods and services will be produced in these sectors and how the associated costs and benefits will be distributed among the members of the

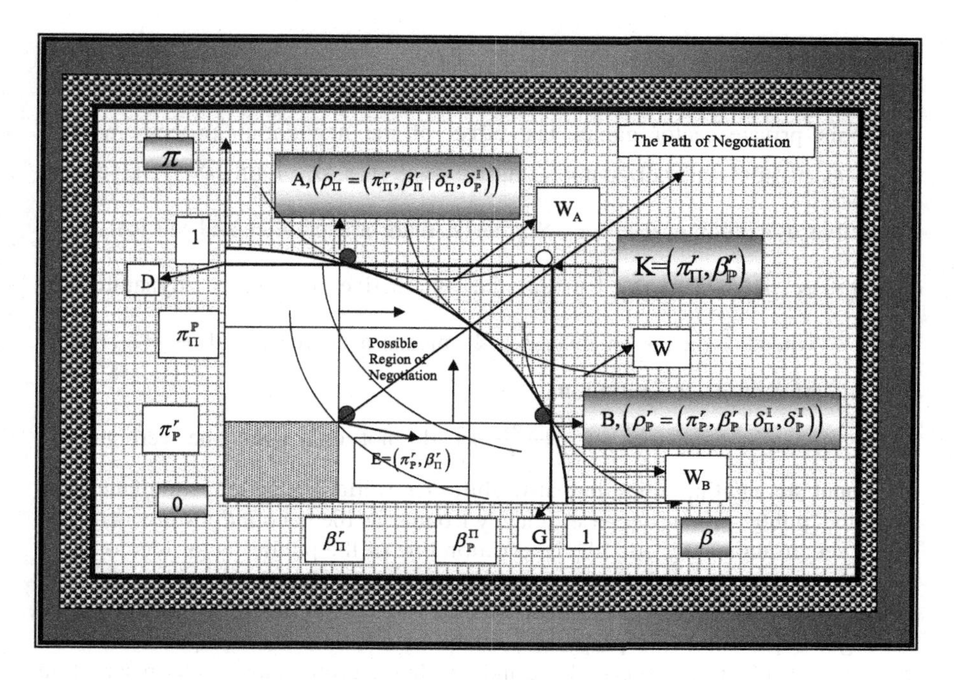

Fig. 5.12 The Negotiation Space and the path of Private-Public Sector Combinations

society. The distribution of the production of these goods and services in support of the elements in the goal-objective set is the production-decision problem and the distribution of the benefits of the production is the income-distribution problem. Both problems are simply net cost-benefit distribution problem in the political economy.

5.5.1 The Adjustment Mechanism to a Temporary Equilibrium

The reservation proportions for the public sector advocates are defined by $\left(\pi_{\Pi}^{r},\beta_{\Pi}^{r} \mid \delta_{\Pi}^{\mathbb{I}},\delta_{\mathbb{P}}^{\mathbb{I}}\right)$ and fixed by the vertical line ρ_{Π} that passes though the point A on the institutional efficiency curve. The point A is the most preferred private-public sector combination for the public-sector ideologues where $\left(\pi_{\Pi}^{r} \gg \beta_{\Pi}^{r}\right)$ in Figure 5.12 with an optimal welfare given on W_A. Similarly, the reservation proportions for the private sector advocates are defined by $\left(\pi_{\mathbb{P}}^{r},\beta_{\mathbb{P}}^{r} \mid \delta_{\mathbb{P}}^{\mathbb{I}},\delta_{\Pi}^{\mathbb{I}}\right)$ and fixed by a horizontal line $\rho_{\mathbb{P}}$ that passes though point B on the institutional efficiency curve. The point B is the most preferred private-public sector combination for the private-sector ideologues where $\left(\beta_{\mathbb{P}}^{r} \gg \pi_{\mathbb{P}}^{r}\right)$ in Figure 5.12 with an optimal welfare given on W_B. The game space for negotiation is restricted to the area defined by points ABE. The rectangular areas $DO\beta_{\Pi}^{r}A$ and $GO\pi_{\mathbb{P}}^{r}B$ are non-negotiable zones and hence outside the area of the public-private sector game. The effect of the negotiation process may be shown as parallel and horizontal shifts of $\rho_{\Pi}(\bullet)$ as $\delta_{\Pi}^{\mathbb{I}}\downarrow$ and $\delta_{\mathbb{P}}^{\mathbb{I}}\uparrow$ given the interactive dynamics of $\overset{\delta_{\Pi}^{\mathbb{I}}}{\underset{\delta_{\mathbb{P}}^{\mathbb{I}}}{\longleftrightarrow}}$. Similarly, there are parallel and vertical shifts of $\rho_{\mathbb{P}}(\bullet)$ as $\delta_{\mathbb{P}}^{\mathbb{I}}\downarrow$ and $\delta_{\Pi}^{\mathbb{I}}\uparrow$ given the interactive preference dynamics of $\overset{\delta_{\mathbb{P}}^{\mathbb{I}}}{\underset{\delta_{\Pi}^{\mathbb{I}}}{\longleftrightarrow}}$ where $\uparrow$ means an increase, $\downarrow$ means a decrease and $\overset{\bullet}{\underset{\bullet}{\longleftrightarrow}}$ means dynamic interactions in continuum.

The dynamic behaviors of $\left(\overset{\delta_{\Pi}^{\mathbb{I}}}{\underset{\delta_{\mathbb{P}}^{\mathbb{I}}}{\longleftrightarrow}}\right)$ do not only shift the vertical and horizontal negotiating lines but also shift the welfare function simultaneously along the public-private sector efficiency frontier as the negotiations proceed. The public sector augmenter sifts the reservation line of the public sector ideologue to the right as the proportion of supporters from the independence increases and similarly the public sector augmenter sifts the reservation line of the private sector ideologues upwards as its proportion of the members in the independence class increases. If all the members of the independence group support the reservation position of the public sector advocates with their votes such that $\left(\delta_{\Pi}+\delta_{\mathbb{I}}>\delta_{\mathbb{P}}\right)$, then the point A will be sustained, since $\rho_{\mathbb{P}}(\bullet)$ will be vertically shifted to coincide with K-line and intersect with $\rho_{\Pi}(\bullet)$ at point A. Similarly, if all the members of the independence support the reservation position of the private sector advocates with their votes such that $\left(\delta_{\mathbb{P}}+\delta_{\mathbb{I}}>\delta_{\Pi}\right)$ then the point B will be sustained since $\rho_{\Pi}(\bullet)$ will be horizontally shifted to cross with $\rho_{\mathbb{P}}(\bullet)$ and coincide with K-line . The shifting of the reservation lines may be seen and translated into the changing social net cost-benefit constraint that defines the social budget constraint on the private-public sector decision-choice actions which then affect the private-public sector institutional configuration.

The decision dynamics of the shifting of the reservation lines and positions are through successful solicitation of votes from the independence class of the decision-making core while maintaining the members of the ideologues. At the point A, the proportion of the voting majority is $\delta_{\Pi}+\delta_{\Pi}^{\mathbb{I}}$ with $\delta_{\mathbb{P}}^{\mathbb{I}}=0$ where $\delta_{\Pi}+\delta_{\Pi}^{\mathbb{I}}>\delta_{\mathbb{P}}$ and $\left(\delta_{\mathbb{P}}\big/\left(\delta_{\Pi}+\delta_{\Pi}^{\mathbb{I}}\right)\right)<1$. Similarly., at the point B, the proportion of the voting majority is $\delta_{\mathbb{P}}+\delta_{\mathbb{P}}^{\mathbb{I}}$ with $\delta_{\Pi}^{\mathbb{I}}=0$ where

$\delta_{\mathbb{P}} + \delta_{\mathbb{P}}^{\mathbb{I}} > \delta_{\Pi}$ and $\left(\delta_{\Pi} / \delta_{\mathbb{P}} + \delta_{\mathbb{P}}^{\mathbb{I}} \right) < 1$. As the majority is formed in accordance with the individual preferences induced by cost-benefit assessments, the indifference map reflecting the social preferences rotates vertically to the left or horizontally to the right. The point of equal proportion is the point where $\left(\pi_{\Pi}^{\mathbb{P}} / \beta_{\mathbb{P}}^{\Pi} \right) = 1$ along the $45^0 - \text{line}$. To the left of the $45^0 - \text{line}$ collective preferences are in favor of greater public-sector size while the collective preferences are in favor with greater size of the private sector to the right of $45^0 - \text{line}$. If the negotiation leads to the settlement on the institutional efficiency frontier then the private-public sector combination is optimized in relation to the social preferences of the decision-making core as the members estimate the social preferences of the voting public at a particular political decision period and the corresponding generation. The case of inefficient public-private sector combination is where the negotiations are stalled and the system finds itself in the triangle ABE, A dictatorial situation is present if the set $\mathbb{D}_{\Pi}$ or $\mathbb{D}_{\mathbb{P}}$ constitutes an effective majority or supermajority in the decision-making core. In this case, the game is uninteresting since there is always one dominating outcome of either $\left(\pi_{\mathbb{P}}^{r}, \beta_{\mathbb{P}}^{r} \mid \right)$ or $\left(\pi_{\Pi}^{r}, \beta_{\Pi}^{r} \right)$ depending on where the supermajority falls. The case where filibuster rule is present leads to the presence of possible decision obstructionism in the favor of the minority who comes to hijack the system into non-decision space.

We shall assume in this negotiation game that neither $\mathbb{D}_{\Pi}$ or $\mathbb{D}_{\mathbb{P}}$ constitutes an effective decision-choice majority and hence the members must work for votes from $\mathbb{D}_3 = \mathbb{D}^{\mathbb{I}} = \left(\mathbb{D}_1 \cap \mathbb{D}_2 \right)$. Furthermore we shall assume that $\delta_{\Pi}^{\mathbb{I}} \neq 0$ and $\delta_{\mathbb{P}}^{\mathbb{I}} \neq 0$ in the voting dynamics. The adjustment process is such that the members in $\mathbb{D}^{\mathbb{I}}$ must be convinced to support either $\mathbb{D}_{\Pi}$ or $\mathbb{D}_{\mathbb{P}}$ in other to establish a majority. From the concept of vote solicitation, two adjustment functions for dynamic changes may be established by eqns. (5.12) and (5.13).

$$\rho_{\Pi} = \pi_{\Pi}^{r} + \alpha_{\Pi} \left(\pi_{\mathbb{P}}^{r} - \pi_{\Pi}^{r} \right), \ \pi_{\Pi}^{r} > \pi_{\mathbb{P}}^{r} \text{ and } \alpha_{\Pi} = \alpha_{\Pi} \left(\delta_{\Pi}^{\mathbb{I}}, \delta_{\mathbb{P}}^{\mathbb{I}} \right) \in (0,1) \quad (5.12)$$

$$\rho_{\mathbb{P}} = \beta_{\mathbb{P}}^{r} + \alpha_{\mathbb{P}} \left(\beta_{\Pi}^{r} - \beta_{\mathbb{P}}^{r} \right), \ \beta_{\mathbb{P}}^{r} > \beta_{\Pi}^{r} \text{ and } \alpha_{\mathbb{P}} = \alpha_{\mathbb{P}} \left(\delta_{\Pi}^{\mathbb{I}}, \delta_{\mathbb{P}}^{\mathbb{I}} \right) \in (0,1) \quad (5.13)$$

The terms $\alpha_{\mathbb{P}} = \alpha_{\mathbb{P}} \left(\delta_{\Pi}^{\mathbb{I}}, \delta_{\mathbb{P}}^{\mathbb{I}} \right)$ and $\alpha_{\Pi} = \alpha_{\Pi} \left(\delta_{\Pi}^{\mathbb{I}}, \delta_{\mathbb{P}}^{\mathbb{I}} \right)$ are adjustment coefficients for the advocators of private and public sectors respectively. They depend on the relative behavior of $\delta_{\Pi}^{\mathbb{I}}$ and $\delta_{\mathbb{P}}^{\mathbb{I}}$ with the structure that as $\delta_{\Pi}^{\mathbb{I}} \uparrow\downarrow \rightleftarrows \delta_{\mathbb{P}}^{\mathbb{I}} \downarrow\uparrow$ respectively and $\alpha_{\mathbb{P}}$ rises or falls relative to α_{Π} to alter the speed of adjustments.

The system's adjustment dynamics will continue until equilibrium is established by the majority votes of the decision-making core. The collection of all points in the area ABE is the *negotiable set*, $\mathcal{N}$, of private-public sector combinations. The set $\mathcal{N}$ contains all the possible equilibrium points where such equilibrium points are temporary with continual changes of the institutional combinations that depend on the changing social preferences of the general population if such changing preferences are reflected in the preferences of the majority of the members of the decision making core. Some of the temporary equilibrium points are temporarily sustained within the inefficient region. Some may be sustained on the efficiency frontier as shown in Figure (5.12).

At the point C, the negotiated temporary equilibrium is inefficient but in favor of greater public sector proportion. The social welfare depends on the negotiated position that presents socially preferred private-public sector institutional combination in the political economy. The position that the system will settle through the democratic decision-making process will depend on the role that the general population (the members of the non-decision making core) plays in influencing decision-choice actions of the decision-making core. If the temporary equilibrium falls on the right of the $45^0 - \text{line}$ then we conclude that the temporary social preferences are in favor of greater private-sector proportion by democratic decision-choice process and vice versa when the temporary equilibrium is on the left of the $45^0 - \text{line}$ as presented in Figure 5.12.

The conditions of the dynamics of the negotiated equilibrium must be linked with the formation of the social goal-objective set and its maintenance through a continual process of negotiation in the political economy. Every negotiated equilibrium is a temporary one waiting to be reconfirmed or for a change. The objective of negotiation is to reduce conflict in the collective decision-choice space through trade-offs of the cost-benefit characteristics as assessed by the opponents regarding elements in the subset that define the zones of possible agreements and disagreements in terms of governance, goal-objective decision and budgetary decisions given the social vision and national interest. The dynamics of the negotiation on private-public-sector combination is a search for optimal mix of private-public sector proportions in terms of institution building and provision of social goods and services and cost-benefit distribution. The analytical structure as presented allows us to explain the concepts of nationalization and privatization in the institutional arrangements in the political economy. Ideally, decision to nationalize and decision to privatize should be based on social cost-benefit rationality but in practice the social cost-benefit rationality, ideological positions and individual cost-benefit rationality, seen in terms of rent-seeking, interact to shape qualitatively and quantitatively the institutional arrangements of private and public sectors.

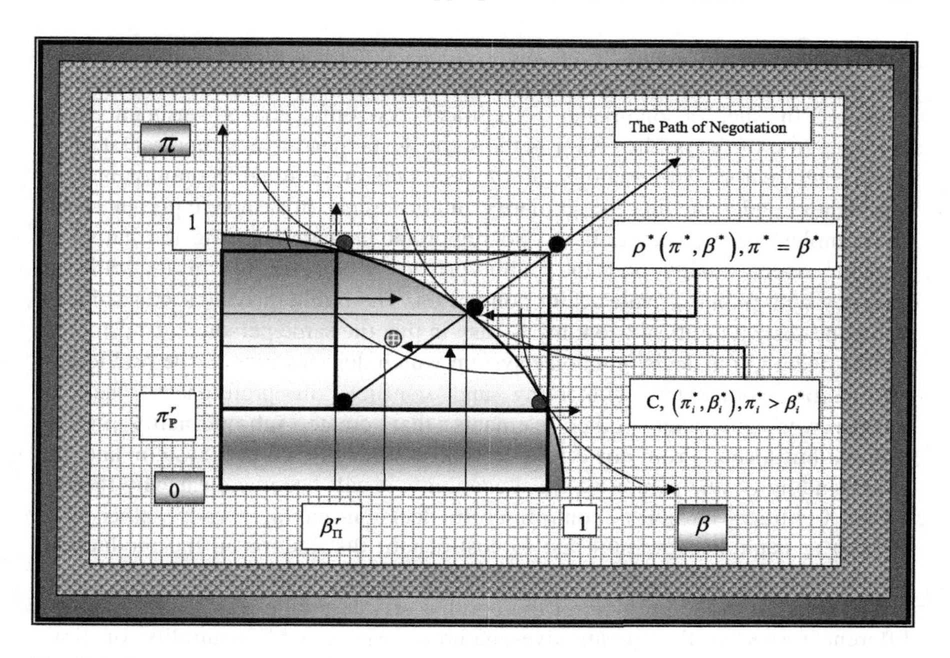

Fig. 5.13 The Negotiation Space and the Path of Private-Public Sector Combination Under Social Welfare Indifference Map

5.6 A Reflection on the Debate on the Appropriate Size of a Government

The debate involves the answer to the question of what is the appropriate size of any government for any social formation and for any given generation. In most cases, reasoning on the basis of ideological principles, rather than scientific rationality, tends to shape the accepted answers and positions of the acceptable size of the government and public sector. The size of the government is equated to the size of the public sector. The arguments center on big and small government without explicit measure of what constitutes big and small governments. The concepts of big and small are qualitative variables whose values may be represented by linguistic numbers. Such linguistic quantities are defined in a fuzzy space and may be viewed in terms of small-big duality with a fuzzy continuum. The effect of social cost-benefit rationality is completely neglected with little regard to the synergetic nature of social systems in terms of inter-supportive mode of the two sectors. In the social organism there is no private sector sitting in some place separate from the public sector. As it has been argued previously that both the private sector and the public sector live in unity and in synergic relation in that every public-sector proportion must have a private-sector support and vice versa if the political economy is to be organize and managed.

The condition for complete private sector or complete public sector is a one-person political economy where the person is both the government and the private in unity. In such a one-person political economy, the government as a representation of the public sector is the same as the person as a representation of the private sector. The analogy of the one-person political economy is very important in understanding the similarity and difference of complete private sector and complete public sector. It is also important in understanding the nature of principal agent duality and the public-private sector duality in the political economy with public-private sector continuum. The public-private sector solution in the one-person political economy is easy in that the same person is the principal and the agent in the decision-choice system which does away with the problem of information asymmetry, competing and conflicts in preferences, lack of transparency and many other elements that create sub-optimality in the information space. From the viewpoint of democratic choice system and the understanding of one-person political economy, it will become clear why there is always a political tension in controlling the decision-making core by either labor (public) or capitalist (private) due to cost-benefit distribution in the production-consumption space as it relates to private-public-sector optimal mix.

The solution to the problem of the optimal private-public-sector mix may take different forms in the quantitative-qualitative space with neutrality of time. Generally, the scientific solution to the problem of the optimal mix of private-public sectors to define an efficient configuration is traded in favor of ideological solution on the basis of simple principles such as high taxes without reference to provision of service or social welfare position of the society where such welfare is specified in terms of provision of services in the three structures of politics, law and economics. The pursuant of this ideological solution is complemented with dubious and unscientific claims that the government is inefficient in institutional construct and management. So far, however, the numerous studies on efficiency of private-public management have no settled position of better managed sector. The current events of 2007 in the global financial sector and other manufacturing sectors of united States point to the private sector inefficient management and synergetic relationships between private and public sectors that lead to financial bailouts of some private sector institutions as well as government financial stimulation of the private real sector. The whole notion of public sector subsidies and tax-breaks to the private sector is completely neglected in the debate.

5.6.1 Fuzzy-Theoretic Reflections on the Optimal Public-Private Sector Mix and the Conditions of Measurement

The debate on the size of governments may be viewed in terms of goods and services that may be provided in the society for creating the path of national welfare measured in some meaningful sense. The government size is seen as the size of the public sector. Let us conceive the political economy in terms of social provision of goods and services in all structures of economy, politics and law.

Definition 5.4: The Political Economy

A political economy $\mathfrak{S}$ is composed of triplet of economic structure, $\mathfrak{E}$, political structure $\mathfrak{P}$ and legal structure $\mathfrak{L}$ and hence we may specify it as:

$$\mathfrak{S} = \mathfrak{E} \otimes \mathfrak{P} \otimes \mathfrak{L} = \left\{ \mathfrak{s} = (\mathfrak{e}, \mathfrak{p}, \mathfrak{l}) \mid \mathfrak{s} \in \mathfrak{S}, \mathfrak{e} \in \mathfrak{E}, \mathfrak{p} \in \mathfrak{P}, \mathfrak{l} \in \mathfrak{L} \right\}$$

The terms $\mathfrak{e}$, $\mathfrak{p}$ and $\mathfrak{l}$ are economic, political and legal configurations respectively that a society may hold at any moment of time. Each of these configurations involves a composite structure of institutions and corresponding goods and services that may be produced in the political, legal and economic structures. The term $\mathfrak{S} = \mathfrak{E} \otimes \mathfrak{P} \otimes \mathfrak{L}$ defines possible institutional arrangements in the organizational structure of a society and $\mathfrak{s} = (\mathfrak{e}, \mathfrak{p}, \mathfrak{l}) \in \mathfrak{S}$ is one of them. The nature of $\mathfrak{s} = (\mathfrak{e}, \mathfrak{p}, \mathfrak{l}) \in \mathfrak{S}$ is ideologically and culturally determined and hence varies from a society to society. Such variations determine the social and individual preferences that affect the decision-choice actions on public-private sector combination and any decision time with regard to distributional structure of provision of goods and services and for whom. They also determine the individual-community freedom relations in terms of collective freedom and individual freedom. The morphology of the political economy is always affected by the nature of the national interest and social vision while the production-consumption configuration is affected by the social-goal-objective set created to support the national interest and social vision.

Given the notion of provision of goods and services, broadly defined, in the society, such provision may be completely undertaken collectively through the government as the public sector in what has been branded as communism or perfect socialist state. Alternatively, the provision may be completely undertaken by the individuals or the group of individuals through the private sector in terms of business organizations in what is called perfect capitalism. There is a third arrangement of the provision that may be undertaken jointly by the public and private sectors in a creative combination in what is called a mixed economy. The provision of goods and services through the pure private or pure public sectors has one important thing in common and that is, there is no taxation and hence the discussion on the size of the government, in terms of taxation and expenditure, is mute and decision-choice irrelevant. Similarly, the discussion on public expenditure and taxation, and collective decision-choice actions, the size of the private sector is also mute. In both cases the principal and the agent are one and the same thing and hence asymmetry in information and decision-choice actions does not arise.

When goods and services are jointly provided by the government and the private sectors then a number of questions tend to arise. One question deals with the decision-choice problem of what goods and services must be supply by the government and what goods and services must be supply by the private sector, given the national resource endowment and its limitations. The second question involves the problem of how are these goods and services going to be provided and in what relative size should the private and public structures assume. The third question involves the problem of whether the government representing the public sector should be allowed to own productive capital and go into production in the same way as the private sector If the answer is no, then a fourth question arises. How should the government's provision of its share of goods and services to the society be financed and should it be through taxation or through user fees? The next question is: should the government accumulate surplus value in the same way that the private sector accumulates surplus value in terms of profits. All these questions and more involve solving a number of decision-choice problems in any political economy of a nation. The answers provided through the problem-solving process help to define the evolving institutions of private and public sectors as well as their relative sizes and hence the sizes of governments. At this juncture, it may be pointed out that there is no one relative size of the government that fits all nations and their political economies. Even for the same nation this relative size of government to private sectors is not one size over time and the same size for different generations. It evolves under different conditions of ideological shifts, cultural dynamics, social preferences, global conditions, resource constraints and the set of socio-economic problems that a nation faces at any time point. The small-big duality of governments exists in policy continuum whose points of equilibrium are temporary and in forward-back motions that must be related to the social provision of goods and services to satisfy needs and wants of the members of the society. In this respect, the linguistic quantity of big-small duality is contextual and must be interpreted in social and cultural dynamics.

By distinguishing the private and government sectors in terms of provision of goods and services to the society, we may see the size of the government to be measured in terms of value of goods and services provided relative to the total value of provision of goods and services for the society. We may assume that these values are directly or indirectly measurable. Alternatively, the size of the government may be seen in terms of the proportion of the total value of the government expenditure $\mathbb{G}$ in the gross national income $\mathbb{Q}$ where such expenditure is the sum of government tax revenue, other fees and deficit spending. The tax burden in this respect should be seen as a relative value to the value of the government's provision of goods and services. The provision of goods and services by the government is related to the government expenditures while the value of the provision of goods and services by the private sector $\mathbb{P}$ is related to private sector expenditures. In this way, the total value of the system's provision of goods and services is the sum of the values of provisions by the government and private sectors such that $\mathbb{Q} = (\mathbb{G} + \mathbb{P})$ which implies that $\left(\mathbb{G}/\mathbb{Q}\right) + \left(\mathbb{P}/\mathbb{Q}\right) = 1$ as we have previously discussed. The measures of $\mathbb{G}$ and $\mathbb{P}$ at any time require us to define economic configuration and the form it might take. It is noted, here , that services in all institutions of law , law enforcement, including religious, philanthropy and others are included in the general economic production.

Definition 5.5: The Economic Structure

The economic structure, $\mathfrak{E}$, of the political economy, $\mathfrak{S}$, is composed of configuration of Cartesian product of a set of decision-choice actions, $\mathfrak{D}$ and a set of commodity elements $\mathfrak{C}$ in the form:

$$\mathfrak{E} = \mathfrak{D} \otimes \mathfrak{C} = \{\mathfrak{e} = (\delta, \mathfrak{c}) \mid \delta \in \mathfrak{D} \text{ and } \mathfrak{c} \in \mathfrak{C}\}$$

The set $\mathfrak{D}$ includes production, income-distribution and consumption decisions. The element $\mathfrak{e} = (\delta, \mathfrak{c})$ is a decision-choice configuration in the economic space. At any time point, if $\aleph$ is an index set of configurations, then the economic structure may be seen as composing of a non-unique set of configurations in the form:

$$\mathfrak{E} = \{\mathfrak{e}_1, \mathfrak{e}_2, \mathfrak{e}_3 \cdots \mathfrak{e}_\ell \cdots \mid \ell \in \aleph \text{ and } (\#\aleph) \in (0, \infty]\} \tag{5.14}$$

The debate on the size of government representing the public sector involves being able to create a crisp partition of the economic structure into goods and services whose provision falls into the domain of the government acting on the behave of the public sector on one hand, and goods and services whose provision is the responsibility of the private sector on the other hand. The facts on the ground and the history of human experiences point to the notion that such crisp partition is impossible since the social system is defined by public-private sector duality that satisfies the principle of fuzzy continuum. What we have is a fuzzy partition through decision-choice actions that reconcile the conflicts in individual preferences in the collective decision-choice space with the collective preferences acting as collective constraint. The fuzzy partitioning allows a creation of a fuzzy economic structure, $\tilde{\mathfrak{E}}$ with a membership characteristic function, $\mu_{\tilde{\mathfrak{E}}}(\mathfrak{e}) \in [0,1]$ that indicates an increasing degree of preference for private sector provision in the form:

$$\tilde{\mathfrak{E}} = \{(\mathfrak{e}_\ell, \mu_{\tilde{\mathfrak{E}}}(\mathfrak{e}_\ell)) \mid \mathfrak{e}_\ell \in \mathfrak{E}, \ell \in \aleph \text{ and } (\#\aleph) \in (0, \infty]\} \tag{5.15}$$

With the fuzzy partition, we decompose the fuzzy economic structure into three crisp economic structures through a fuzzy decision-choice action where $\mathfrak{E}_1$ is the substructure of the private sector's provision, $\mathfrak{E}_2$ is the substructure of joint private-public sector's provision and $\mathfrak{E}_3$ is the substructure of public sector's provision of goods and services. This is done through the application of the fuzzy decomposition theorem where $\alpha_1 > \alpha_2$ [R.1.23]. The partition may be written as in equations (5.16) – (5.18)

$$\mathfrak{E}_1 = \left\{ \left(\mathfrak{e}_\ell, \mu_{\tilde{\mathfrak{E}}}(\mathfrak{e}_\ell) \right) \mid \mathfrak{e}_\ell \in \mathfrak{E},\ \ell \in \aleph \text{ and } (\#\aleph) \in (0,\infty], \mu_{\tilde{\mathfrak{E}}}(\mathfrak{e}_\ell) > \alpha_1 \right\} \quad (5.16)$$

$$\mathfrak{E}_2 = \left\{ \left(\mathfrak{e}_\ell, \mu_{\tilde{\mathfrak{E}}}(\mathfrak{e}_\ell) \right) \mid \mathfrak{e}_\ell \in \mathfrak{E},\ \ell \in \aleph \text{ and } (\#\aleph) \in (0,\infty], \alpha_2 \le \mu_{\tilde{\mathfrak{E}}}(\mathfrak{e}_\ell) \le \alpha_1,\ \alpha_1 > \alpha_2 \right\} \quad (5.17)$$

$$\mathfrak{E}_3 = \left\{ \left(\mathfrak{e}_\ell, \mu_{\tilde{\mathfrak{E}}}(\mathfrak{e}_\ell) \right) \mid \mathfrak{e}_\ell \in \mathfrak{E},\ \ell \in \aleph \text{ and } (\#\aleph) \in (0,\infty], \mu_{\tilde{\mathfrak{E}}}(\mathfrak{e}_\ell) < \alpha_2 \right\} \quad (5.18)$$

The individual preferences regarding private-public sector provisions are shaped by ideological position around some important elements of the nature of income distribution (inequality, equality, poverty and others), ownership of means of production (private or public), size of the government (spending, cost of governance, taxes, interference of private enterprise, power of the nation, military expenditures , welfare conditions, (employment, unemployment, joblessness, stable prices, law and order, domestic stability, health, education, justice and others). All these are related to the individual and community and the shared platform of freedom, responsibility and fairness, and the role that governments play to ensure freedom, responsibility, justice and fairness of all members regarding the cost-benefit structure of the political economy. It is here that the concepts of individual-community duality, individual-community continuum and individual-community unity find their analytical strength in the organizational structure of the political economy involving the people, their institutions and their government.

The size of the government that may rule at any time for any given generation may be seen as a political game to established temporary relative size. A permanent relative size of the government is impossible and inconsistent with the changing culture and political dynamics. Thus there will always be conflicts in the preference space in the public-private sector duality. The dynamics within the private-public-sector duality in the continuum is through the activities in the political sector in terms of political games. Let us look at the structure of the game in a political duopoly like that of United States with Republican ($\mathscr{R}$)and Democratic($\mathscr{D}$) parties in terms of big government, $\mathbb{G}_B$ and small government $\mathbb{G}_S$ that define the state of nature for the negotiated size through a democratic decision-choice action. Since the size of the government is measured in terms of the value of the provision of goods and services, the negotiation decisions are centered on the budget size and distribution. The game of small-big government and the prevailing size may be seen in terms of prisoner's dilemma involving deception and trust by the two parties. The structure of the game in terms of negotiation is presented in Figure 5.14.

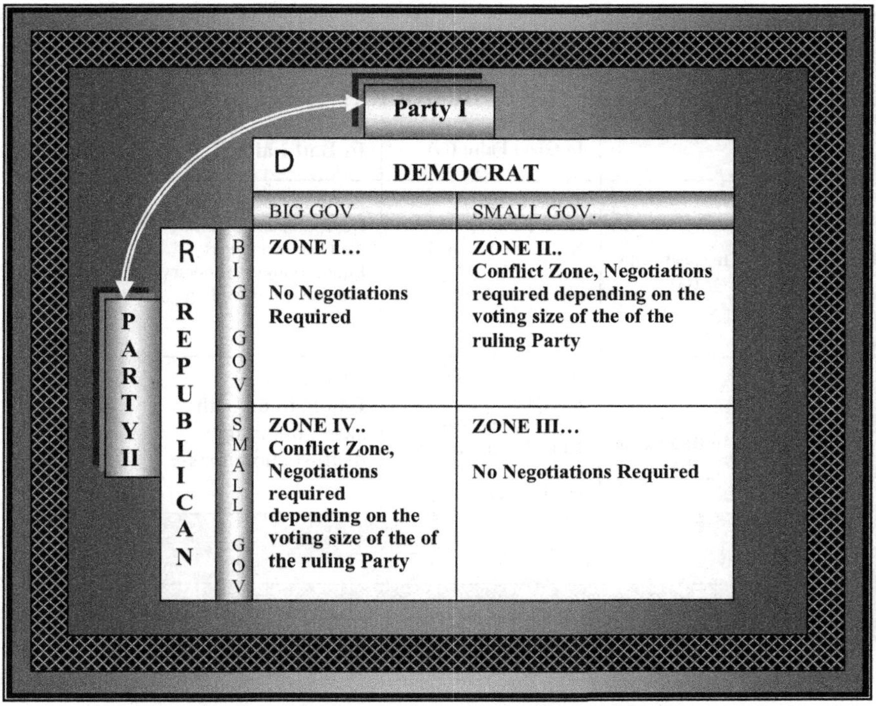

Party I

D DEMOCRAT

R REPUBLICAN		BIG GOV	SMALL GOV.
PARTY II	BIG GOV	**ZONE I...** **No Negotiations Required**	**ZONE II.. Conflict Zone, Negotiations required depending on the voting size of the of the ruling Party**
	SMALL GOV	**ZONE IV.. Conflict Zone, Negotiations required depending on the voting size of the of the ruling Party**	**ZONE III...** **No Negotiations Required**

Fig. 5.14 Small-Big Government Game and Negotiation Zones

The possible outcomes in the continuum of big-small government duality are shown in Figure 5.14 as zones of no negotiations and possible negotiations on the basis of implicit cost-benefit values in relation to the welfares of the nation and party. In Zones I and III, both parties agree on either big or small government size as measured in terms of the value of provision of goods and services and hence determine the position of the public-private sector efficiency frontiers in Figure 5.14. In other words they agree on the budget size in support of the government size and public sector. They may, however, disagree on the budget distribution over programs which may require negotiations on the budget allocations. In Zones II and IV both parties disagree on the size of the government. In zone II the Democratic Party prefers small government; the Republican Party prefers big government. The preferences are reversed in the Zone IV where the Democratic Party prefers big government and the Republican Party prefers small government. The democratic collective decision-choice process becomes more complicated by two sequential decision-choice modes for determining the size of the government and the distribution of budget. The negotiation process involves the use of information to convince some opposing members to change their position and accept your social preference ordering. It is here that the creation and manipulation of deceptive information structure enter into the politico-economic space to affect preference ordering. The deceptive information structure is seen in terms of good-bad faith duality in the negotiations.

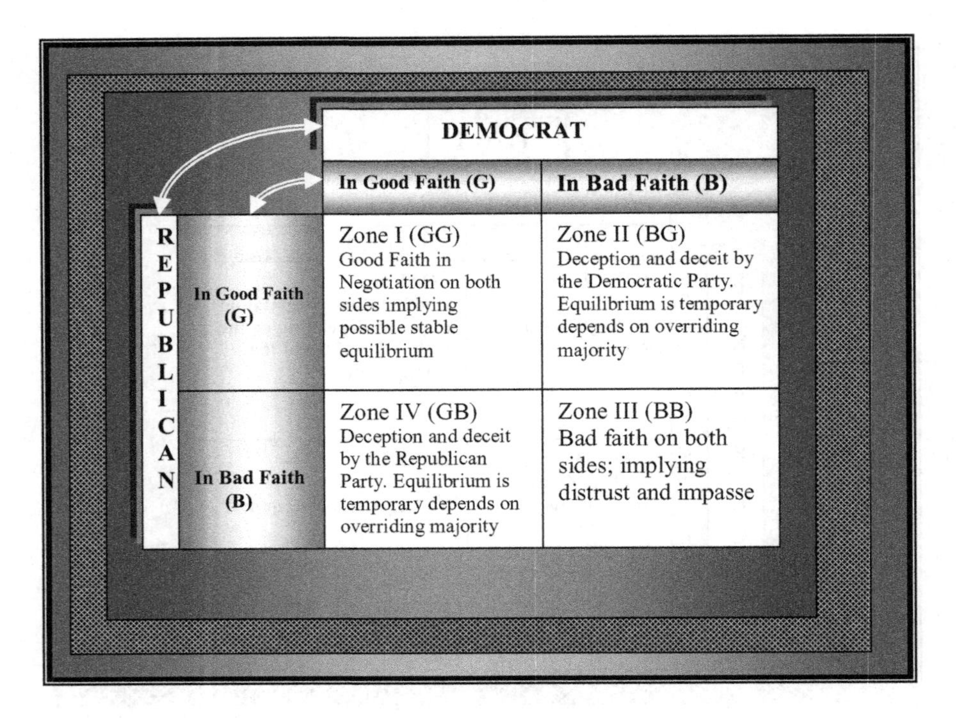

Fig. 5.15 Good-Bad Faith Game in Negotiations

In this respect all the negotiation processes are then mapped into the trust space of good-bad duality of faith with a fuzzy continuum where the size of the fuzzy residual is defined in terms of good-faith residual and determined by the relative negotiation strength of the parties. The measure of this fuzzy residual may be conceptualized from the view point of fuzzy risk in addition to stochastic risk [R15.16]. The temporally possible decision-choice equilibria in Zones II and IV of Figure 5.15 will depend on the relative size of the majority party and the size of the socioeconomic power which the parties have. Such equilibria are then mapped onto Zone II and IV of the accountability game in Figure 5.16. Zone I of Figure 5.15 is mapped onto Zone I of Figure 5.16 of the accountability game. Zone III is unstable and cannot be maintain since it leads to the collapse of governance with an increasing social cost that reduces the net benefit of governance. This zone is an extreme partisan sociopolitical economy that creates popular distrust in the overall government leading to a blame game. Zone III of Figure 5.15 is mapped onto Zone III of Figure 5.16 of party-nation accountability game.

It is in the good-bad faith game in negotiation of Figure 5.15 that problems of accountability and corruption tend to arise and their respective zones become mapped onto the corresponding zones of accountability of the accountability game. The Zones of the accountability game are mapped onto the respective zones of corruption game. The zones in which the party members will fall will depend on their assessments of the corresponding cost-benefit configurations and the socioeconomic incentive structure built into the political economy. The zone in which the party will fall in the small-big government game will depend on the party ideology and platform that will shape the cognitive computations of relative importance of cost-benefit attributes and socioeconomic incentive elements in the political economy. The cost-benefit configurations and the incentive structures may be composed of measurable and non-measurable attributes that will directly or indirectly influence the direction of preferences leading to a particular game zone that the party will end. In this party game, the individual in the decision-making core is assumed to follow the party ideology and platform as we have discussed. It may be pointed out that the analytical structure of the political economy, the explanation of its behavior and the prescriptive process for its change must be seen in quantity-quality-time space. This quantity-quality-time space may be partitioned into a) quantity-quality space, b) quantity-time space with quantitative equation of motion, and c) quality-time space with qualitative motion. These sub-spaces may be accounted for if we are to understand social dynamics.

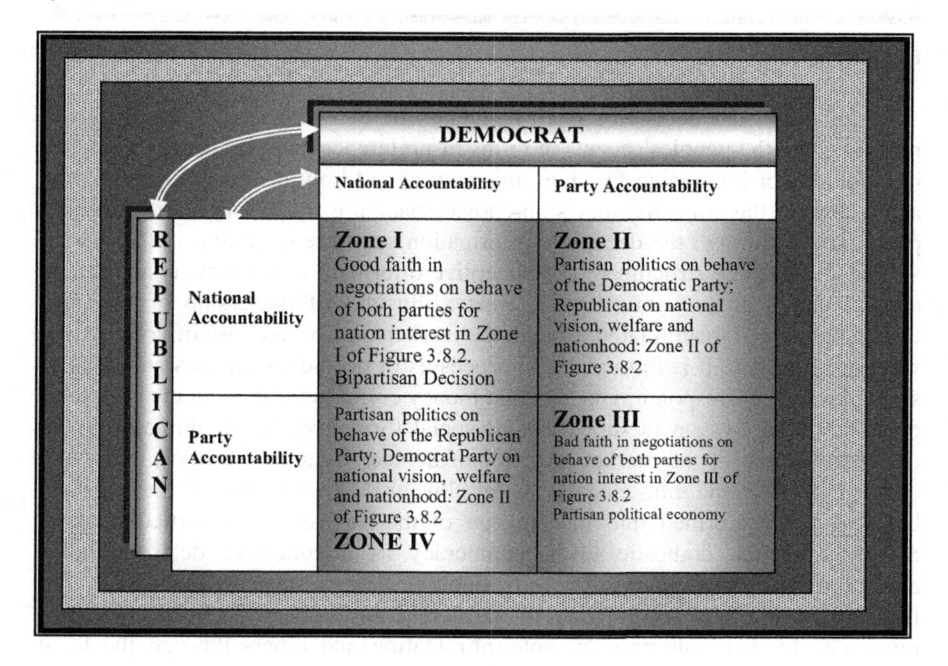

Fig. 5.16 The Party-Nation Accountability Game

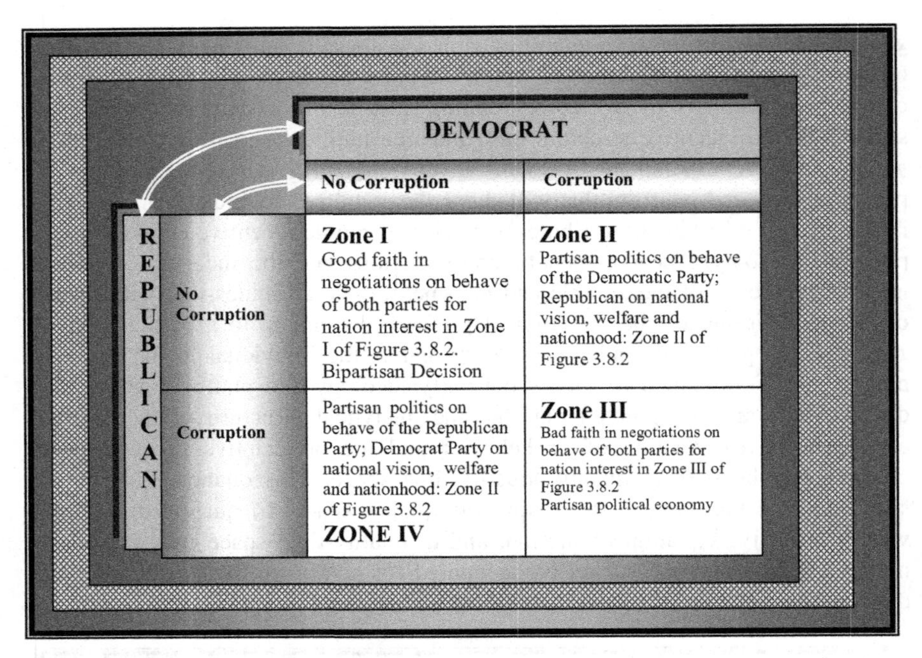

Fig. 5.17 The Party-Nation Corruption Game

In all these, the decision-choice behaviors of the members of the decision-making core (the peoples' agent) in terms of preferences will define the nature of individual decision-choice freedom and power, and how the individual freedom and power in the collective decision-choice space are transformed into phantom phenomenon through the deceptive information structure operating in the market system in the three structures. The deceptive information structure, composed of disinformation and misinformation, creates illusions of ownership of decision-choice power and manipulated through the markets for the exercise of vote-casting without substance. It is through these processes of political strategies and counter strategies with corresponding tactics that money corrupts the democratic collective decision-choice system in support of its social formation. At the domestic level, the phenomenon of the market mockery of democracy arises. In most cases, the concept of free and fair election is an illusion. The concept of the peoples' business is transformed into the concept of the business of a particular class. Similarly, at the domestic level, democracy as a collective decision-choice algorithm is simply transformed into ideological instrument of class war. The illusion of individual sovereignty in the democratic collective-choice system is reinforced by vote suppression, vote-intimidation and others through the legal structure or indirect force. At the global level, the democracy is transformed into an instrument of intimidation for justifying terror and territorial invasions.

5.7 Epistemic Reflections on the Private-Public Sector Duality and the Demecratic Decision-Choice Process

The discussions on collective decision-choice process that we have carried on in this political economic framework are to advance the principle that all individual and collective decisions are made in reference to the private-public duality. Social goal-objective set is established by reconciling individual and community freedoms in relation to a number of basic distributions of power, resource, justice and income as seen from the conflicts of the individual-community cost-benefit balances. Here, the community is broadly defined to cover society, social and collective entities whose decision-choice actions produce cost-benefit balances for all in the economic space. The concept of individual absolute freedom in any political economy or social thought is illusion and chaotic. Similarly the concept of absolute collective freedom in any political economy is an illusion and unattainable. These **two freedoms exist in social tension** and antagonistic relation in terms of opposites in a continuum. Between the two absoluteness of individual and the collective freedoms lies **a set of** sustainable social formations that reconciles the two antagonistic freedoms in a dynamic socioeconomic equilibrium in a continuum. This socioeconomic equilibrium is always temporary and constantly reversing itself in relation to the culture and generation of the political economy composed of political, legal and economic structures with connecting institutions.

An important question arises concerning the integrity and stability of the political economy. What should be the balance between individual and the collective freedoms in relation to power, justice fairness and resources in the community in relation to cost-benefit balances in the national decision-choice space? Power, its distribution and allocation relate to the political structure, justice and its distribution and allocation relate to the legal structure; and resource, its distribution and allocation relate to the economic structure where the three structure constitute a trinity whose unity is established by the social institutions for fairness. The search for points of equilibrium and balance is the core of political decisions and practices that generate antagonistic social relations which then affect the distribution of cost-benefit balances that motivate constantly evolving decision-choice actions in the legal and economic structures. Freedom as used in this analytical work relates to decision-choice power and the ability to exercise it over the national choice space. Notice that through the political structure the society collectively establishes the nature and the structure of sovereignty of the individual and collective decision-choice system in terms of power distribution in the individual-community duality. The exercise of the power distribution may be constrained by the code of conduct of the legal structure as well as made inoperative by the lack of resource capacity of an individual in the economic structure.

The essential social conflict is expressed in a question as to which sector of the public or private \ can bring about the fulfillment of the elements in the goal-objective set in support of the national interest and social aspirations as well as maintain individual welfare perceptions of social states. It is this social conflict in the spaces of political, legal and the economic structures that one can appreciate the analytical contributions of Marx and Schumpeter. The Marxian analytical process leads to the establishment of workers' socialism with worker's government where the means of production is under public ownership; and the control of all the political, legal and economic structures is used to serve the public and the people, at least in theory. In the Marxian political economy, the community rights and freedom are prior to those of the individuals within the individual-community duality. Both profits and costs or the cost and benefit configurations are socialized [R4.71] [R4.73]. The Schumpeterian analytical process leads to the establishment of capitalists' socialism with the establishment of capitalist government where the means of production is under private ownership; and the control of all the political, legal and economic structures is used to serve the capitalist class at least in theory. In the Schumpeterian political economy, the private rights and freedom are prior to those of the community within the individual-community duality. Here, profits or benefit configurations are privatized while the cost configurations are socialized [R4.94] [R4.95].

The social formation process is such that the public-private-sector duality and its internal conflicts are transformed into worker-capitalist duality and its internal conflicts. The process of resolving these conflicts becomes never-ending game process that generates a continual energy for socio-economic transformations. Thus in Marxian political economy the public and the workers form one social polarity where the social decision-making power is vested in the workers (labor). In this situation, the social collectivity constitutes the principal and the agent as in one person political economy where every person works for the society to generate surplus value for continual social improvement. In the Schumpeterian political economy the private and capitalists form the other polarity where the social decision-making power is vested in the capitalists. In this situation, the capitalists as a class constitutes the agent while the workers constitute the principal in the political where the workers work for the capitalist to generate a surplus value for the capitalists where the continual social improvement is at the mercy of the capitalist. The conditions of the decision-choice system for the control of the three structures are reflected under the conditions of the control of the political structure in which the social decision-making power is vested. Thus by capturing the political structure one comes to control the destiny and the welfare of the individuals and the nation by controlling the nature of the power distributions in the three structures. The destiny and welfare control may be tilt in favor of the capital-owing class or working class depending on the dominant ideology held by the members of the social decision-making core in relation to the mode of the individual conflicting preferences in the collective decision-choice space.

The individual preferences relate to the cost-benefit balances which in turn relate to the social goal-objective set in the democratic decision-choice system. The elements in the goal-objective set are established by preferences over elements between complete private freedom and complete collective freedom over the political and economic spaces. By the democratic decision-choice process, as established by a simple majority (non-weighted majority) rule, the greater the size of the private sector (measured in some meaningful sense), the greater is the intensity of the majority preference toward individual freedom and the smaller is the intensity to collective freedom as seen in terms of the relative size of the private to public sectors. These statements hold under a particular information regime that excludes deceptive information structure composed of disinformation that corrupts truths and misinformation that incorporates propaganda and lies to affect the preference ordering.

These decided true outcomes of the voting preferences are not realizable if the rule of the non-weighted majoritarian democracy is not practiced or allowed to take effect. This might be the case where the collective preference of the decision-making core as the agent is incompatible with the majoritarian rule, where the masses constitutes the principal, and where the practices of the decision-choice actions violate the majoritarian constraint. The persistence of this violation in the political economy gives rise to a dictatorship not of the majority over the minority of the voting public, but rather a dictatorship of the decision-making core whose aggregate preference shapes the path of the collective and individual welfare along the transformational dynamics of the national history. It is at this decision-choice juncture that a mockery of democracy under the majoritarian principle tends to arise where the majority of the decision-making core defines the path of the national welfare and history and where the voting public becomes disfranchised from the social decision-choice actions involving the three structures of politics, law and economics. In this case, the voting public is moved to a field of zombies that just follow the dictatorship of the decision-making core and silenced by a regime of secrecy. This situation in a democratic decision-choice system is indirectly controlled with deceptive information structure and information limitations through powerful instruments of the principle of national secrete to generate a lack of transparency in the democratic choice system.

The conflicts in the private-public duality and the relativity of the public-private-sector size in the political economy are resolved purely by the members of the decision-making core in accordance with their preferences over the political, economic and legal structures. Here, the members of the general public are reduced to servants of the members of the decision making core. The consolation in this process is that the masters are willingly selected by the servants to assume power over them through the institutional role of governance. Operationally, this is done in two ways of using brute force or ideological strategies that distort the public information in the collective decision-choice space. It is here that the character of social information structure becomes important in the efficient working mechanism of the democratic decision-choice system. The strategy of information distortions relative to social decisions is the most powerful way to push the members of the voting public into sub-optimal decision-choice space and hence into a temporary or permanent submission with their acceptance. The integrity and stability of the social system, with its democratic collective decision-choice system, depend on the use of combination of ideology and enlightened force on the part of the decision-making core to maintain the stability of its decision-choice structure by the manipulation of the information and motivation structures.

The preservation and the efficient functioning of the political economy depend on the relative independence and unity of the private (individual) and the public (collective) sectors. The interdependence and unity depend on how the decision-making core manages the integrity of the socio-political economy that harbors both the individual and the collective decision-choice structure by manipulating the information and motivation structures in the decision-choice space in other to create the elements of the goal-objective set that will affect the relative size of the private-public sector through the distribution of their provision in the production-consumption space. The private and public sectors exist in a multiplicity of give-and-take relation where the public sector supplies to the private sector something that is wanting and needed for completeness of its existence. Similarly, the private sector also supplies to the public sector something that is wanting and needed for completeness of its existence. This is the inter-dependency principle that maintains the private-public sector unity. The give-and-take process proceeds through decision-choice agents that are part of both sectors and exercise differential preferences over how the sectors are to be managed to define the areas of participation of the individual and collective decisions in the political, legal and economic structures through the construct of the elements in the goal-objective set. The individual and collective preferences are always conflicting and antagonistic requiring negotiations to resolve the conflicting preferences over the goal-objective elements that will shape the path of the relative size of the private-public sector division of the socio-political economy. It must be emphasized that the whole social decision-choice space is viewed in terms of conflicts within the individual preferences and between the individual and collective freedoms in the political, legal and economic structures. The statement may be seen in terms of individual-community duality that resides in the public-private polarity to shape the path of public-private-sector relative size irrespective of the accepted measurement. In fact, the whole of the government budgetary disputes must be seen in relation to conflicts in the private-public-sector duality that must shape the vision of the society. The negotiated equilibrium position at any time is temporary for any given political time. The conflicts in private-public relativity are always ideologically based where each of the negotiated equilibrium may sub-optimal in terms of the efficient functioning of the social organism. It is also these conflicts that provide the conversion moment for continual social transformations. The rewards in the private-public sector game are defined in terms of the gains realized in an increase in either the proportion of the public sector or the proportion in the private sector where the social cost-benefit constraint is under relative ideological influences.

References: Interdisplanary

R1 Classical Game Theory, Information, and Decision-Choice Conflicts

[R1.1] Aumann, R.: Correlated Equilibrium as an Expression of Bayesian Rationality. Econometrica 55, 1–18 (1987)

[R1.2] Aumann, R., Hart, S.: Handbook of Game Theory with Economic Applications. North-Holland, New York (1992)

[R1.3] Border, K.: Fixed Point Theorems with Applications to Economics and Game Theory. Cambridge University Press, Cambridge (1985)

[R1.4] Brams, S., Marc Kilgour, D.: Game Theory and National Security. Basil Blackwell, Oxford (1988)

[R1.5] Brandenburger, A.: Knowledge and Equilibrium Games. Journal of Economic Perspectives 6, 83–102 (1992)

[R1.6] Campbell, R., Sowden, L.: Paradoxes of Rationality and Cooperation: Prisoner's Dilemma and Newcomb's Problem. University of British Columbia Press, Vancouver (1985)

[R1.7] Crawford, V., Sobel, J.: Strategic Information Transmission. Econometrica 50, 1431–1452 (1982)

[R1.8] Dresher, M., et al. (eds.): Contributions to the Theory of Games. Annals of Mathematics Studies, vol. III(39). Princeton University Press, Princeton (1957)

[R1.9] Scott, G., Humes, B.: Games, Information, and Politics: Applying Game Theoretic Models to Political Science. University of Michigan Press, Ann Arbor (1996)

[R1.10] Gjesdal, F.: Information and Incentives: The Agency Information Problem. Review of Economic Studies 49, 373–390 (1982)

[R1.11] Harsanyi, J.: Games with Incomplete Information Played by 'Bayesian' Players I: The Basic Model. Management Science 14, 159–182 (1967)

[R1.12] Harsanyi, J.: Games with Incomplete Information Played by 'Bayesian' Players II: Bayesian Equilibrium Points. Management Science 14, 320–334 (1968)

[R1.13] Harsanyi, J.: Games with Incomplete Information Played by 'Bayesian' Players III: The Basic Probability Distribution of the Game. Management Science 14, 486–502 (1968)

[R1.14] Harsanyi, J.: Rational Behavior and Bargaining Equilibrium in Games and Social Situations. Cambridge University Press, New York (1977)

[R1.15] Haussmann, U.G.: A Stochastic Maximum Principle for Optimal Control of Diffusions. Longman, Essex (1986)

[R1.16] Haywood, O.: Military Decisions and Game Theory. Journal of the Operations Research Society of America 2, 365–385 (1954)

[R1.17] Krasovskii, N.N., Subbotin, A.I.: Game-theoretical Control Problems. Springer, New York (1988)

[R1.18] Kuhn, H. (ed.): Classics in Game Theory. Princeton University Press, Princeton (1997)

[R1.19] Lagunov, V.N.: Introduction to Differential Games and Control Theory. Heldermann Verlag, Berlin (1985)

[R1.20] Luce, D.R., Raiffa, H.: Games and Decisions. John Wiley and Sons, New York (1957)

[R1.21] Maynard Smith, J.: The Theory of Games and the Evolution of Animal Conflicts. Journal of Theoretical Biology 47, 209–221 (1974)

[R1.22] Maynard Smith, J.: Evolution and the Theory of Games. Cambridge University Press, Cambridge (1982)

[R1.23] Milgrom, P., Roberts, J.: Rationalizablility, Learning and Equilibrium in Games with Strategic Complementarities. Econometrica 58, 1255–1279 (1990)

[R1.24] Myerson, R.: Game Theory: Analysis of Conflict. Harvard University Press, Cambridge (1991)

[R1.25] Nalebuff, B., Riley, J.: Asymmetric Equilibrium in the War of Attrition. Journal of Theoretical Biology 113, 517–527 (1985)

[R1.26] Pauly, M.V.: Clubs Commonality, and the Core: An Integration of Game Theory and the Theory of Public Goods. Economica 35, 314–324 (1967)

[R1.27] Rapoport, A., Chammah, A.: Prisoner's Dilemma: A Study in Conflict and Cooperation. University of Michigan Press, Ann Arbor (1965)

[R1.28] Roth, A.E.: The Economist as Engineer: Game Theory, Experimentation, and Computation as Tools for Design Economics. Econometrica 70, 1341–1378 (2002)

[R1.29] Shubik, M.: Game Theory in the Social Sciences: Concepts and Solutions. MIT Press, Cambridge (1982)

[R1.30] Von Neumann, J., Morgenstern, O.: The Theory of Games in Economic Behavior. John Wiley and Sons, New York (1944)

R2 Collective Rationality, Public Choice and Social Decision-Choice Process

[R2.1] Aaron, H.J., et al.: Efficiency and Equity in the Optimal Supply of a Public Good. Rev. of Econ. Stat. 51, 31–39 (1969)

[R2.2] Aaron, H.J., et al.: Public Goods and Income Distribution. Econometrica 38, 907–920 (1970)

[R2.3] Apter, D.E.: Choice and politics of allocation. Yale University Press, New Haven (1971)

[R2.4] Archibald, G.C.: Welfare Economics Ethics and Essentialism. Economica (New Series) 26, 316–327 (1959)

[R2.5] Arrow, K.J.: Social Choice and Individual Values. Wiley, New York (1951)

[R2.6] Arrow, K.J.: Behavior under Uncertainty and Its Implication for Policy, Technical Report #399, Center for Research on Organizational Efficiency, Stanford, Stanford University (1983)

[R2.7] Arrow, K.J.: A Difficulty in the Concept of Social Welfare. Jour. of Political Econ. 58, 328–346 (1950)

[R2.8] Arrow, K.J.: Equality in Public Expenditure. Quarterly Jour. of Econ. 85, 409–415 (1971)

[R2.9] Arrow, K.J., et al. (eds.): Readings in Welfare Economics. Homewood, Irwin (1969)

[R2.10] Atkinson, A.B., et al.: Lecture on Public Economics. McGraw-Hill, New York (1980)

[R2.11] Atkinson, A.B.: Optimal Taxation and the Direct Versus Indirect Tax Controversy. Canadian Jour. of Economics 10, 590–606 (1977)

[R2.12] Atkinson, A.B., et al.: The Structure of Indirect Taxation and Economic Efficiency. Jour. of Public Economics 1, 97–119 (1972)

[R2.13] Atkinson, A.B.: The Design of Tax Structure: Direct Versus Indirect Taxation. Journal of Public Econ. 6, 55–75 (1977)

[R2.14] Axelrod, R.: Conflict of Interest, Chicago, Markham (1970)

[R2.15] Balassa, B., et al.: Economic Progress, Private Values and Public Policy. North-Holland, New York (1977)

[R2.16] Baldwin, D.A.: Foreign Aid, Interventions and Influence. World Politics 21, 425–447 (1969)

[R2.17] Baumol, W.J.: Welfare Economics and the Theory of the State. Harvard University Press, Cambridge (1967)

[R2.18] Becker, G.S.: A Theory of Competition among pressure Groups for Political Influence. Quarterly Jour. of Econ. 97(XCVII), 371–400 (1983)

[R2.19] Becker, G.S.: A Theory of Social Interactions. Jour. of Political Economy 82, 1063–1093 (1974)

[R2.20] Bergson, A.: A Reformation of Certain Aspects of Welfare Economics. Quart. Jour. of Econ. 52, 314–344 (1938)

[R2.21] Bergson, A.: On the Concept of Social Welfare. Quarterly Jour. of Econ. 68, 233–253 (1954)

[R2.22] Bernholz, P.: Economic Policies in a Democracy. Kyklos 19, fasc. 1, 48–80 (1966)

[R2.23] Black, D.: On the Rationale of Group Decision Making. Jour. of Political Economy 56, 23–34 (1948)

[R2.24] Blackorby, C., et al.: Utility vs. Equity. Jour. Public Economics 7, 365–381 (1977)

[R2.25] Blau, J.H., et al.: Social Decision Functions and the Veto. Econometrica 45, 871–879 (1977)

[R2.26] Bowen, H.R.: The Interpretation of Voting in the Allocation of Economic Resources. Quart. Jour. of Econ. 58, 27–48 (1943)

[R2.27] Brown, D.J.: Aggregation of Preferences. Quart. Jour. of Econ. 89, 456–469 (1975)

[R2.28] Buchanan, J.M.: Individual Choice in Voting and the Market. Jour. of Political Econ. 62, 334–343 (1954)

[R2.29] Buchanan, J.M.: An Economic Theory of Clubs. Economica 32, 1–14 (1965)

[R2.30] Buchanan, J.M.: Notes for an Economic Theory of Socialism. Public Choice 8, 29–43 (1970)

[R2.31] Buchanan, J.M.: The Demand and Supply of Public Goods. Rand McNally, Chicago (1968)

[R2.32] Buchanan, J.M.: Public Finance and Public Choice. National Tax Jour. 28, 383–394 (1975)

[R2.33] Buchanan, J.M., et al. (eds.): Theory of Public Choice. The University of Michigan Press, Ann Arbor (1972)

[R2.34] Buchanan, J.M., et al. (eds.): Toward a Theory of the Rent Seeking Society. Texas A and M Univ. Press, College Station (1980)

[R2.35] Buchanan, J.M., et al.: The Calculus of Consent. The University of Michigan Press, Ann Arbor (1962)

[R2.36] Chipman, J.S.: The Welfare Ranking of Pareto Distributions. Jour. of Econ. Theory 9, 275–282 (1974)

[R2.37] Coleman, J.S.: Foundations for a Theory of Collective Decisions. Amer. Jour. of Sociology 71, 615–627 (1966)

[R2.38] Coleman, J.S.: The Possibility of a Social Welfare Function. Amer. Econ. Rev. 56, 1105–1112 (1966)

[R2.39] Coleman, J.S.: The Possibility of a Social Welfare Function: A Reply. Amer. Econ. Rev. 57, 1311–1317 (1967)

[R2.40] Comanor, W.S.: The Median Voter and the Theory of Political Choice. Jour. of Public Econ. 5, 169–177 (1976)

[R2.41] Corlette, W., et al.: Complementality and Excess Burden of Taxation. Rev. Econ. Studies 21, 21–30 (1953-1954)

[R2.42] D'Aspermont, C., et al.: Equity and the Information Basis of Collective Choice. Rev. of Econ. Stud. 44, 119–209 (1977)

[R2.43] DeMeyer, F., et al.: A Welfare Function Using Relative Intensity of Preferences. Quart. Jour. of Econ. 85, 179–186 (1971)

[R2.44] Diamond, P.: Cardinal Welfare, Individualistic Ethics, and Interpersonal Comparisons of Utility: A Comment. Jour. of Political Econ. 75, 765–766 (1967)

[R2.45] Dobb, M.A.: Welfare Economics and the Economics of Socialism. Cambridge Univ. Press, Cambridge (1969)

[R2.46] Downs, A.: Economic Theory of Democracy. Harper and Row, New York (1957)

[R2.47] Dubins, L.E., et al.: How to Cut a Cake Fairly. Amer. Math. Monthly, 1–17 (1961); Reprinted in Newman, P. (ed.): Readings in Mathematical Economics. Value Theory, vol. 1. John Hopkins Univ. Press, Baltimore (1968)

[R2.48] Ellickson, B.: A Generalization of the Pure Theory of Public Goods. American Econ. Rev. 63, 417–432 (1973)

[R2.49] Farquharson, R.: Theory of Voting. Yale University Press, New Haven (1969)

[R2.50] Fontaine, E.: Economic Principles for Project Evaluation. Organization of American States, Washington, DC (1975)

[R2.51] Green, J., Laffont, J.-J.: Characterization of Satisfactory Mechanisms for Revelation of Preferences for Public Goods. Econometrica 45, 427–438 (1977)

[R2.52] Green, J., et al.: Imperfect Personal Information and the Demand Revealing Process: A Sampling Approach. Public Choice 29, 79–94 (1977)

[R2.53] Green, J.R., et al. (eds.): Incentatives in Public Decision-Making. North-Holland, New York (1979)

[R2.54] Haefele, E.T.: A Utility Theory of Representative Government. Amer. Econ. Rev. 61, 350–367 (1971)

[R2.55] Hammond, P.J.: Why Ethical Measures of Inequality Need Interpersonal Comparisons. Theory and Decision 7, 263–274 (1976)

[R2.56] Harberger, A.C.: The Basic Postulates for Applied Welfare Economics; An Interpretive Essay. Jour. of Econ. Lit. 9(3), 785–793 (1971)

[R2.57] Hardin, R.: Collective Action as an Agreeable n-Prisoner's Dilemma. Behav. Science 16, 471–481 (1971)

[R2.58] Hause, J.C.: The Theory of Welfare Cost Measurement. Jour. of Polit. Econ. 83, 1145–1182 (1975)

[R2.59] Head, J.G., et al.: Public Goods, Private Goods and Ambiguous Goods. Econ. Journal 79, 567–572 (1969)

[R2.60] Heineke, J.M.: Economic Model of Criminal Behavior. North-Holland, New York (1978)

[R2.61] Hicks, J.R.: The Four Consumer's Surpluses. Rev. of Econ. Stud. 13, 68–73 (1944)
[R2.62] Hicks, J.R.: The Foundations of Welfare Economics. Economic Journal 59, 696–712 (1949)
[R2.63] Intrilligator, M.D.: A Probabilistic Model of Social Choice. Rev. Econ. Stud. 40, 553–560 (1973)
[R2.64] Jakobsson, U.: On the Measurement of the Degree of Progression. Jour. of Public Economics 14, 161–168 (1976)
[R2.65] Kaldor, N.: Welfare Propositions of Economics and Interpersonal Comparison. Economic Jour. 49, 549–552 (1939)
[R2.66] Kaldor, N. (ed.): Conflicts in Policy objectives. Oxford Univ. Press, Oxford (1971)
[R2.67] Kemp, M.C., et al.: More on Social Welfare Functions: The Incompatibility of Individualism and Ordinalism. Economica 44, 89–90 (1977)
[R2.68] Kemp, M.C., et al.: On the Existence of Social Welfare Functions: Social Orderings and Social Decision Function. Economica 43, 59–66 (1976)
[R2.69] Laffont, J.-J. (ed.): Aggregation and Revelation of Preferences. North-Holland, New York (1979)
[R2.70] Lancaster, K.J.: A New Approach to Consumer Theory. Jour. of Political Economy 74, 132–157 (1966)
[R2.71] Lancaster, K.J.: Variety, Equity, and Efficiency. Columbia University Press, New York (1979)
[R2.72] Lin, S.A.Y. (ed.): Theory and Measurement of Economic Externalities. Academic Press, New York (1976)
[R2.73] Krueger, A.: The Political Economy of Rent Seeking Society. American Econ. Rev. 64, 291–302 (1974)
[R2.74] Lindsay, C.M.: A Theory of Government Enterprise. Jour. of Political Econ. 84, 1061–1077 (1976)
[R2.75] Lipsey, R.G., et al.: The General Theory of Second Best. Rev. of Econ. Stud. 24, 11–32 (1957)
[R2.76] Little, I.M.D.: Social Choice and Individual Values. Jour. of Political Econ. 60, 422–432 (1952)
[R2.77] Little, I.M.D.: A Critique of Welfare Economics. Clarendon Press, Oxford (1957)
[R2.78] McFadden, D.: The Revealed Preferences of a Public Bureaucracy Theory. Bell Jour. of Econ. 6(2), 55–72 (1975)
[R2.79] McFadden, D.: The Revealed Preferences of a Government Bureaucracy: Empirical Evidence. Bell Jour. of Econ. 7(1), 55–72 (1976)
[R2.80] McGuire, M.: Private Good Clubs and Public Good Clubs: Economic Model of Group Formation. Swedish Jour. of Econ. 74, 84–99 (1972)
[R2.81] McGuire, M.: Group Segregation and Optimal Jurisdiction. Jour. of Political Econ. 82, 112–132 (1974)
[R2.82] Mirkin, B.: Group Choice. John Wiley, New York (1979)
[R2.83] Millerson, J.C.: Theory of Value with Public Goods: A Survey Article. Jour. Econ. Theory 5, 419–477 (1972)
[R2.84] Mishan, E.J.: Survey of Welfare Economics: 1939-1959. In: Surveys of Economic Theory, vol. 1, pp. 156–222. Macmillan, New York (1968)
[R2.85] Mishan, E.J.: Welfare Criteria: Resolution of a Paradox. Economic Journal 83, 747–767 (1973)
[R2.86] Mishan, E.J.: Flexibility and Consistency in Cost Benefit Analysis. Economica 41, 81–96 (1974)

[R2.87] Mishan, E.J.: The Use of Compensating and Equivalent Variation in Costs-Benefit Analysis. Economica 43, 185–197 (1976)

[R2.88] Mishan, E.J.: Introduction to Normative Economics. Oxford Univ. Press, New York (1981)

[R2.89] Mueller, D.C.: Voting Paradox. In: Rowley, C.K. (ed.) Democracy and Public Choice, pp. 77–102. Basil Blackwell, New York (1987)

[R2.90] Mueller, D.C.: Public Choice: A Survey. Jour. of Econ. Lit. 14, 396–433 (1976)

[R2.91] Mueller, D.C.: The Possibility of a Social Welfare Function: Comment. Amer. Econ. Rev. 57, 1304–1311 (1967)

[R2.92] Mueller, D.C.: Allocation, Redistribution and Collective Choice. Public Finance 32, 225–244 (1977)

[R2.93] Mueller, D.C.: Voting by Veto. Jour. of Public Econ. 10, 57–75 (1978)

[R2.94] Mueller, D.C.: Public Choice. Cambridge Univ. Press, New York (1979)

[R2.95] Mueller, D.C., et al.: Solving the Intensity Problem in a Representative Democracy. In: Leiter, R.D., et al. (eds.) Economics of Public Choice, pp. 54–94. Cyro Press, New York (1975)

[R2.96] Musgrave, R.A., et al. (eds.): Classics in the Theory of Public Finance. St. Martin's Press, New York (1994)

[R2.97] Musgrave, R.A.: Public Finance in a Democratic Society: Collected Papers. New York Univ. Press, New York (1986)

[R2.98] Newbery, D., et al. (eds.): The Theory of Taxation for Developing Countries. Oxford University Press, New York (1987)

[R2.99] Olson, M.: The Logic of Collective Action. Harvard Univ. Press, Cambridge (1965)

[R2.100] Park, R.E.: The Possibility of a Social Welfare Function: Comment. Amer. Econ. Rev. 57, 1300–1304 (1967)

[R2.101] Pattanaik, P.K.: Voting and Collective Choice. Cambridge Univ. Press, New York (1971)

[R2.102] Pauly, M.V.: Cores and Clubs. Public Choice 9, 53–65 (1970)

[R2.103] Plott, C.R.: Ethics, Social Choice Theory and the Theory of Economic Policy. Jour. of Math. Soc. 2, 181–208 (1972)

[R2.104] Plott, C.R.: Axiomatic Social Choice Theory: An Overview and Interpretation. Amer. Jour. Polit. Science 20, 511–596 (1976)

[R2.105] Rae, D.W.: Decision-Rules and Individual Values in Constitutional Choice. Amer. Polit. Science Rev. 63, 40–56 (1969)

[R2.106] Rae, D.W.: The Limit of Consensual Decision. Amer. Polit. Science Rev. 69, 1270–1294 (1975)

[R2.107] Rapport, A., et al.: Prisoner's Dilemma. Michigan University Press, Ann Arbor (1965)

[R2.108] Rawls, J.A.: A Theory of Justice. Harvard Univ. Press, Cambridge (1971)

[R2.109] Rawls, J.A.: Concepts of Distributional Equity: Some Reasons for the maximum Criterion. Amer. Econ. Rev. 64, 141–146 (1974)

[R2.110] Ray, P.: Independence of Irrelevant Alternatives. Econometrica 41, 987–991 (1973)

[R2.111] Reimer, M.: The Case for Bare Majority Rule. Ethics 62, 16–32 (1951)

[R2.112] Reimer, M.: The Theory of Political Coalition. Yale Univ. Press, New Haven (1962)

[R2.113] Roberts, F.S.: Measurement Theory: with Applications to Decision Making, Utility and the Social Science. Addison-Wesley, Reading (1979)

[R2.114] Roberts, K.W.S.: Voting Over Income Tax Schedules. Jour. of Public Econ. 8, 329–340 (1977)

[R2.115] Rothenberg, J.: The Measurement of Social Welfare. Prentice-Hall, Englewood (1961)
[R2.116] Satterthwaite, M.A.: Strategy-Proofness and Arrows Conditions: Existence and Correspondence Theorem for Voting Procedures andSocial Welfare Functions. Jour. of Econ. Theory 10, 187–217 (1975)
[R2.117] Schneider, H.: National Objectives and Project Appraisal in Developing Countries. Development Centre of OECD, Paris (1975)
[R2.118] Sen, A.K.: A Possibility Theorem on Majority Decisions. Econometrica 34, 491–499 (1966)
[R2.119] Sen, A.K.: Quasi-transitivity, Rational Choice and Collective Decisions. Rev. Econ. Stud. 36, 381–394 (1969)
[R2.120] Sen, A.K.: Rawls versus Benthan: An Axiomatic Examination of the Pure Distribution Problem. Theory and Decision 4, 301–310 (1974)
[R2.121] Sen, A.K.: Informational Basis of Alternative Welfare Approaches Aggregation and Income Distribution. Jour. Public Econ. 3, 387–403 (1974)
[R2.122] Sen, A.K.: Liberty, Unanimity and Rights. Economica 43, 217–245 (1976)
[R2.123] Sen, A.K.: Social Choice Theory: A Re-examination. Econometrica 45, 43–89 (1977)
[R2.124] Sen, A.K.: On Weight and Measures Informational Constraints in Social Welfare Analysis. Econometrica 45 (October 1977)
[R2.125] Sen, A.K.: Collective Choice and Social Welfare. Holden-Day, San Francisco (1970)
[R2.126] Siegan, B.H.: Economic Liberties and the Constitution. Univ. of Chicago Press, Chicago (1980)
[R2.127] Stone, A.H., et al.: Generalized Sandwich Theorems. Duke Math. Jour. 9, 356–359 (1942)
[R2.128] Taylor, M.J.: Graph Theoretical Approach to the Theory of Social Choice. Public Choice 4, 35–48 (1968)
[R2.129] Taylor, M.J.: Proof of a Theorem on Majority Rule. Behavioral Science 14, 228–231 (1969)
[R2.130] Tideman, J.N., et al.: A New and Superior Process for Making Social Choice. Jour. of Polit. Economy 84, 1145–1159 (1976)
[R2.131] Tollison, R.M., et al.: Information and Voting: An Empirical Note. Public Choice 24, 43–49 (1975)
[R2.132] Tullock, G.: Some Problems of Majority Voting. Jour. of Polit. Econ. 67, 571–579 (1959)
[R2.133] Tullock, G.: The Politics of Bureaucracy. Public Affairs Press, Washington, D.C. (1965)
[R2.134] Varian, H.R.: Equity, Envy and Efficiency, Jour. Econ. Theory 9, 63–91 (1974)
[R2.135] Varian, H.R.: Two Problems in the Theory of Fairness. Jour. of Public Econ. 5, 249–260 (1976)
[R2.136] Williamson, O.E., et al.: Social Choice: A Probabilistic Approach. Econ. Jour. 77, 797–813 (1967)
[R2.137] Wilson, R.: A Game-Theoretic Analysis of Social Choice. In: Liebermann, B. (ed.) Social Choice, pp. 393–407. Gordon and Breach, New York (1971)
[R2.138] Wilson, R.A.: Stable Coalition Proposals in Majority-RuleVoting. Jour. of Econ. Theory 3, 254–271 (1971)
[R2.139] Wingo, L., et al. (eds.): Public Economics and Quality of Life. John Hopkins Univ. Press, Baltimore (1977)
[R2.140] Yaari, M.E., et al.: On Dividing Justly. Social Choice and Welfare 1, 1–24 (1984)

R3 Cost-Benefit Foundations For Decision-Choice Systems

R.3.1 Cost-Benefit Rationality and Contingent Valuation Method (CVM)

[R3.1.1] Ajzen, I., Fishbein, M.: Understanding Attitudes and Predicting Social Behavior. Prentice-Hall, Inc., Englewood Cliffs (1980)

[R3.1.2] Arrow, K., et al.: Repeat of NOAA Panel on Contingent Valuation. Federal Register 58, 4601–4614 (1993)

[R3.1.3] Bateman, I., Willis, K. (eds.): Valuing Environmental Preference: Theory and Practice of the Contingent Valuation Method in the US, EC and Developing Countries. Oxford University Press, Oxford (2002)

[R3.1.4] Batie, S.S., et al.: Valuing Non-Market Goods-Conceptual and Empirical Issues: Discussion. Amer. Jour. of Agricultural Economics 61(5), 931–932 (1979)

[R3.1.5] Bentkover, J.D., et al. (eds.): Benefit Assessment: The State of the Art. D. Reidel, Boston (1986)

[R3.1.6] Bishop, R.C., Heberlein, T.A.: The Contingent Valuation Method. In: Johnson, R.L., Johnson, G.V. (eds.) Economic Valuation of Natural Resources: Issues, Theory, and Applications, pp. 81–104. Westview Press, Boulder (1990)

[R3.1.7] Bishop, R.C., et al.: Contingent Valuation of Environmental Assets: Comparison with a Simulated Market. National Resources Jour. 23(3), 619–634 (1983)

[R3.1.8] Brookshire, D.S., et al.: The Advantage of Contingent Valuation Methods for Benefit Cost Analysis. Public Choice 36(2), 235–252 (1981)

[R3.1.9] Brookshire, D.S., et al.: Valuing Public Goods: A Compromise of Survey and Hedonic Approaches. Amer. Economic Review 72(1), 165–177 (1982)

[R3.1.10] Burness, H.S., et al.: Valuing Policies which Reduce Environmental Risk. Natural Resources, Jour. 23(3), 675–682 (1983)

[R3.1.11] Carson, R.T., et al.: Temporal Reliability of Estimates from Contingent Valuations, Discussion Paper 95-37. Resource for the Future, Washington, D.C. (August 1995)

[R3.1.12] Carson, R., et al.: A Contingent Valuation Study of Lost Passive Use Values Resulting from the Exxon Valdex Oil Spill. Report to the Attorney General of Alaska. Natural Resource Damage Assessment, Inc., La Jolla (1992)

[R3.1.13] Cummings, R.G., Brookshire, D.S., Schulze, W.D., et al. (eds.): Valuing Environmental Goods: An Assessment of the Contingent Valuation Method. Rowman and Allanheld, Totowa (1996)

[R3.1.14] Diamond, P.A., Hausman, J.A.: On Contingent Valuation Measurement of Nonuse Values. In: Hausman, J. (ed.) Contingent Valuation: A Critical Assessment, pp. 3–38. North- Holland Press, Amsterdam (1993)

[R3.1.15] Diamond, P.A., et al.: Contingent Valuation: Is Some Number Better then no Number? Jour. Economics Perspectives 8, 45–64 (1994)

[R3.1.16] Dompere, K.K.: The Theory of Fuzzy Decisions, Cost Distribution Principle in Social Choice and Optimal Tax Distribution. Fuzzy Sets and Systems 53, 253–274 (1993)

[R3.1.17] Dompere, K.K.: A Fuzzy-Decision Theory of Optimal Social Discount Rate: Collective-Choice-Theoretic. Fuzzy Sets and Systems 58, 269–301 (1993)

[R3.1.18] Dompere, K.K.: The Theory of Social Costs and Costing for Cost-Benefit Analysis in a Fuzzy Decision Space. Fuzzy Sets and Systems 76, 1–24 (1995)

[R3.1.19] Dompere, K.K.: The Theory of Approximate Prices: Analytical Foundations of Experimental Cost-Benefit Analysis in a Fuzzy-Decision Space. Fuzzy Sets and Systems 87, 1–26 (1997)

[R3.1.20] Dompere, K.K.: Cost-Benefit Analysis, Benefit Accounting and Fuzzy Decisions I: Theory. Fuzzy Sets and Systems 92, 275–287 (1987)

[R3.1.21] Dompere, K.K.: Cost-Benefit Analysis, Benefit Accounting and Fuzzy Decisions: Part II, Mental Illness in Hypothetical Community. Fuzzy Sets and Systems 100, 101–116 (1998)

[R3.1.22] Dompere, K.K.: Cost-Benefit Analysis of Information Technology. In: Kent, A., et al. (eds.) Encyclopedia of Computer Science and Technology, vol. 41(suppl. 26), pp. 27–44. Marcal Dekker, New York (1999)

[R3.1.23] Dompere, K.K. (ed.): Cost-Benefit Analysis and the Theory of Fuzzy Decision: Identification and Measurement Theory. STUDFUZZ, vol. 158. Springer, Heidelberg (2004)

[R3.1.24] Dompere, K.K. (ed.): Cost-Benefit Analysis and the Theory of Fuzzy Decision: The Fuzzy Value Theory. STUDFUZZ, vol. 160. Springer, Heidelberg (2004)

[R3.1.25] Freeman, A.M.: The Measurement of Environment and Resource Values: theory and Method. Resources for the Future, Washington, D.C. (1993)

[R3.1.26] Gregory, R.: Interpreting Measures of Economic Loss: Evidence from Contingent Valuation and Experimental Studies. Jour. of Environmental Economics and Management 13, 325–337 (1986)

[R3.1.27] Hanemann, W.M.: Welfare Evaluations in Contingent Valuation Experiments with Discrete Responses. American Journal of Agricultural Economics 66, 332–341 (1984)

[R3.1.28] Harrison, G.W.: Valuing Public Goods with the Contingent Valuation Method: A Critique of Kahneman and Knetsch. Journal of Environmental Economics and Management 23, 248–257 (1992)

[R3.1.29] Hausman, J.A. (ed.): Contingent Valuation: A Critical Assessment. North-Holland, New York (1993)

[R3.1.30] Hoehn, J.P., Randall, A.: Too Many Proposals Pass the Benefit Cost Test. American Economic Review 79, 544–551 (1989)

[R3.1.31] Kahneman, D.: Comments on the Contingent Valuation Method. In: Cummings, R.G., Brookshire, D.S., Schulze, W.D. (eds.) Valuing Environmental Goods: A State of the Arts Assessment of the Contingent Valuation Method, pp. 185–194. Rowman and Allanheld, Totowa (1986)

[R3.1.32] McNeal, B.J.: On the elicitation of Preferences for alternative Therapies. New England Jour. of Medicine 306, 1259–1262 (1982)

[R3.1.33] Portney, P.R.: The Contingent Valuation Debate: Why Economists Should Care. Jour. of Economic Perspectives 8, 3–17 (1994)

[R3.1.34] Randall, A., et al.: Contingent Valuation Survey for Evaluating Environmental Assets. Natural Resources Jour. 23, 635–648 (1983)

[R3.1.35] Smith, V.L.: Experiments with a Decentralized Mechanism for Public Good Decision. American Economic Review 70, 584–599 (1980)

[R3.1.36] Smith, V.K.: An Experimental Comparison of Three Public Good Decision Mechanisms. Scandinavian Jour. of Economics 81, 198–215 (1979)

[R3.1.37] Smith, V.K.: Indirect Revelation of demand for Public Goods: AnOverview and Critique. Scot. Jour. of Political Econ. 26, 183–189 (1979)

[R3.1.38] Tversky, A., et al.: The Framing of Decisions and the Psychology of Choice. Science 211, 453–458 (1981)

R3.2 Cost-Benefit Rationality and the Revealed Preference Approach (RPA)

[R3.2.1] Bain, J.S.: Criteria for Undertaking Water-Resource Development. Amer. Econ. Rev. Papers and Proceedings 50(2), 310–320 (1960)

[R3.2.2] Barsb, S.L.: Cost-Benefit Analysis and Manpower Programs. D.C. Heath & Co., Toronto (1972)

[R3.2.3] Benefit-Cost and Policy Analysis 1973, An Aldine Annual. Aldine Pub. Co., Chicago (1974)

[R3.2.4] Benefit-Cost Analysis and Policy Analysis 1974, An Aldine Annual. Aldine Pub. Co., Chicago (1975)

[R3.2.5] Benefit-Cost Analysis and Policy Analysis 1971, An Aldine Annual. Aldine Pub. Co., Chicago (1972)

[R3.2.6] Brookings Institution, Applying Benefit-Cost Analysis to Public Programs. Brookings Research Report #79, Washington, D.C. (1968)

[R3.2.7] Devine, E.T.: The Treatment of Incommensurables in Cost-Benefit Analysis. Land Economics 42(3), 383–387 (1966)

[R3.2.8] Gramlich, E.M.: Benefit-Cost Analysis of Government Programs. Prentice-Hall, Inc., Englewood Cliffs (1981)

[R3.2.9] Haveman, R.H.: Benefit-Cost Analysis: Its Relevance to Public Investment Decisions: Comment. Quarterly, Jour. of Econ. 81(4), 695–702 (1967)

[R3.2.10] Knesse, A.V.: Research Goals and Progress Toward them. In: Jarrett, H. (ed.) Environmental Quality in a Growing Economy, pp. 69–87. John Hopkins Press, Washington, DC (1966)

[R3.2.11] Layard, R. (ed.): Cost-Benefit Analysis. Penguin, Baltimore (1972)

[R3.2.12] Lesourne, J.: Cost-Benefit Analysis and Economic theory. North-Holland, New York (1975)

[R3.2.13] Mass, A.: Benefit-Cost Analysis: Its Relevance to Public Investment Decisions. Quarterly Jour. of Economics 80(2), 208–226 (1966)

[R3.2.14] MacDonald, J.S.: Benefits and Costs: Theoretical and Methodological Issues: Discussion. In: Somers, G.G., et al. (eds.) Cost-Benefit Analysis of Manpower Policies, Proceedings of a North American Conference, pp. 30–37. Kingston, Ontario (1969)

[R3.2.15] Marciariello, J.A.: Dynamic Benefit-Cost Analysis. Heath and Co., Toronto (1975)

[R3.2.16] Mishan, E.J.: Cost-Benefit Analysis. Praeger, New York (1976)

[R3.2.17] Musgrave, R.A.: Cost-Benefit Analysis and the Theory of Public Finance. Journal of Econ. Literature 7, 797–806 (1967)

[R3.2.18] Prest, A.R., et al.: Cost-Benefit Analysis: A Survey. In: Survey in Economic Theory, vol. III. St. Martin Press, New York (1966), Also in Economic Jour. 75, 685–705 (1965)

[R3.2.19] Raynauld, A.: Benefits and Costs: Theoretical and Methodological Issue: Discussion. In: Somers, G.G., et al. (eds.) Cost-Benefit Analysis of Manpower Policies, Proceedings of a North American Conference, pp. 37–41. Kingston, Ontario (1969)

[R3.2.20] Schwartz, H., et al. (eds.): Social and Economic Dimensions of Project Evaluation, Symposium on the Use of Socioeconomic Investment Criteria. Inter-American Dev. Bank, Washington, D.C. (1973)

[R3.2.21] Solo, R.A.: Benefit-Cost Analysis and Externalities in Private Choice: Comment. Southern Econ. Jour. 34(4), 569–570 (1968)

[R3.2.22] Somers, G.G., et al. (eds.): Cost-Benefit Analysis of Manpower Policies, Proceedings of a North American Conference. Kingston, Ontario (1969)
[R3.2.23] United States of America Committee on Interstate and Foreign Commerce House of Representatives, Use of Cost-Benefit Analysis By Regulatory Agencies, Joint and Subcommittee on Consumer Protection and Finance, Serial #96-157 (1979)
[R3.2.24] Water Resource Council of USA, Procedures for evaluation of national economic development (NED) benefits and costs in water resources planning (level C) final rule. Fed. Register 44, 72892–72977 (1979)

R4 Democracy, Governance and Political Markets

[R4.1] Alexander, K.J.W.: Political Economy of Change. Blackwell, Oxford (1975)
[R4.2] Arrow, K.: The Limits of Organization. Norton, New York (1974)
[R4.3] Barber, J.D.: Power in Committees. Rand McNally, Chicago IL (1966)
[R4.4] Beard, C.A.: The Economic Basis of Politics and Related Writings. Random House, New York (1957)
[R4.5] Bentley, A.F.: The Process of Government. University of Chicago Press, Chicago (1907)
[R4.6] Berle, A.A.: The Twentieth Century Capitalist Revolution. Harcourt Brace Jovanovich, New York (1954)
[R4.7] Berle, A.A.: Power without Property: A New Development in American Political Economy. Harcourt Brace and Co., New York (1959)
[R4.8] Bernholz, P.: Economic Policies in a Democracy. Kyklos 19, 48–80 (1966)
[R4.9] Black, D.: The Theory of Committees and Elections. Cambridge Univ. Press, Cambridge (1958)
[R4.10] Blaisdell, D.C.: Unofficial Government: Pressure Groups and Lobbies, Philadelphia, The Annals, vol. 319. The American Academy of Political and Social Science (1958) Sellin, T., Lambert, R.D. (Gen. eds.)
[R4.11] Bowles, S., Gintis, H.: Property, Community, and Contradictions of Modern Social Thought. Basic Books, New York (1986)
[R4.12] Breton, A.: The Economic Theory of Representative Government. Aldine Pub. Co., Chicago (1974)
[R4.13] Brenton, A., Galeotti, G., Salmon, P., Wintrobe, R. (eds.): Rational Foundations of Democratic Politics. Cambridge University Press, Cambridge (2003)
[R4.14] Brenton, A., Galeotti, G., Salmon, P., Wintrobe, R. (eds.): Understanding Democracy. Cambridge University Press, New York (1997)
[R4.15] Brams, S.J.: Measuring the Concentration of Power in Political Systems. American Political Science Review 62, 461–475 (1968)
[R4.16] Breton, A.: The Economic Constitution of Federal States. University of Toronto Press, Toronto (1978)
[R4.17] Barry, B.M.: Sociologists, Economists and Democracy. Collier- Macmillan, London (1970)
[R4.18] Buchanan, J.M.: Individual Choice in Voting and the Market. Journal of Political Economy 62, 334–343 (1954)
[R4.19] Buchanan, J.M.: Public Finance in Democratic Process. North Carolina. University of North Carolina Press, Chapel Hill (1967)
[R4.20] Buchanan, J.M., Tullock, G.: The Calculus of Consent. University of Michigan Press, Ann Arbor (1962)

[R4.21] Carson, R.B.: Economic Issues Today: Alternative Approaches. St. Martin's Press, New York (1991)

[R4.22] Carson, R.B., et al.: Government in the American Economy. D.C. Heath and Co., Lexington (1973)

[R4.23] Champlin, J.: On the Study of Power. Politics and Society 1, 91–111 (1971)

[R4.24] Clark, J.M.: Alternative to Selfdom. Random House/Vintage Books, New York (1960)

[R4.25] Clark, J.M.: Social Control of Business. McGraw-Hill, New York (1939)

[R4.26] Cohen, C. (ed.): Communism, Fascism and Democracy: The Theoretical Foundations I&II. Random House, New York (1962)

[R4.27] Cornforth, M.: The Open Philosophy and the Open Society. International Publishers, New York (1968)

[R4.28] Crossman, R.H.S.: The Politics of Socialism. Athenaeum Pub., New York (1965)

[R4.29] Dahl, R.A.: The Concept of Power. Behavioral Science 2, 201–215 (1957)

[R4.30] Dahl, R.A.: Who Governs? Democracy and Power in an American City. Yale University Press, New Haven (1961)

[R4.31] Dell, E.: Political Responsibility and Industry. Allen & Unwin, London (1973)

[R4.32] Diermier, D., Merlo, A.: Government Turnover in Parliamentary Democracies. Journal of Economic Theory 94, 46–79 (2000)

[R4.33] Dobb, M.: Studies in the Development of Capitalism. International Publishers, New York (1970)

[R4.34] Domhoff, W.: Who Rules America? Prentice-Hall, Englewood Cliffs (1967)

[R4.35] Downs, A.: An Economic Theory of Democracy. Harper and Row, New York (1957)

[R4.36] Downs, A.: Inside Bureaucracy. Little Brown, Boston (1966)

[R4.37] Easton, D. (ed.): Varieties of Political Theory. Prentice-Hall, Englewood Cliffs (1968)

[R4.38] Farquharson, R.: Theory of Voting. Yale University Press, New Haven (1969)

[R4.39] Fiorina, M.P.: Majority Rule Models and Legislative Elections. Journal of Politics 41, 1081–1104 (1979)

[R4.40] Fishkin, J.S.: Democracy and Deliberation: New Direction for Democratic Reform. Yale University Press, New Haven (1991)

[R4.41] Friedman, M.: Capitalism and Freedom. University of Chicago Press, Chicago (1962)

[R4.42] Frisch, H. (ed.): Schumpeterian Economics. Praeger, New York (1982)

[R4.43] Fulkerson, D.R.: Networks, Frames, Blocking Systems. In: Mathematics of Decision Sciences, vol. 1. American Mathematical Society, Providence (1968)

[R4.44] Galbraith, J.K.: The Affluent Society. Houghton Mifflin, Boston (1971)

[R4.45] Galbraith, J.K.: Economics in Perspective. Houghton Mifflin, Boston (1987)

[R4.46] Galbraith, J.K.: The New Industrial State. Houghton Mifflin, Boston (1967)

[R4.47] Galbraith, J.K.: Economics and Public Purpose. Houghton Mifflin, Boston (1967)

[R4.48] Galbraith, J.K.: American Capitalism: The Concept of Countervailing Powers. Houghton Mifflin, Boston (1956)

[R4.49] Gamson, W.A.: Coalition Formation at Presidential Nominating Conventions. American Journal of Sociology 68, 157–171 (1962)

[R4.50] Gamson, W.A.: The Theory of Coalition Formation. American Sociological Review 26, 373–382 (1961)

[R4.51] Gibbard, A.: Manipulation of Voting Schemes: A General Result. Econometrica 41, 587–601 (1973)

[R4.52] Ginsberg, B.: Elections and Public Policy. American Political Science Review 68, 41–49 (1976)
[R4.53] Graham, F.D.: Social Goals and Economic Institutions. Princeton University Press, Princeton (1942)
[R4.54] Hahn, E.L.: Revival of Political Economy: The Wrong Issues and the Wrong Argument. Economic Record 51, 360–364 (1975)
[R4.55] Harris, S.E.: The Economics of the Two Political Parties. Macmillan, New York (1962)
[R4.56] Hayek, F.: The Road to Serfdom. University of Chicago Press, Chicago (1944)
[R4.57] Heertje, A.: Schumpeter's Vision: Capitalism Socialism and Democracy after 40 Years. Praeger, New York (1981)
[R4.58] Heilbroner, R.: The Nature and Logic of Capitalism. Norton, New York (1985)
[R4.59] Heller, W.W.: New Dimensions of Political Economy. Harvard University Press, Cambridge (1966)
[R4.60] Hibbs Jr., D.: Political Parties and Macroeconomic Policy. American Political Science Review 71, 1467–1487 (1977)
[R4.61] Ingberman, D.E.: Running Against the Status Quo: Institutions for Direct Democracy Referenda and Allocation over Time. Public Choice 46, 19–43 (1985)
[R4.62] Key Jr., V.O.: Politics, Parties, and Pressure Groups. Thomas Y. Cromwell, New York (1964)
[R4.63] Kirk, R.: The Conservative Mind. Regnery, Chicago (1954)
[R4.64] Kitschelt, H.: Linkages between Citizens and Politicians in Democratic Politics. Comparative Political Studies 33(6/7), 845–879
[R4.65] Kitschelt, H., Wilkinson, S. (eds.): Citizen-Politician Linkages in Democratic Politics. Cambridge University Press, Cambridge (2006)
[R4.66] Knight, F.: Freedom and Reform. Harper & Row, New York (1947)
[R4.67] Knoke, D.: Change and Continuity in American Politics: The SocialBasis of Political Parties. The Johns Hopkins University Press, Baltmore (1976)
[R4.68] Kollman, K., Miller, J.H., Page, S.E. (eds.): Computational Models in Political Economy. MIT Press, Cmbridge (2003)
[R4.69] La Palombra, J.: Bureaucracy and Political Development. Princeton University Press, Princeton (1963)
[R4.70] Lindbeck, A.: The Political Economy of the New Left. Harper and Row, New York (1977)
[R4.71] Lohmann, S.: An Information Rationale for the Power of Special Interest. American Political Science Review 92(4), 809–827 (1998)
[R4.72] Malthus, T.R.: Definitions in Political Economy. Augustus M. Kelley, New York (1963)
[R4.73] March, J.G.: An Introduction to the Theory and Measurement of Influence. American Political Science Review 49, 431–451 (1955)
[R4.74] March, J.G.: The Power of Power. In: Easton, D. (ed.) Varieties of Political Theory, pp. 39–70. Prentice-Hall, Englewood Cliffs (1968)
[R4.75] Martindale, D. (ed.): National Character in the Perspective of the Social Sciences, Philadelphia. The Annals, vol. 370. The American Academy of Political and Social Science (1967); Sellin, T., Lambert, R.D. (Gen. eds.)
[R4.76] Marx, K.: Contribution to the Critique of Political Economy. Charles H. Kerr and Co., Chicago (1904)
[R4.77] Marx, K.: Capital, vol. 1, 2 & 3. Progress Publishers, Moscow (1887)

[R4.78] Marx, K.: Theories of Surplus–Value. Progress Publishers, Moscow, Part 1 (1963), Part II (1968), Part III (1971)

[R4.79] Maschler, M.: The Power of Coalition. Management Science 10, 8–29 (1963)

[R4.80] Mayhew, D.: Congress: The Electoral Connection. Yale University Press, New Haven (1974)

[R4.81] Midgaard, K.: Strategy and Ethics in international Politics. Cooperation and Conflict 4, 224–240 (1770)

[R4.82] Mueller, D.C.: Constitutional Democracy and Social Welfare. Quart. Jour. of Econ. 87, 60–80 (1973)

[R4.83] Niskanen, W.A.: Bureaucracy and Representative Government. Aldine, Chicago (1971)

[R4.84] Noll, R., Fiorina, M.: Voters, Bureaucrates and Legislators. Journal of Public Economics 7(32), 239–254 (1978)

[R4.85] Normanton, E.L.: The Accountability and Audit of Governments. Praeger, New York (1966)

[R4.86] Phelps, E.S. (ed.): Private Wants and Public Needs. W.W.Norton, New York (1957)

[R4.87] Piattoni, S. (ed.): Clientelism, Interests, and Democratic Representation. Cambridge University Press, Cambridge (2001)

[R4.88] Popkin, S.L.: The Reasoning Voter: Communication and Persuasion in Presidential Campaigns. University of Chicago Press, Chicago (1991)

[R4.89] Preston, N.S.: Politics, Economics and Power: Ideology and Practice Under Capitalism, Socialism, Communism and Fascism. Macmillan, London (1967)

[R4.90] Richter, M.K.: Coalitions, Core, and Competition. Journal of Economic Theory 3, 323–334 (1971)

[R4.91] Riker, W.H.: A Test of the Adequacy of the Power Index. Behavioral Science 4, 120–131 (1959)

[R4.92] Riker, W.H.: The Theory of Political Coalitions. Yale University Press, New Haven (1962)

[R4.93] Riker, W.H.: Some Ambiguities in the Notion of Power. American Political Science Review 58, 341–349 (1964)

[R4.94] Robbins, L.O.: Political Economy Past and Present. Macmillan, London (1976)

[R4.95] Rostow, E.: Planning for Freedom: The Public Law of American Capitalism. Yale University Press, New Haven (1959)

[R4.96] Rowley, C.K. (ed.): Democracy and Public Choice. Basil Blackwell, New York (1987)

[R4.97] Salamon, L.M., Siegfried, J.J.: Economic Power and PoliticalInfluence: The impact of Industry Structure on Public Policy. American Political Science Review 71, 1026–1043 (1977)

[R4.98] Schotter, A.: The Economic Theory of Social Institutions. Cambridge University Press, Cambridge (1981)

[R4.99] Schumpter, J.A.: The Theory of Economic Development. Harvard University Press, Cambridge (1934)

[R4.100] Schumpter, J.A.: Capitalism, Socialism and Democracy. Harper & Row, New York (1950)

[R4.101] Schumpeter, J.A.: March to Socialism. American Economic Review 40, 446–456 (1950)

[R4.102] Schumpeter, J.A.: Theoretical Problems of Economic Growth. Journal of Economic History 8, 1–9 (1947)

[R4.103] Schumpeter, J.A.: The Analysis of Economic Change. Review of Economic Statistics 17, 2–10 (1935)

[R4.104] Shapley, L.S.: Pure Competition, Coalitional Power, and Fair Division. International Economic Review 10, 337–362 (1969)
[R4.105] Shepsle, K., Weingast, B.: When Do Rules of Procedure Matter? The Journal of Politics 46, 206–221
[R4.106] Sigel, R. (ed.): Political Socialization: Its Role in the Political Process, Philadelphia. The Annals, vol. 361. The American Academy of Political and Social Science (1965); Sellin, T., Lambert, R.D. (Gen. eds.)
[R4.107] Stigler, G.J.: General Economic Conditions and National Elections. American Economic Review 63, 160–167 (1973)
[R4.108] Stokes, D.E., Campbell, A., Miller, W.E.: Components of Electoral Decision. American Political Science Review 52, 367–387 (1958)
[R4.109] Thurow, L.C.: Dangerous Currents. Random House, New York (1983)
[R4.110] Tufte, E.R.: The Relationship between Seats and Votes in Two-Party System. American Political Science Review 67, 540–554 (1973)
[R4.111] Tullock, G.: The Politics of Bureaucracy. Public Affairs Press, Washington, D.C (1965)
[R4.112] Tullock, G.: Bureaucracy: The Selected Works, vol. 6. Liberty Fund, Indianapolis (2005)
[R4.113] Tullock, G.: The Economics of Politics: The Selected Works, vol. 4. Liberty Fund, Indianapolis (2004)
[R4.114] Tullock, G.: Virginia Political Economy: The Selected Works, vol. 1. Liberty Fund, Indianapolis (2005)
[R4.115] Tullock, G.: The Calculus of Consent:Logical Foundations of Constitutional Democracy, The Selected Works, vol. 2. Liberty Fund, Indianapolis (2005)
[R4.116] Tullock, G.: Some Problems of Majority Voting. Journal of Political Economy 67, 571–579 (1959)
[R4.117] Windmuller, J.P. (ed.): Industrial Democracy in International Perspective, Philadelphia. The Annals, vol. 431. The American Academy of Political and Social Science (1977); Lambert, R.D., Heston, A.W. (Gen.eds.)
[R4.118] Winters, T.: Party Control and Policy Change. American Journal of Political Science 20, 597–636 (1976)
[R4.119] Yergin, D., Stanislaw, J.: The Commanding Heights: The Battle Between Government and the Marketplace that is Remaking the Modern Word. Simon & Schuster, New York (1999)
[R4.120] Young, H.P.: Power, Prices and Income in Voting Systems. Mathematical Programming 14, 129–148 (1978)
[R4.121] Young, H.P.: The Allocation of Funds in Lobbying and Campaigning. Behavioral Science 23, 21–31 (1978)

R5 Fuzzy Game Theory

[R5.1] Aubin, J.P.: Cooperative Fuzzy Games. Mathematics of Operations Research 6, 1–13 (1981)
[R5.2] Aubin, J.P.: Mathematical Methods of Game and Economics Theory. North-Holland, New York (1979)
[R5.3] Butnaria, D.: Fuzzy Games: A description pf the concepts. Fuzzy Sets and Systems 1, 181–192 (1978)
[R5.4] Butnaria, D.: Stability and shapely value for a n – persons Fuzzy Games. Fuzzy Sets and Systems 4(1), 63–72 (1980)

[R5.5] Nurmi, H.: A Fuzzy Solution to a Majority Voting Game. Fuzzy Sets and Systems 5, 187–198 (1981)

[R5.6] Regade, R.K.: Fuzzy Games in the Analysis of Options. Jour. of Cybernetics 6, 213–221 (1976)

[R5.7] Spillman, B., et al.: Coalition Analysis with Fuzzy Sets. Kybernetes 8, 203–211 (1979)

[R5.8] Wernerfelt, B.: Semifuzzy Games. Fuzzy Sets and Systems 19, 21–28 (1986)

R6 Fuzzy Mathematics and Optimal Rationality

[R6.1] Bandler, W., et al.: Fuzzy Power Sets and Fuzzy Implication Operators. Fuzzy Sets and Systems 4(1), 13–30 (1980)

[R6.2] Banon, G.: Distinction between Several Subsets of Fuzzy Measures. Fuzzy Sets and Systems 5(3), 291–305 (1981)

[R6.3] Bellman, R.E.: Mathematics and Human Sciences. In: Wilkinson, J., et al. (eds.) The Dynamic Programming of Human Systems, pp. 11–18. MSS Information Corp., New York (1973)

[R6.4] Bellman, R.E., Glertz, M.: On the Analytic Formalism of the Theory of Fuzzy Sets. Information Science 5, 149–156 (1973)

[R6.5] Brown, J.G.: A Note On Fuzzy Sets. Information and Control 18, 32–39 (1971)

[R6.6] Butnariu, D.: Fixed Points For Fuzzy Mapping. Fuzzy Sets and Systems 7(2), 191–207 (1982)

[R6.7] Butnariu, D.: Decompositions and Range For Additive Fuzzy Measures. Fuzzy Sets and Systems 10(2), 135–155 (1983)

[R6.8] Cerruti, U.: Graphs and Fuzzy Graphs. In: Fuzzy Information and Decision Processes, pp. 123–131. North-Holland, New York (1982)

[R6.9] Chakraborty, M.K., et al.: Studies in Fuzzy Relations Over Fuzzy Subsets. Fuzzy Sets and Systems 9(1), 79–89 (1983)

[R6.10] Chang, C.L.: Fuzzy Topological Spaces. J. Math. Anal. and Applications 24, 182–190 (1968)

[R6.11] Chang, S.S.L.: Fuzzy Mathematics, Man and His Environment. IEEE Transactions on Systems, Man and Cybernetics SMC-2, 92–93 (1972)

[R6.12] Chang, S.S.L., et al.: On Fuzzy Mathematics and Control. IEEE Transactions, System, Man and Cybernetics SMC-2, 30–34 (1972)

[R6.13] Chang, S.S.: Fixed Point Theorems for Fuzzy Mappings. Fuzzy Sets and Systems 17, 181–187 (1985)

[R6.14] Chapin, E.W.: An Axiomatization of the Set Theory of Zadeh, Notices. American Math. Society 687-02-4 754 (1971)

[R6.15] Chaudhury, A.K., Das, P.: Some Results on Fuzzy Topology on Fuzzy Sets. Fuzzy Sets and Systems 56, 331–336 (1993)

[R6.16] Cheng-Zhong, L.: Generalized Inverses of Fuzzy Matrix. In: Gupta, M.M., et al. (eds.) Approximate Reasoning In Decision Analysis, pp. 57–60. North Holland, New York (1982)

[R6.17] Chitra, H., Subrahmanyam, P.V.: Fuzzy Sets and Fixed Points. Jour. of Mathematical Analysis and Application 124, 584–590 (1987)

[R6.18] Cohn, D.L.: Measure Theory. Birkhauser, Basel (1980)

[R6.19] Cohen, P.J., Hirsch, R.: Non-Cantorian Set Theory. Scientific America, 101–116 (December 1967)

[R6.20] Czogala, J., et al.: Fuzzy Relation Equations On a Finite Set. Fuzzy Sets and Systems 7(1), 89–101 (1982)

[R6.21] Das, P.: Fuzzy Topology on Fuzzy Sets: Product Fuzzy Topology and Fuzzy Topological Groups. Fuzzy Sets and Systems 100, 367–372 (1998)

[R6.22] DiNola, A., et al. (eds.): The Mathematics of Fuzzy Systems. Verlag TUV Rheinland, Koln (1986)

[R6.23] DiNola, A., et al.: On Some Chains of Fuzzy Sets. Fuzzy Sets and Systems 4(2), 185–191 (1980)

[R6.24] Dombi, J.: A General Class of Fuzzy Operators, the DeMorgan Class of Fuzzy Operators and Fuzzy Measures Induced by Fuzzy Operators. Fuzzy Sets and Systems 8(2), 149–163 (1982)

[R6.25] Dubois, D., Prade, H.: Towards Fuzzy Differential Calculus, Part I: Integration of Fuzzy Mappings. Fuzzy Sets and Systems 8(1), 1–17 (1982)

[R6.26] Dubois, D., Prade, H.: Towards Fuzzy Differential Calculus, Part 2: Integration On Fuzzy Intervals. Fuzzy Sets and Systems 8(2), 105–116 (1982)

[R6.27] Dubois, D., Prade, H.: Towards Fuzzy Differential Calculus, Part 3: Differentiation. Fuzzy Sets and Systems 8(3), 225–233 (1982)

[R6.28] Dubois, D., Prade, H.: Fuzzy Sets and Systems. Academic Press, New York (1980)

[R6.29] Dubois, D.: Fuzzy Real Algebra: Some Results. Fuzzy Sets and Systems 2(4), 327–348 (1979)

[R6.30] Dubois, D., Prade, H.: Gradual Inference rules in approximate reasoning. Information Sciences 61(1-2), 103–122 (1992)

[R6.31] Dubois, D., Prade, H.: On the combination of evidence in various mathematical frameworks. In: Flamm, J., Luisi, T. (eds.) Reliability Data Collection and Analysis, pp. 213–241. Kluwer, Boston (1992)

[R6.32] Dubois, D., Prade, H. (eds.): Readings in Fuzzy Sets for Intelligent Systems. Morgan Kaufmann, San Mateo (1993)

[R6.33] Dubois, D., Prade, H.: Fuzzy sets and probability: Misunderstanding, bridges and gaps. In: Proc. Second IEEE Intern. Conf. on Fuzzy Systems, San Francisco, pp. 1059–1068 (1993)

[R6.34] Dubois, D., Prade, H.: A survey of belief revision and updating rules in various uncertainty models. Intern. J. of Intelligent Systems 9(1), 61–100 (1994)

[R6.35] Erceg, M.A.: Functions, Equivalence Relations, Quotient Spaces and Subsets in Fuzzy Set Theory. Fuzzy Sets and Systems 3(1), 79–92 (1980)

[R6.36] Feng, Y.-J.: A Method Using Fuzzy Mathematics to Solve the Vectormaximum Problem. Fuzzy Sets and Systems 9(2), 129–136 (1983)

[R6.37] Filev, D.P., et al.: A Generalized Defuzzification Method via Bag Distributions. Intern. Jour. of Intelligent Systems 6(7), 687–697 (1991)

[R6.38] Foster, D.H.: Fuzzy Topological Groups. Journal of Math. Analysis and Applications 67, 549–564 (1979)

[R6.39] Goetschel Jr., R., et al.: Topological Properties of Fuzzy Number. Fuzzy Sets and Systems 10(1), 87–99 (1983)

[R6.40] Goguen, J.A.: Mathematical Representation of Hierarchically Organized System. In: Attinger, E.O. (ed.) Global System Dynamics, pp. 111–129. S. Karger, Berlin (1970)

[R6.41] Goodman, I.R.: Fuzzy Sets As Random Level Sets: Implications and Extensions of the Basic Results. In: Lasker, G.E. (ed.) Applied Systems and Cybernetics. Fuzzy Sets and Systems, vol. VI, pp. 2756–2766. Pergamon Press, New York (1981)

[R6.42] Goodman, I.R.: Fuzzy Sets As Equivalence Classes of Random Sets. In: Yager, R.R. (ed.) Fuzzy Set and Possibility Theory: Recent Development, pp. 327–343. Pergamon Press, New York (1992)

[R6.43] Gupta, M.M., et al. (eds.): Fuzzy Antomata and Decision Processes. North-Holland, New York (1977)

[R6.44] Gupta, M.M., Sanchez, E. (eds.): Fuzzy Information and Decision Processes. North-Holland, New York (1982)

[R6.45] Higashi, M., Klir, G.J.: On measure of fuzziness and fuzzy complements. Intern. J. of General Systems 8(3), 169–180 (1982)

[R6.46] Higashi, M., Klir, G.J.: Measures of uncertainty and information based on possibility distributions. International Journal of General Systems 9(1), 43–58 (1983)

[R6.47] Higashi, M., Klir, G.J.: On the notion of distance representing information closeness: Possibility and probability distributions. Intern. J. of General Systems 9(2), 103–115 (1983)

[R6.48] Higashi, M., Klir, G.J.: Resolution of finite fuzzy relation equations. Fuzzy Sets and Systems 13(1), 65–82 (1984)

[R6.49] Higashi, M., Klir, G.J.: Identification of fuzzy relation systems. IEEE Trans. on Systems, Man, and Cybernetics 14(2), 349–355 (1984)

[R6.50] Ulrich, H.: A Mathematical Theory of Uncertainty. In: Yager, R.R. (ed.) Fuzzy Set and Possibility Theory: Recent Developments, pp. 344–355. Pergamon Press, New York (1982)

[R6.51] Jin-Wen, Z.: A Unified Treatment of Fuzzy Set Theory and Boolean Valued Set theory: Fuzzy Set Structures and Normal Fuzzy Set Structures. Jour. Math. Anal. and Applications 76(1), 197–301 (1980)

[R6.52] Kandel, A.: Fuzzy Mathematical Techniques with Applications. Addison-Wesley, Reading (1986)

[R6.53] Kandel, A., Byatt, W.J.: Fuzzy Processes. Fuzzy Sets and Systems 4(2), 117–152 (1980)

[R6.54] Kaufmann, A., Gupta, M.M.: Introduction to fuzzy arithmetic: Theory and applications. Van Nostrand Rheinhold, New York (1991)

[R6.55] Kaufmann, A.: Introduction to the Theory of Fuzzy Subsets, vol. 1. Academic Press, New York (1975)

[R6.56] Kaufmann, A.: Theory of Fuzzy Sets. Merson Press, Paris (1972)

[R6.57] Kaufmann, A., et al.: Fuzzy Mathematical Models in Engineering and Management Science. North-Holland, New York (1988)

[R6.58] Kim, K.H., et al.: Generalized Fuzzy Matrices. Fuzzy Sets and Systems 4(3), 293–315 (1980)

[R6.59] Klement, E.P.: Fuzzy and Fuzzy Measurable Functions. Fuzzy Sets and Systems 4, 83–93 (1980)

[R6.60] Klement, E.P.: Characterization of Finite Fuzzy Measures Using Markoff-kernels. Journal of Math. Analysis and Applications 75, 330–339 (1980)

[R6.61] Klement, E.P.: Construction of Fuzzy Using Triangular Norms. Journal of Math. Analysis and Applications 85, 543–565 (1982)

[R6.62] Klement, E.P., Schwyhla, W.: Correspondence Between Fuzzy Measures and Classical Measures. Fuzzy Sets and Systems 7(1), 57–70 (1982)

[R6.63] Klir, G., Yuan, B.: Fuzzy Sets and Fuzzy Logic. Prentice Hall, Upper Saddle River (1995)

[R6.64] Kokawa, M., et al.: Fuzzy-Theoretical Dimensionality Reduction Method of Multi-Dimensional Quality. In: Gupta, M.M., Sanchez, E. (eds.) Fuzzy Information and Decision Processes, pp. 235–250. North-Holland, New York (1982)

[R6.65] Kramosil, I., et al.: Fuzzy Metrics and Statistical Metric Spaces. Kybernetika 11, 336–344 (1975)

[R6.66] Kruse, R.: On the Construction of Fuzzy Measures. Fuzzy Sets and Systems 8(3), 323–327 (1982)

[R6.67] Kruse, R., et al.: Foundations of Fuzzy Systems. John Wiley and Sons, New York (1994)

[R6.68] Lasker, G.E. (ed.): Applied Systems and Cybernetics. Fuzzy Sets and Systems, vol. VI. Pergamon Press, New York (1981)

[R6.69] Lake, L.: Fuzzy Sets, Multisets and Functions I. London Math. Soc. 12(2), 323–326 (1976)

[R6.70] Lientz, B.P.: On Time Dependent Fuzzy Sets. Inform. Science 4, 367–376 (1972)

[R6.71] Lowen, R.: On the Existence of Natural Non-Topological Fuzzy Topological Space. Haldermann Verlag, Berlin (1986)

[R6.72] Martin, H.W.: Weakly Induced Fuzzy Topological Spaces. Jour. Math. Anal. and Application 78, 634–639 (1980)

[R6.73] Michalek, J.: Fuzzy Topologies. Kybernetika 11, 345–354 (1975)

[R6.74] Mizumoto, M., Tanaka, K.: Some Properties of Fuzzy Numbers. In: Gupta, M.M., et al. (eds.) Advances in Fuzzy Sets Theory and Applications, Amsterdam, North Holland, pp. 153–164. North-Holland, Amsterdam (1979)

[R6.75] Negoita, C.V., et al.: Applications of Fuzzy Sets to Systems Analysis. Wiley and Sons, New York (1975)

[R6.76] Negoita, C.V.: Representation Theorems For Fuzzy Concepts. Kybernetes 4, 169–174 (1975)

[R6.77] Negoita, C.V., et al.: On the State Equation of Fuzzy Systems. Kybernetes 4, 231–241 (1975)

[R6.78] Negoita, C.V.: Fuzzy Sets in Topoi. Fuzzy Sets and Systems 8(1), 93–99 (1982)

[R6.79] Netto, A.B.: Fuzzy Classes. Notices, American Mathematical Society 68T-H28, 945 (1968)

[R6.80] Nguyen, H.T.: Possibility Measures and Related Topics. In: Gupta, M.M., et al. (eds.) Approximate Reasoning in Decision Analysis, pp. 197–202. North-Holland, New York (1982)

[R6.81] Nowakowska, M.: Some Problems in the Foundations of Fuzzy Set Theory. In: Gupta, M.M., et al. (eds.) Approximate Reasoning in Decision Analysis, pp. 349–360. North-Holland, New York (1982)

[R6.82] Ovchinnikov, S.V.: Structure of Fuzzy Binary Relations. Fuzzy Sets and Systems 6(2), 169–195 (1981)

[R6.83] Pedrycz, W.: Fuzzy Relational Equations with Generalized Connectives and Their Applications. Fuzzy Sets and Systems 10(2), 185–201 (1983)

[R6.84] Raha, S., et al.: Analogy Between Approximate Reasoning and the Method of Interpolation. Fuzzy Sets and Systems 51(3), 259–266 (1992)

[R6.85] Ralescu, D.: Toward a General Theory of Fuzzy Variables. Jour. of Math. Analysis and Applications 86(1), 176–193 (1982)

[R6.86] Rao, M.B., et al.: Some Comments On Fuzzy Variables. Fuzzy Sets and Systems 6(2), 285–292 (1981)

[R6.87] Rodabaugh, S.E.: Fuzzy Arithmetic and Fuzzy Topology. In: Lasker, G.E. (ed.) Applied Systems and Cybernetics. Fuzzy Sets and Systems, vol. VI, pp. 2803–2807. Pergamon Press, New York (1981)

[R6.88] Rodabaugh, S., et al. (eds.): Application of Category Theory to Fuzzy Subsets. Kluwer, Boston (1992)

[R6.89] Roubens, M., et al.: Linear Fuzzy Graphs. Fuzzy Sets and Systems 10(1), 798–806 (1983)
[R6.90] Rosenfeld, A.: Fuzzy Groups. Jour. Math. Anal. Appln. 35, 512–517 (1971)
[R6.91] Rosenfeld, A.: Fuzzy Graphs. In: Zadeh, L.A., et al. (eds.) Fuzzy Sets and Their Applications to Cognitive and Decision Processes, pp. 77–95. Academic Press, New York (1974)
[R6.92] Rubin, P.A.: A Note on the Geometry of Reciprocal Fuzzy Relations. Fuzzy Sets and Systems 7(3), 307–309 (1982)
[R6.93] Ruspini, E.H.: Recent Developments In Mathematical Classification Using Fuzzy Sets. In: Lasker, G.E. (ed.) Applied Systems and Cybernetics. Fuzzy Sets and Systems, vol. VI, pp. 2785–2790. Pergamon Press, New York (1981)
[R6.94] Sanchez, E.: Resolution of Composite Fuzzy Relation Equations. Information and Control 3, 39–47 (1976)
[R6.95] Santos, E.S.: Maximin, Minimax and Composite Sequential Machines. Jour. Math. Anal. and Appln. 24, 246–259 (1968)
[R6.96] Santos, E.S.: Maximin Sequential Chains. Jour. of Math. Anal. and Appln. 26, 28–38 (1969)
[R6.97] Santos, E.S.: Fuzzy Algorithms. Inform. and Control 17, 326–339 (1970)
[R6.98] Sarkar, M.: On Fuzzy Topological Spaces. Jour. Math. Anal. Appln. 79, 384–394 (1981)
[R6.99] Skala, H., et al. (eds.): Aspects of Vagueness. D. Reidel, Boston (1984)
[R6.100] Slowinski, R., Teghem, J. (eds.): Stochastic versus Fuzzy Approaches to Multiobjective Mathematical Programming Under Uncertainty. Kluwer, Dordrecht (1990)
[R6.101] Stein, N.E., Talaki, K.: Convex Fuzzy Random Variables. Fuzzy Sets and Systems 6(3), 271–284 (1981)
[R6.102] Sugeno, M.: Inverse Operation of Fuzzy Integrals and Conditional Fuzzy Measures. Transactions SICE 11, 709–714 (1975)
[R6.103] Taylor, J.G. (ed.): Mathematical Approaches to Neural Networks. North-Holland, New York
[R6.104] Wright, C.: On the Coherence of Vague Predicates. Synthese 3, 325–365 (1975)
[R6.105] Yager, R.R., Filver, D.P.: Essentials of Fuzzy Modeling and Control. John Wiley and Sons, New York (1994)
[R6.106] Perano, T., et al.: Fuzzy Systems Theory and its Applications. Academic Press, New York (1992)
[R6.107] Terano, T., et al.: Applied Fuzzy Systems. AP Professional, New York (1994)
[R6.108] Triantaphyllon, E., et al.: The Problem of Determining Membership Values in Fuzzy Sets in Real World Situations. In: Brown, D.E., et al. (eds.) Operations Research and Artificial Intelligence: The Integration of Problem-Solving Strategies, pp. 197–214. Kluwer, Boston (1990)
[R6.109] Tsichritzis, D.: Participation Measures. Jour. Math. Anal. and Appln. 36, 60–72 (1971)
[R6.110] Tsichritzis, D.: Approximation and Complexity of Functions on the Integers. Inform. Science 4, 70–86 (1971)
[R6.111] Turksens, I.B.: Four Methods of Approximate Reasoning with Interval-Valued Fuzzy Sets. Intern. Journ. of Approximate Reasoning 3(2), 121–142 (1989)
[R6.112] Turksen, I.B.: Measurement of Membership Functions and Their Acquisition. Fuzzy Sets and Systems 40(1), 5–38 (1991)

[R6.113] Verdegay, J., et al.: The Interface Between Artificial Intelligence and Operations Research in Fuzzy Environment. Verlag TUV Rheinland, Koln (1989)

[R6.114] Wang, L.X.: Adaptive Fuzzy Sets and Control: Design and Stability Analysis. Prentice Hall, Englewood Cliffs (1994)

[R6.115] Wang, P.P. (ed.): Advances in Fuzzy Sets, Possibility Theory, and Applications. Plenum Press, New York (1983)

[R6.116] Wang, P.P. (ed.): Advances in Fuzzy Theory and Technology, vol. 1. Bookwright Press, Durham (1992)

[R6.117] Wang, Z., Klir, G.: Fuzzy Measure Theory. Plenum Press, New York (1992)

[R6.118] Wang, P.Z., et al. (eds.): Between Mind and Computer: Fuzzy Science and Engineering. World Scientific Press, Singapore (1993)

[R6.119] Wang, P.Z.: Contactability and Fuzzy Variables. Fuzzy Sets and Systems 8(1), 81–92 (1982)

[R6.120] Wang, S.: Generating Fuzzy Membership Functions: A Monotonic Neural Network Model. Fuzzy Sets and Systems 61(1), 71–82 (1994)

[R6.121] Whalen, T., et al.: Usuality, Regularity, and Fuzzy Set Logic. Intern. Jour. of Approximate Reasoning 6(4), 481–504 (1992)

[R6.122] Wierzchon, S.T.: An Algorithm for Identification of Fuzzy Measure. Fuzzy Sets and Systems 9(1), 69–78 (1983)

[R6.123] Wong, C.K.: Fuzzy Topology: Product and Quotient Theorems. Journal of Math. Analysis and Applications 45, 512–521 (1974)

[R6.124] Wong, C.K.: Fuzzy Points and Local Properties of Fuzzy Topology. Jour. Math. Anal. and Appln. 46, 316–328 (1987)

[R6.125] Wong, C.K.: Fuzzy Topology. In: Zadeh, L.A., et al. (eds.) Fuzzy Sets and Their Applications to Cognitive and Decision Processes, pp. 171–190. Academic Press, New York (1974)

[R6.126] Wong, C.K.: Categories of Fuzzy Sets and Fuzzy Topological Spaces. Jour. Math. Anal. and Appln. 53, 704–714 (1976)

[R6.127] Wygralak, M.: Fuzzy Inclusion and Fuzzy Equality of two Fuzzy Subsets. Fuzzy Sets and Systems 10(2), 157–168 (1983)

[R6.128] Yager, R.R.: On the Lack of Inverses in Fuzzy Arithmetic. Fuzzy Sets and Systems 4(1), 73–82 (1980)

[R6.129] Yager, R.R. (ed.): Fuzzy Set and Possibility Theory: Recent Development. Pergamon Press, New York (1992)

[R6.130] Yager, R.R.: Fuzzy Subsets with Uncertain Membership Grades. IEEE Transactions on Systems, Man and Cybernetics 14(2), 271–275 (1984)

[R6.131] Yager, R.R., et al.: Essentials of Fuzzy Modeling and Control. John Wiley, New York (1994)

[R6.132] Yager, R.R., et al. (eds.): Fuzzy Sets, Neural Networks, and Soft Computing. Nostrand Reinhold, New York (1994)

[R6.133] Yager, R.R.: On the Theory of Fuzzy Bags. Intern. Jour. of General Systems 13(1), 23–37 (1986)

[R6.134] Yager, R.R.: Cardinality of Fuzzy Sets via Bags. Mathematical Modelling 9(6), 441–446 (1987)

[R6.135] Zadeh, L.A.: A Computational Theory of Decompositions. Intern. Jour. of Intelligent Systems 2(1), 39–63 (1987)

[R6.136] Zadeh, L.A.: The Birth and Evolution of Fuzzy Logic. Intern. Jour. of General Systems 17(2-3), 95–105 (1990)

[R6.137] Zadeh, L.A., et al.: Fuzzy Logic for the Management of Uncertainty. John Wiley, New York (1992)

[R6.138] Zimmerman, H.J.: Fuzzy Set Theory and Its Applications. Kluwer, Boston (1985)

R7 Fuzzy Optimization and Decision-Choice Rationality

[R7.1] Bose, R.K., Sahani, D.: Fuzzy Mappings and Fixed Point Theorems. Fuzzy Sets and Systems 21, 53–58 (1987)

[R7.2] Buckley, J.J.: Fuzzy Programming and the Pareto Optimal Set. Fuzzy Set and Systems 10(1), 57–63 (1983)

[R7.3] Butnariu, D.: Fixed Points for Fuzzy Mappings. Fuzzy Sets and Systems 7, 191–207 (1982)

[R7.4] Carlsson, G.: Solving Ill-Structured Problems Through Well Structured Fuzzy Programming. In: Brans, J.P. (ed.) Operation Research 1981, pp. 467–477. North-Holland, Amsterdam (1981)

[R7.5] Carlsson, C.: Tackling an AMDM - Problem with the Help of Some Results From Fuzzy Set Theory. European Journal of Operational Research 10(3), 270–281 (1982)

[R7.6] Cerny, M.: Fuzzy Approach to Vector Optimization. Intern. Jour. of General Systems 20(1), 23–29

[R7.7] Chang, C.L.: Interpretation and Execution of Fuzzy Programs. In: Zadeh, L.A., et al. (eds.) Fuzzy Sets and Their Applications to Cognitive and Decision Processes, pp. 191–218. Academic Press, New York (1975)

[R7.8] Chang, S.K.: On the Execution of Fuzzy Programs Using Finite State Machines. IEEE, Trans. Comp. C-12, 214–253 (1982)

[R7.9] Chang, S.S.: Fixed Point Theorems for Fuzzy Mappings. Fuzzy Sets and Systems 17, 181–187 (1985)

[R7.10] Chang, S.S.L.: Fuzzy Dynamic Programming and the Decision Making Process. In: Proc. 3rd Princeton Conference on Information Science and Systems, Princeton, pp. 200–203 (1969)

[R7.11] Chang, S.Y., et al.: Modeling To Generate Alternatives: A Fuzzy Approach. Fuzzy Sets and Systems 9(2), 137–151 (1983)

[R7.12] Dompere, K.K.: Fuzziness, Rationality, Optimality and Equilibrium in Decision and Economic Theories. In: Lodwick, W.A., Kacprzyk, J. (eds.) Fuzzy Optimization. STUDFUZZ, vol. 254, pp. 3–32. Springer, Heidelberg (2010)

[R7.13] Dompere, K.K.: On Epistemology and Decision-Choice Rationality. In: Trappl, R. (ed.) Cybernetics and System Research, pp. 219–228. North-Holland, New York (1982)

[R7.14] Dompere, K.K.: Fuzzy Rationality: Methodological Critique and Unity of Classical, Bounded and Other Rationalities. STUDFUZZ, vol. 235. Springer, New York (2009)

[R7.15] Dompere, K.K.: Epistemic Foundations of Fuzziness: Unified Theories on Decision-Choice Processes. STUDFUZZ, vol. 236. Springer, New York (2009)

[R7.16] Dompere, K.K.: Fuzziness and Approximate Reasoning: Epistemics on Uncertainty, Expectations and Risk in Rational Behavior. STUDFUZZ, vol. 237. Springer, Heidelberg (2009)

[R7.17] Dubois, D., et al.: Systems of Linear Fuzzy Constraints. Fuzzy Sets and Systems 3(1), 37–48 (1980)
[R7.18] Dubois, D.: An Application of Fuzzy Arithmetic to the Optimization of Industrial Machining Processes. Mathematical Modelling 9(6), 461–475 (1987)
[R7.19] Edwards, W.: The Theory of Decision Making. Psychological Bulletin 51, 380–417 (1954)
[R7.20] Eaves, B.C.: Computing Kakutani Fixed Points. Journal of Applied Mathematics 21, 236–244 (1971)
[R7.21] Feng, Y.J.: A Method Using Fuzzy Mathematics to Solve the Vector Maxim Problem. Fuzzy Set and Systems 9(2), 129–136 (1983)
[R7.22] Hamacher, H., et al.: Sensitivity Analysis in Fuzzy Linear Programming. Fuzzy Sets and Systems 1, 269–281 (1978)
[R7.23] Hannan, E.L.: On the Efficiency of the Product Operator in Fuzzy Programming with Multiple Objectives. Fuzzy Sets and Systems 2(3), 259–262 (1979)
[R7.24] Hannan, E.L.: Linear Programming with Multiple Fuzzy Goals. Fuzzy Sets and Systems 6(3), 235–248 (1981)
[R7.25] Heilpern, S.: Fuzzy Mappings and Fixed Point Theorem. Journal of Mathematical Analysis and Applications 83, 566–569 (1981)
[R7.26] Ignizio, J.P., et al.: Fuzzy Multicriteria Integer Programming via Fuzzy Generalized Networks. Fuzzy Sets and Systems 10(3), 261–270 (1983)
[R7.27] Jarvis, R.A.: Optimization Strategies in Adaptive Control: A Selective Survey. IEEE Trans. Syst. Man. Cybernetics SMC-5, 83–94 (1975)
[R7.28] Jakubowski, R., et al.: Application of Fuzzy Programs to the Design of Machining Technology. Bulleting of the Polish Academy of Science 21, 17–22 (1973)
[R7.29] Kabbara, G.: New Utilization of Fuzzy Optimization Method. In: Gupta, M.M., et al. (eds.) Approximate Reasoning In Decision Analysis, pp. 239–246. North Holland, New York (1982)
[R7.30] Kacprzyk, J., et al. (eds.): Optimization Models Using Fuzzy Sets and Possibility Theory. D. Reidel, Boston (1987)
[R7.31] Kakutani, S.: A Generalization of Brouwer's Fixed Point Theorem. Duke Mathematical Journal 8, 416–427 (1941)
[R7.32] Kaleva, O.: A Note on Fixed Points for Fuzzy Mappings. Fuzzy Sets and Systems 15, 99–100 (1985)
[R7.33] Kandel, A.: On Minimization of Fuzzy Functions. IEEE Trans. Comp. C-22, 826–832 (1973)
[R7.34] Kandel, A.: Comments on Minimization of Fuzzy Functions. IEEE Trans. Comp. C-22, 217 (1973)
[R7.35] Kandel, A.: On the Minimization of Incompletely Specified Fuzzy Functions. Information, and Control 26, 141–153 (1974)
[R7.36] Lai, Y., et al.: Fuzzy Mathematical Programming. Springer, New York (1992)
[R7.37] Leberling, H.: On Finding Compromise Solution in Multcriteria Problems, Using the Fuzzy Min-Operator. Fuzzy Set and Systems 6(2), 105–118 (1981)
[R7.38] Lee, E.S., et al.: Fuzzy Multiple Objective Programming and Compromise Programming with Pareto Optimum. Fuzzy Sets and Systems 53(3), 275–288 (1993)
[R7.39] Lowen, R.: Connex Fuzzy Sets. Fuzzy Sets and Systems 3, 291–310 (1980)
[R7.40] Luhandjula, M.K.: Compensatory Operators in Fuzzy Linear Programming with Multiple Objectives. Fuzzy Sets and Systems 8(3), 245–252 (1982)

[R7.41] Luhandjula, M.K.: Linear Programming Under Randomness and Fuzziness. Fuzzy Sets and Systems 10(1), 45–54 (1983)

[R7.42] Negoita, C.V., et al.: Fuzzy Linear Programming and Tolerances in Planning. Econ. Group Cybernetic Studies 1, 3–15 (1976)

[R7.43] Negoita, C.V., Stefanescu, A.C.: On Fuzzy Optimization. In: Gupta, M.M., et al. (eds.) Approximate Reasoning In Decision Analysis, pp. 247–250. North Holland, New York (1982)

[R7.44] Negoita, C.V.: The Current Interest in Fuzzy Optimization. Fuzzy Sets and Systems 6(3), 261–270 (1981)

[R7.45] Negoita, C.V., et al.: On Fuzzy Environment in Optimization Problems. In: Rose, J., et al. (eds.) Modern Trends in Cybernetics and Systems, pp. 13–24. Springer, Berlin (1977)

[R7.46] Orlovsky, S.A.: On Programming with Fuzzy Constraint Sets. Kybernetes 6, 197–201 (1977)

[R7.47] Orlovsky, S.A.: On Formulation of General Fuzzy Mathematical Problem. Fuzzy Sets and Systems 3, 311–321 (1980)

[R7.48] Ostasiewicz, W.: A New Approach to Fuzzy Programming. Fuzzy Sets and Systems 7(2), 139–152 (1982)

[R7.49] Pollatschek, M.A.: Hieranchical Systems and Fuzzy-Set Theory. Kybernetes 6, 147–151 (1977)

[R7.50] Ponsard, G.: Partial Spatial Equilibra With Fuzzy Constraints. Journal of Regional Science 22(2), 159–175 (1982)

[R7.51] Prade, M.: Operations Research with Fuzzy Data. In: Want, P.P., et al. (eds.) Fuzzy Sets, pp. 155–170. Plenum, New York (1980)

[R7.52] Ralescu, D.: 0ptimization in a Fuzzy Environment. In: Gupta, M.M., et al. (eds.) Advances in Fuzzy Set Theory and Applications, pp. 77–91. North-Holland, New York (1979)

[R7.53] Ralescu, D.A.: Orderings, Preferences and Fuzzy Optimization. In: Rose, J. (ed.) Current Topics in Cybernetics and Systems. Springer, Berlin (1978)

[R7.54] Sakawa, M.: Fuzzy Sets and Interactive Multiobjective Optimization. Plenum Press, New York (1993)

[R7.55] Sakawa, M., et al.: Feasibility and Pareto Optimality for Multi-objective Nonlinear Programming Problems with Fuzzy Parameters. Fuzzy Sets and Systems 43(1), 1–15

[R7.56] Tanaka, K., et al.: Fuzzy Programs and Their Execution. In: Zadeh, L.A., et al. (eds.) Fuzzy Sets and Their Applications to Cognitive and Decision Processes, pp. 41–76 (1974)

[R7.57] Tanaka, K., et al.: Fuzzy Mathematical Programming. Transactions of SICE, 109-115 (1973)

[R7.58] Tanaka, H., et al.: On Fuzzy-Mathematical Programming. Journal of Cybernetics 3(4), 37–46 (1974)

[R7.59] Vira, J.: Fuzzy Expectation Values in Multistage Optimization Problems. Fuzzy Sets and Systems 6(2), 161–168 (1981)

[R7.60] Verdegay, J.L.: Fuzzy Mathematical Programming. In: Gupta, M.M., et al. (eds.) Fuzzy Information and Decision Processes, pp. 231–238. North-Holland, New York (1982)

[R7.61] Warren, R.H.: Optimality in Fuzzy Topological Polysystems. Jour. Math. Anal. 54, 309–315 (1976)

[R7.62] Weiss, M.D.: Fixed Points, Separation and Induced Topologies for Fuzzy Sets. Jour. Math. Anal. and Appln. 50, 142–150 (1975)

[R7.63] Wiedey, G., Zimmermann, H.J.: Media Selection and Fuzzy Linear Programming. Journal Oper. Res. Society 29, 1071–1084 (1978)

[R7.64] Wilkinson, J.: Archetypes, Language, Dynamic Programming and Fuzzy Sets. In: Wilkinson, J., et al. (eds.) The Dynamic Programming of Human Systems, MSS New York, Information Corp., pp. 44–53 (1973)
[R7.65] Yager, R.R.: Mathematical Programming with Fuzzy Constraints and Preference on the Objective. Kybernetes 8, 285–291 (1979)
[R7.66] Zadeh, L.A.: Outline of a New Approach to the Analysis of Complex Systems and Decision Process. In: Cochrane, J.L., et al. (eds.) Multiple Criteria Decision Making. Univ. of South Carolina Press, Columbia (1973); Also in IEEE Transactions on System, Man and Cybernetics 1, 28–44
[R7.67] Zadeh, L.A.: The Role of Fuzzy Logic in the Management of Ucertainty in expert Systems. Fuzzy Sets and Systems 11, 199–227 (1983)
[R7.68] Zimmerman, H.-J.: Description and Optimization of Fuzzy Systems. Intern. Jour. Gen. Syst. 2(4), 209–215 (1975)
[R7.69] Zimmerman, H.-J.: Fuzzy Programming and Linear Programming with Several Objective Functions. Fuzzy Sets and Systems 1(1), 45–56 (1978)
[R7.70] Zimmerman, H.J.: Applications of Fuzzy Set Theory to Mathematical Programming. Information Science 36(1), 29–58 (1985)

R8 National Interest and Foreign Policy

[R8.1] Almond, G.A.: The American people and foreign policy. Harcourt, Brace, New York (1950)
[R8.2] Allen, J.S.: Atomic Imperialism: The State, Monopoly and the Bomb. International Publishers, New York (1952)
[R8.3] Beard, C.A.: The Idea of National Interest. Macmillan, New York (1934)
[R8.4] Beloff, M.: The future of British foreign policy. Secker & Warburg, London (1969)
[R8.5] Boulding, K.E.: The Image. University of Michigan Press, Ann Arbor (1956)
[R8.6] Burton, J.W.: International Relations; a General Theory. CambridgeUniversity Press, London (1965)
[R8.7] Butterfield, H.: The Scientific v. the Moralistic Approach in International Affairs. International Affairs 27, 411–422 (1951)
[R8.8] Butwell, R. (ed.): Foreign policy and the developing nation: Papers by Henry Bienen. University of Kentucky Press, Lexington (1969)
[R8.9] Cherlesworth, J. (ed.): American Foreign Policy Challenged, Philadelphia,The Annals, Vol. 342 (1962); Sellin, T., Lambert, R.D.(Gen. eds.). The American Academy of Political and Social Science
[R8.10] Cohen, B.C.: The public's impact on foreign policy. Little, Brown, Boston (1972)
[R8.11] Cook, T.I., Moos, M.: Power through purpose; the realism of idealism as a basis for foreign policy. Johns Hopkins Press, Baltimore (1954)
[R8.12] Crabb Jr., C.V. (ed.): Nonalignment in Foreign Affairs, vol. 362. The Annals, Philadelphia (1965); Sellin, H., Lambert, R.D. (Gen. eds.). The American Academy of Political and Social Science
[R8.13] Dahl, R.A.: Congress and foreign policy. Norton, New York (1964)
[R8.14] Davenant, C.: Essays Upon the Balance of Power: The Right of Making War, Peace and Alliances; and Univeral Monarchy. Knapton, London (1701)
[R8.15] Dennett, R., Turner, R.K. (eds.): Documents on American Foreign Relations, vol. XII. Princeton University Press, Princeton (1952)

[R8.16] Derber, M.: The American idea of industrial democracy, 1865-1965. University of Illinois Press, Urbana (1970)

[R8.17] Destler, I.M.: Making Foreign Economic Policy. The Brookings Institutions, Washington, D.C. (1980)

[R8.18] Dougherty, J.E., Faltzgraff, P. (eds.): Contending Theories of International Relations. Lippincott, Philadelphia (1971)

[R8.19] Evans, P.B., et al.: Double – Edge Diplomacy. International Bargaining and Domestic politics. University of Californian Press, Berkeley (1993)

[R8.20] Fischman, L.L.: World mineral trends and U.S. supply problems. Resources for the Future. Johns Hopkins University Press, Washington, D.C., Baltimore (1980)

[R8.21] Galtung, J.: A Structural Theory of Imperialism. Journal of Peace Research 2, 81–117 (1971)

[R8.22] Goldwin, R.A. (ed.): Readings in American Foreign policy. Oxford University Press, New York (1971)

[R8.23] Govett, G.J.S., Govett, M.H. (eds.): World mineral supplies: assessment and perspective. Elsevier Scientific Pub. Co., Amsterdam / New York (1976)

[R8.24] Hanrieder, W.F. (ed.): Comparative foreign policyl theoretical essays. McKay, New York (1971)

[R8.25] Hargreaves, D., Fromson, S.: World index of strategic minerals:production, exploitation, and risk. Facts on File, New York (1983)

[R8.26] Harrod, J.: International Relations: Perceptions and Neo-realism. Year Book of World Affairs 31, 289–305 (1977)

[R8.27] Hartmann, F.H. (ed.): Basic documents of international relations. Kennikat Press, Port Washington (1969)

[R8.28] Hill, M.: A Goals – Achievement Matrix for Evaluating Alternative Plans. Journal of Amer. Institute of Planners 34(1), 19–29

[R8.29] Hinsley, F.H.: Power and the Pursuit of Peace. Cambridge University Press, London (1963)

[R8.30] Hollingsworth, J.R.: Social Theory and Public Policy, vol. 434. The Annals, Philadelphia (1977); Laambert, R.D., Heston, A. W., (Gen. eds.). The American Academy of Political and Social Science

[R8.31] Hoskins, H.L. (ed.): Aiding Underdeveloped Areas Abroad, vol. 268. The Annals, Philadelphia (1950); Sellin, T., Charlesworth, J.C. (Gen. eds.).The American Academy of Political and Social Science

[R8.32] Houghton, N.D.: The Challenge to Political Scientists in Recent American Foreign Policy: Scholarship or Ideology? American Political Science Review 52, 678–688 (1958)

[R8.33] Johnson, R.A.: The administration of United States foreign policy. University of Texas Press, Austin (1971)

[R8.34] Kaldor, N. (ed.): Conflicts in Policy objective. Oxford University. Press, Oxford (1971)

[R8.35] Kalijarvi, T.V., Merrow, C.E.: Congress and Foreign Relations, vol. 289. The Annals, Philadelphia (1953); The American Academy of Political and Social Science

[R8.36] Kindleberger, C. (ed.): The International Corporation. M.I.T.Press, Cambridge (1970)

[R8.37] Kelman, H.C.: International Behavior: A Socail-Psychological Analysis. Hol, Rinehart and Winston, New York (1963)

[R8.38] Kissinger, H.: American foreign policy; three essays. Norton, New York (1969)

[R8.39] Lauterpacht, H.: The Function of Law in the International Community. Clarendon, Oxford (1933)
[R8.40] Lieth, C.K., Furness, J.W., Lewis, C.: World Minerals and World Peace. The Brookings Institution, Washington, D.C. (1943)
[R8.41] Lowe, C.J.: The reluctant imperialists; British foreign policy. Macmillan, New York (1969)
[R8.42] Marshall, C.B.: The exercise of sovereignty: papers on foreignpolicy. Johns Hopkins Press, Baltimore (1965)
[R8.43] Merton, R.K.: Social Theory and Social Structure. Free Press, New York (1949)
[R8.44] Nielsen, W.A.: The great powers and Africa (Published for the Council on Foreign Relations). Praeger Publishers, New York (1969)
[R8.45] Northedge, F.S.: The International Political System. Faber, London (1976)
[R8.46] Ornstein, N.: Changing Congress: The Committee System. The Annals, Philadelphia (1974); Sellin, T., Lambert, R.D. (General eds.), vol. 411. The American Academy of Political and Social Science
[R8.47] Palmer, N. (ed.): The National Interest- - - - - -Alone or with Others? The American Academy of Political and Social Science; Selling, T. (General eds.), vol. 282. The Annals (July 1952)
[R8.48] Paterson, T.G. (ed.): Major Problems in American Foreign Policy: Documents and Essays. Heath and Co., Lexington (1978)
[R8.49] Patterson, E.M.: World Government, vol. 264. The Annals, Philadelphia (1949); Sellin, T., Lambert, R.D. (eds.), The American Academy of Political and Social Science
[R8.50] Patterson, E.M.: Formulating a Point Four Program, vol. 270. The Annals, Philadelphia (1950); Sellin, T., Charlesworth, J.C. (eds.). The American Academy of Political and Social Science
[R8.51] Dorner, P., El-Shafie, M.A.: Resources and development: natural resource policies and economic development in an interdependent world. University of Wisconsin Press, Croom Helm, Madison, London (1980)
[R8.52] Petras, J.: Critical Perspectives on Imperialism and Social Class in the Third World. Monthly Review Press, New York (1978)
[R8.53] Pfaltzgraff, R.L.: National Security Policy for the 1980s, vol. 457. The Annals, Philadelphia (1981); Lambert, R.D., Heston, A.W. (Gen. eds.). The American Academy of Political and Social Science
[R8.54] Prest, A.R.: The Budget and Interpersonal Distribution. Public Finance 23, 80–98 (1968)
[R8.55] Presthus, R.: Interest Groups in International Perspective, vol. 413. The Annals, Philadelphia (1974); Thorsten, S., Lambert, R.D. (Gen. eds.). The American Academy of Political and Social Science
[R8.56] Raushenbush, S.: The Future of Natural Resources, vol. 281. The Annals, Philadelphia (1952); Thorsten, S., Lambert, R.D. (Gen. eds.). The American Academy of Political and Social Science
[R8.57] Russett, B.M., Hanson, E.C.: Interest and Ideology; The Foreign Policy Beliefs of American Businessmen. W.H.Freeman, San Francisco (1975)
[R8.58] Rosenau, J.N. (ed.): International politics and foreign policy; a reader in research and theory. Free Press, New York (1969)
[R8.59] Rosenau, J.N.: National leadership and foreign policy; a case study in the mobilization of public support. Princeton University Press, Princeton (1963)
[R8.60] Rourke, F.E.: Bureaucracy and foreign policy. Johns Hopkins University Press, Baltimore (1972)

[R8.61] Rummel, R.J.: Understanding Conflict and War, vol. I. Sage Publication, New York (1975)

[R8.62] Schneider, H.: National Objectives and project Appraisal in Developing countries, Paris, Development Center, OECD (1975)

[R8.63] Seelig, L.C.: Resource management in peace and war. National Defense University Press, Washington, DC (1990)

[R8.64] Sherif, M.: Group Conflict and Cooperation. Routledge and Kegan Paul, London (1966)

[R8.65] Simon, H.A.: On the concept of organizational goal. Administrative Science Quarterly 9(1), 1–22 (1964)

[R8.66] Thompson, K.W.: Beyond National Interest: A Critical Evaluation of Reinhold Neibuh's Theory of International Politics. Review of Politics 13, 379–391 (1956)

[R8.67] Van Rensburg, W.C.J.: Strategic minerals. Prentice-Hall, Englewood Cliffs (1986)

[R8.68] Leontief, W., et al.: The Future of nonfuel minerals in the U.S. and world economy: input-output projections, pp. 1980–2030. Lexington Books, Lexington (1983)

[R8.69] Westerfield, H.B.: Foreign policy and party politics: Pearl Harbor to Korea. Octagon Books, New York (1972)

[R8.70] Wilson, J.H.: American business & foreign policy. Beacon Press, Boston (1973)

[R8.71] Young, R.G.: Goals and Goal-Setting. Jour of Amer. Institute of Planners, 76–85 (March 1966)

R9 Negotiations and Resolutions in Conflicts

[R9.1] Deutsch, M.: The Resolution of conflict: constructive and Destructive process. Conn. Yale University Press, New Haven (1977)

[R9.2] Dixit, A., Nalebeff, B.: Thinking Strategically. Norton, New York (1991)

[R9.3] Raiffa, H., et al.: The Science and Art of collaborative Decision. The Belknap – Harvard University Press, Cambridge, Mass. (2002)

[R9.4] Raiffa, H., et al.: Negotiation Analysis: The Science and Art of collaborative

[R9.5] Mnookin, R.H.: Beyond Winning. Negotiate to create value in Deals and Disputes. Belknap, Harnerd Press, Cambridge, Mass. (2000)

[R9.6] Mnookin, R.H.: Beyond Winning. Negotiate to create value" in Deals and Disputes, vol. 1979, pp. 203–211. Belknap, Harnerd Press, Cambridge, Mass. (1979)

[R9.7] Zartman, W.I. (ed.): The Negotiation process: Theories and applications. Calif. Sage Pub., Beverly Hill (1978)

[R9.8] Zeckhouser, R.J., et al. (eds.): Wise Choices: Decisions. Harvard Business School Pub., Harvard University Press, Cambridge, Mass. (1996)

R10 Political Decisions, Games and Rationality

[R10.1] Adelman, I., Cynthia: Society, Politics and Economic Development. John Hopkins Press, Baltimore (1967)

[R10.2] Austen-Smith, D.: Sophisticated Sincerity: Voting over EndogenousAgendas. American Political Science Review 81, 1323–1330 (1987)

[R10.3] Banks, J.S.: Sophisticated Voting Outcomes and Agenda Control. Social Choice and Welfare 1, 295–306 (1985)
[R10.4] Baron, D.P.: Electoral Competition with Informed and UninformedVoters. American Political Science Review 88, 1–14 (1994)
[R10.5] Baron, D.P.: A Dynamic Theory of Collective Goods Provision. American Political Science Review 90, 216–330 (1996)
[R10.6] Baron, D.P., Ferejohn, J.: Bargaining in Legislatures. American Political Science Review 83, 1181–1206 (1989)
[R10.7] Bernard, J.: The Theory of Games of Strategy as a Modern Sociology of Conflict. American Journal of Sociology 59, 411–424
[R10.8] Douglas Bernheim, B., Rangel, A., Rayo, L.: The Power of theLast Word in Legislative Policy Making. Econometrica 74(5), 1161–1190 (2006)
[R10.9] Brams, S.J.: Game Theory and Politics. Free Press, New York (1975)
[R10.10] Brams, S.J.: The Presidential Election Game. Yale University Press, New Haven (1978)
[R10.11] Cowart, A.T., Brofass, K.E.: Decision, Politics, and Change. Universitetsforlaget, Oslo (1979)
[R10.12] Dompere, K.K.: The Social Goal-objective Formation and National Interest in Political Economies Under Fuzzy Rationality, A Working Monograph. Department of Economics, Howard University, Washington D.C. (2009)
[R10.13] Dompere, K.K.: Fuzziness,and the Market Mockery of Democracy: ThePolitical Economy of Rent-Seeking and Profit-Harvesting,, A Working Monograph. Department of Economics, Howard University, Washington D.C. (2010)
[R10.14] Dompere, K.K.: African Union: Pan-African Analytical Foundations. Adonis& Abbey Publishers, London (2006)
[R10.15] Dompere, K.K.: Polyrhythmicity: Foundations of African Philosophy. Adonis& Abbey Publishers, London (2006)
[R10.16] Epple, D., Riordan, M.H.: Cooperation and Punishment Under Repeated Majority Voting. Public Choice 55, 41–73 (1987)
[R10.17] Mc Kelvey, R.: Intransitivities in Multidimensional Voting Models and Some Implication of Agenda Control. Journal of Economic Theory 12, 472–482 (1976)
[R10.18] Noll, R.G., Owen, B.M.: The Political Economy of Deregulation: Interest Groups in the Regulatory Process. American Enterprise Institute for Public Policy Research, Washington, D.C. (1983)
[R10.19] Rapoport, A.: Fights, Games and Debates. Michigan University Press, Lansing (1963)
[R10.20] Romer, R., Rosenthal, H.: Political Resource Allocation, Controlled Agendas and the Status quo. Public Choice 33, 27–43

R11 Public-Private Sector Ralative Size

[R11.1] Arrow, K.J., Boskin, M.J. (eds.): The Economics of Public Debt. Macmillan Press (1988)
[R11.2] Aschauer, D.A.: Is Public Expenditure Productive? Journal of Monetary Economics 23, 177–200 (1989)
[R11.3] Baumol, W.J. (ed.): Public and Private Enterprise in a MixedEconomy' (Proceedings of Conference held by the International Economic Association in Mexico City). St. Martin's Press, New York (1980)

[R11.4] Beck, M.: Public Sector Growth: A Real Perspective. Public Finance 34, 14–27 (1979)

[R11.5] Baird, C.W.: On Profits and Hospitals. Journal of Economic Issues 5, 57–66 (1971)

[R11.6] Balkan, E., Greene, K.V.: On Democracy and Debt. Public Choice 67, 201–211 (1990)

[R11.7] Baumol, W.J. (ed.): Public and Private Enterprise in a Mixed Economy (Proceedings of Conference held by the International Economic Associationin Mexico City). St. Martin's Press, New York (1980)

[R11.8] Barro, R.: On the Determination of the Public Debt. Journal of Political Economy 87, 940–971 (1979)

[R11.9] Borcherding, T.E.: Budgets and Bureaucrats: The Sources of Government Growth. Duke University Press, Durham (1977)

[R11.10] Buchanan, J.M., Rowley, C.K., Tollison, R.D. (eds.): Deficits. Basil Blackwell, Oxford (1987)

[R11.11] Butler, S.M.: Privatizing Federal Spending: A Strategy to Eliminate the Deficit. Universal Books, New York (1985)

[R11.12] Cameron, D.R.: The Expansion of the Public Economy: A Comparative Analysis. American Political Science Review 72, 1243–1261 (1978)

[R11.13] Clarke, S.R.: The Management of the Public Sector of the National Economy. Athlone Press, London (1964)

[R11.14] Coombes, D.: State Enterprise: Business or Politics? Allen and Unwin, London (1970)

[R11.15] Davies, D.G.: The Efficiency of Public Versus Private Firms, The Case of Australia's Two Airlines. Journal of Law and Economics 14(31), 149–165 (1971)

[R11.16] Eisner, R.: How real Is the Federal Deficits. The Free Press, New York (A division of Macmillan)

[R11.17] Eisner, R.: Budget Deficits: Rhetoric and Reality. Journal of Economic Perspectives 3, 72–93 (1989)

[R11.18] Eisner, R., Pieper, P.J.: Deficits, Monetary Policy and Real Economic Activity. In: Arrow, K.J., Boskin, M.J. (eds.) The Economics of Public Debt., pp. 3–40. Macmillan Press, New York (1988)

[R11.19] Eisner, R., Pieper, P.J.: A New View of the Federal Debt and Budget Deficits. American Economic Review 74, 11–29 (1984)

[R11.20] Feldman, A.M.: A Model of majority Voting and Growth in Government Expenditure. Public Choice 46(#i), 3–17 (1985)

[R11.21] Fitch, L.C.: Increasing the Role of the Private Sector in Providing Public Services. In: Hawley, W.D., Rogers, D. (eds.) Improving the Quality of Urban Management, pp. 501–559. Sage, Beverly Hills Ca. (1974)

[R11.22] Gramlich, E.M.: U. S. Federal Budget Deficits and Gramm- Rudman-Holdings. American Economic Review 80, 75–80 (1990)

[R11.23] Galbraith, J.K.: Economics and Public Purpose. Houghton Mifflin, Boston (1973)

[R11.24] Hanke, S.H.: Prospects for Privatization. Academy of Political Science, New York (1987)

[R11.25] Hanson, A. (ed.): Organization and Administration of Public Enterprise. United Nations, New York (1968)

[R11.26] Hayek, F.A.: New Studies in Philosophy, Politics, Economics and History of Ideas. University of Chicago Press, Chicago (1978)

[R11.27] Heilbroner, R.L., Bernstein, P.L.: The Debt and Deficit; False Alarms/Real Possibilities. Norton, New York (1990)

[R11.28] Heclo, H., Wildavsky, A.: The Private Government of Public Money: Community and Policy Inside British Politics. Macmillan, London (1974)

[R11.29] Hsiao, W.: Public Versus Private Administration of Health Insurance: A Study in Relative Economic Efficiency. Inquiry 15, 379–387 (1978)

[R11.30] Johnson, R.T.: Historical Beginnings ...The Federal Reserve. Federal Reserve Bank of Boston, Boston (1977)

[R11.31] Kaldor, N.: Public or Private Enterprise- The Issues to be Considered. In: Baumol, W.J. (ed.) Public and Private Enterprise in a Mixed Economy (Proceedings of Conference held by the International Economic Association in Mexico City), pp. 1–12. St. Martin's Press, New York (1980)

[R11.32] Kendrick, J.W.: The Formation and Stocks of Total Capital. Columbia University Press, New York (1976)

[R11.33] Kimmel, L.H.: Federal Budget and Fiscal policy, 1789-1958. Brookings Institution, Washington, D.C (1959)

[R11.34] Kotlikoff, L.J.: Deficit Delusion. Public Interest 84, 53–65 (1984)

[R11.35] Kuttner, R.: The Economic Illusion: False Choices Between Prosperityand Social Justices. Houghton Mifflin, Boston (1984)

[R11.36] Kotlikoff, L.J.: The Deficit Is Not a Well-Defined Measure of Fiscal Policy. Science 241, 791–795 (1988)

[R11.37] Lindsay, C.M.: A Theory of Government Enterprise. Journal of Political Economy 84, 1061–1077 (1976)

[R11.38] Meltzer, A.H., Richard, S.F.: Why Government Grows (and Grows) in Democracy. Public Interest 52, 111–118 (1978)

[R11.39] Meyer, R.A.: Publically Owned Versus Privately Owned Utilities: A Policy Choice. Review of Economics and Statistics 57(4), 391–399 (1975)

[R11.40] Naylor, R.T.: Hot Money and the Politics of Debt. Simon and Schuster, New York (1987)

[R11.41] Peacock, A.T., Wiseman, J.: The Growth of Public Expenditure in theUnited Kingdom, 1890-1955. Princeton University Press, Princeton (1961)

[R11.42] Peacock, A., Shaw, G.K.: The economic Theory of Fiscal Policy. St Martin's Press, New York (1976)

[R11.43] Peston, M.: Public Goods and the Private Sector. Macmillan, London (1972)

[R11.44] Phelps, E.S. (ed.): Private Wants and Public Needs. W. W. Norton, New York (1957)

[R11.45] Reich, R.: The Power of Public Ideas. Mass., Ballinger Pub., Cambridge (1987)

[R11.46] Rock, J.M. (ed.): Debt and the Twin Deficits Debate. Mayfield Pub., Mountain View (1991)

[R11.47] Runge, C.F.: The Fallacy of Privatization. Journal of Contemporary Studies, 3–17 (Winter 1984)

[R11.48] Saunders, P.: Public Expenditure and Economic Performance in OECD Countries. Journal of Public Policy 5, 1–21 (1986)

[R11.49] Savas, E.S.: Public vs. Private Refuse Collection: A Critical Review of the Evidence. Journal of Urban Analysis 6, 1–13 (1979)

[R11.50] Savas, E.S.: Privatizing the Public Sector: How to Shrink Government. Chatham House Publishers, Inc., Chatham (1982)

[R11.51] Schick, A.: Congress and Money: Budgeting, Spending and Taxing. The Urban Institute, Washington, D.C (1980)

[R11.52] Shonfield, A., Capitalism, M.: The Changing Balance of Public and Private Power. Oxford University Press, New York (1965)

[R11.53] Spann, R.M.: Public Versus Private Provision of Government Services. In: Borcherding, T.E. (ed.) Budgets and Bureaucrats: The Sources of Government Growth, pp. 82–87. Duke University Press, Durham (1977)
[R11.54] Wagner, R.E.: Revenue Structure, Fiscal Illusion, and Budgetary Choice. Public Choice 25, 45–61 (1976)
[R11.55] Watson, D.S.: Economic Policy: Business and Government. Houghton Mifflin, Boston (1960)
[R11.56] Wilson, J.Q., Rachal, P.: Can the Government Regulate Itself? Public Interest 46, 3–14 (1977)
[R11.57] Yunker, J.E.: Economic Performance of Public and Private Enterprise: The Case of U. S. Electric Utilities. Journal of Economics and Business 28, 60–67 (1975)

R12 Regulation-Deregulation Game

[R12.1] Beesley, M.E.: Privatization, Regulation and Deregulation. Routledge, London (1992)
[R12.2] Buchanan, J.M.: A Contractual Paradigm for Applying EconomicTheory. American Economic Review 65, 225–230 (1975)
[R12.3] Cavaco-Silva, A.: Economic Effects of Public Debt. St. Martin's Press, New York (1977)
[R12.4] Gayle, D.J., Goodrich, J.N. (eds.): Privatization andderegulation in global perspective. Quorum Books, New York (1990)
[R12.5] Fowler, R.B.: Energy & the Deregulated Marketplace: 1998 survey. The Fairmont Press, Lilburn (1998)
[R12.6] Landy, M.K., Levine, M.A., Shapiro, M. (eds.): CreatingCompetitive Markets: The Politics of Regulatory Reform. Brookings Institution Press, Washington, D.C. (2007)
[R12.7] Letwin, W.: Law and Economic Policy in America: The Evolution of Sherman Antitrust Act. Random House, New York (1965)
[R12.8] Milanovic, B.: Liberalization and Entrepreneurship: Dynamics of Reform in Socialism and Capitalism. M. E. Sharp, Inc., New York (1989)
[R12.9] Peltzman, S.: Toward a More General Theory of Regulation. Journal of Law and, Economics, 211–240 (1976)
[R12.10] Phillips Jr., C.F.: The Economics of Regulation. Ill. Irwin, Holmwood (1969)
[R12.11] Posner, R.A.: The Social Cost of Monopoly and Regulation. Journal of Political Economy 83, 507–827 (1975)
[R12.12] Spulber, D.F.: Regulation and Merkets. MIT Press, Cambridge (1989)
[R12.13] Swary, I., Topf, B.: Global Financial Deregulation: CommercialBanking at the Crossroads. Blackwell, Cambridge (1992)

R13 Rent-Seeking and Pork Barreling

[R13.1] Buchanan, J.M., et al. (eds.): Toward a theory of the Rent seeking society, collegestation. Taxas A and M University Press, Taxas (1980)
[R13.2] Ferejohn, J.: Pork Barrel Politics: Rivers and Harbors Legislation, pp. 1947–1968. Stanford University Press, Stanford (1974)
[R13.3] Krueger, A.: The Political Economy of Rent seeking society. American Economic Review 64, 291–302 (1974)

[R13.4] Nti, K.O.: Rent-Seeking with Asymmetric Valuation. Public Choice 98, 415–430 (1999)

[R13.5] Tollison, R.D.: Rent-Seeking: A Survey. Kyklos 35, 575–602 (1982)

[R13.6] Tullock, G.: The Rent-Seeking Society: The Selected Works, vol. 5. Liberty Fund, Indianapolis (2005)

R14 Government Revenues, Taxes and Expenditures

[R14.1] Arrow, K.J.: Equality in Public Expenditure. Quarterly Jour. of Econ. 85, 409–415 (1971)

[R14.2] Dorfman, R.: Measuring Benefits of Government Investments. The Brookings Institution, Washington, D.C. (1965)

[R14.3] Downs, G.W., Larkey, P.D.: The Search for Government Efficiency. Temple University Press, Philadelphia (1986)

[R14.4] Else, P.K., Marshall, G.P.: The Management of Public Expenditure. Policy Studies Institute, London (1979)

[R14.5] Gerwin, D.: Towards a Theory of Public Budgetary Decision Making. In: Byrne, R.F., et al. (eds.) Studies in Budgeting. North-Holland, Amsterdam (1971)

[R14.6] Oates, W.E.: The Political Economy of Fiscal Federalism. Lexington Press, Lexington (1977)

[R14.7] Peacock, A.: The Economic Analysis of Government and Related Themes. St. Martins's Press, New York (1979)

[R14.8] Schultze, C.L.: The Politics and Economics of Public Spending. The Brookings Institution, Washington, D.C. (1968)

R15 Some Relevant Revolutions in Thought and Political Economy

[R15.1] Arrow, K.J.: Limited Knowledge and Economic Analysis. American Economic Review 64, 1–10 (1974)

[R15.2] Arrow, K.J.: General Economic Equilibrium: Purpose, Analytic Techniques, Collective Choice. American Economic Review 64, 253–272 (1974)

[R15.3] Black, M.: The Nature of Mathematics. Littlefield, Adams and Co., Totowa, N.J. (1965)

[R15.4] Blass, A.: The Interaction Between Category and Set Theory. Mathematical Applications of Category Theory 30, 5–29 (1984)

[R15.5] Brouwer, L.E.J.: Intuitionism and Formalism. Bull. of American Math. Soc. 20, 81–96 (1913) Also in Benecerraf, P., Putnam, H. (eds.), Philosophy of Mathematics: Selected Readings, pp. 77-89. Cambridge University Press, Cambridge (1983)

[R15.6] Brouwer, L.E.J.: Consciousness, Philosophy, and Mathematics. In: Benecerraf, P., Putnam, H. (eds.) Philosophy of Mathematics: Selected Readings, pp. 90–96. Cambridge University Press, Cambridge (1983)

[R15.7] Brown, B., Woods, J.: Logical Consequence; Rival Approaches and New Studies in exact Philosophy: Logic, Mathematics and Science, vol. II. Hermes, Oxford (2000)

[R15.8] Campbell, N.R.: What is Science? Dover, New York (1952)
[R15.9] Carnap, R.: Foundations of Logic and Mathematics. In: International Encyclopedia of Unified Science, pp. 143–211. Univ. of Chicago, Chicago (1939)
[R15.10] Carnap, R.: The Two Concepts of Probability. Philosophy and phenomenonological Review 5, 513–5532 (1945)
[R15.11] Carnap, R.: The Methodological Character of Theoretical Concepts. In: Feigl, H., Scriven, M. (eds.) Minnesota Studies in the Philosophy of Science, vol. I, pp. 38–76 (1956)
[R15.12] Carson, R.B., Ingles, J., McLand, D.: Government in the American Economy: Conventional and Radical Studies on the Growth of State Economic Power. Lexington Press, Lexington (1977)
[R15.13] Dompere, K.K.: The Theory of the Knowledge Square: The Fuzzy Rational Foundations of the Knowledge-Production Systems. STUDFUZZ, vol. 289. Springer, New York (2013)
[R15.14] Dompere, K.K.: Fuzziness and Foundations of Exact and Inexact Sciences. STUDFUZZ, vol. 290. Springer, New York (2013)
[R15.15] Dompere, K.K.: Fuzziness, Rationality, Optimality and Equilibrium in Decision and Economic Theories. In: Lodwick, W.A., Kacprzyk, J. (eds.) Fuzzy Optimization: Recent Advances and Applications. STUDFUZZ, vol. 254, pp. 3–32. Springer, Heidelberg (2010)
[R15.16] Dompere, K.K., Ejaz, M.: Epistemics of Development Economics: Toward a Methodological Critique and Unity, pp. 219–228. Greenwood Press, Westport (1995, 1982)
[R15.17] Dompere, K.K.: Fuzziness and the Market Mockery of Democracy: The Political Economy of Rent-Seeking and Profit-Harvesting, A Working Monograph, Department of Economics, Howard University, Washington D.C. (2009)
[R15.18] Dompere, K.K.: Social Goal-Objective Formation, Democracy and National Interest: Political Economy Under Fuzzy Rationality, A Working Monograph, Department of Economics, Howard University, Washington D.C. (2009)
[R15.19] Dretske, F.I.: Knowledge and the Flow of Information. MIT Press, Cambridge (1981)
[R15.20] Harwood, E.C.: Reconstruction of Economics. Amarican Institute of Economic Research, Great Barrington (1955)
[R15.21] Hayek, F.A.: New Studies in Philosophy, Politics, Economics and the Historyof Ideas. The University of Chicago Press, Chicago (1978)
[R15.22] Helmer, O., Resher, N.: On the Epistemology of Inexact Science, P-1513. Rand Corporation (October 13, 1958)
[R15.23] Kay, G.: The Economic Theory of the Working Class. St. Martin's Press, New York (1979)
[R15.24] Keirstead, B.S.: The Conditions of Survival. American Economic Review 40(2), 435–445
[R15.25] Knight, F.H.: On the History and Method of Economics. University of Chicago Press, Chicago (1966)
[R15.26] Kuznets, S.: Toward a Theory of Economic Growth. Norton, New York (1968)
[R15.27] Kühne, K.: Economics and Marxism. The Renaissance of the Marxian System, vol. I. St Martin's Press, New York (1979)
[R15.28] Kühne, K.: Economics and Marxism. The Dynamics of the Marxian System, vol. II. St. Martin's Press, New York (1979)

[R15.29] March, J.G.: Bounded Rationality, Ambiguity and Engineering of Choice. The Bell Journal of Economics 9(2), 587–608 (1978)
[R15.30] Marx, K.: The Poverty of Philosophy. International Publishers, New York (1963)
[R15.31] Mátyás, A.: History of Modern Non-Marxian Economics: From Marginalist Revolution through Keynesian Revolution to Contemporary Monetarist Counter-revolution, Budapest, Akademiai Kiodo (1980)
[R15.32] Niebyl, K.H.: Modern Mathematics and Some Problems of Quantity, Quality and Motion in Economic Analysis. Philosophy of Science 7(1), 103–120 (1940)
[R15.33] Nkrumah, K.: Consciencism. Modern Reader, New York (1970)
[R15.34] Pollock, J.: Knowledge and Justification. Princeton University Press, Princeton (1974)
[R15.35] Polanyi, M.: Personal Knowledge. Routledge and Kegan Paul, London (1958)
[R15.36] Price, H.H.: Thinking and Experience. Hutchinson, London (1953)
[R15.37] Popper, K.R.: Objective Knowledge. Macmillan, London (1949)
[R15.38] Putman, H.: Reason, Truth and History. Cambridge University Press, Cambridge (1981)
[R15.39] Putman, H.: Realism and Reason. Cambridge University Press, Cambridge (1983)
[R15.40] Quiggin, J.: Zombie Economics: How Dead Ideas Still Walk among Us. Princeton University Press, Princeton (2010)
[R15.41] Robinson, J.: Economic Philosophy. Anchor Books, New York (1962)
[R15.42] Robinson, J.: Freedom and Necessity: An Introduction to the Study of Society. Vintage Books, New York (1971)
[R15.43] Robinson, J.: Economic Heresies: Some Old-Fashioned Questions in Economic Theory. Basic Books, New York (1973)
[R15.44] Simon, H.A.: Models of Bounded Rationality, vol. 2. MIT Press, Cambridge (1982)
[R15.45] Tigar, M.E., Levy, M.R.: Law and the Rise of Capitalism. Monthly Review Press, New York (1977)

Subject Index

E

F

G

I

W

Z

Printed in the United States
By Bookmasters